Springer Climate

Series Editor

John Dodson, Menai, Australia

Springer Climate is an interdisciplinary book series dedicated on all climate research. This includes climatology, climate change impacts, climate change management, climate change policy, regional climate, climate monitoring and modeling, palaeoclimatology etc. The series hosts high quality research monographs and edited volumes on Climate, and is crucial reading material for Researchers and students in the field, but also policy makers, and industries dealing with climatic issues. Springer Climate books are all peer-reviewed by specialists (see Editorial Advisory board). If you wish to submit a book project to this series, please contact your Publisher (elodie.tronche@springer.com).

More information about this series at http://www.springer.com/series/11741

Komaragiri Srinivasa Raju
Dasika Nagesh Kumar

Impact of Climate Change on Water Resources

With Modeling Techniques and Case Studies

Springer

Komaragiri Srinivasa Raju
Department of Civil Engineering
Birla Institute of Technology and Science-Pilani
Hyderabad, Telangana
India

Dasika Nagesh Kumar
Department of Civil Engineering, Centre for Earth Sciences, Divecha Centre for Climate Change, Interdisciplinary Centre for Water Research
Indian Institute of Science
Bangalore, Karnataka
India

Additional material to this book can be downloaded from http://extras.springer.com.

ISSN 2352-0698 ISSN 2352-0701 (electronic)
Springer Climate
ISBN 978-981-13-5570-7 ISBN 978-981-10-6110-3 (eBook)
https://doi.org/10.1007/978-981-10-6110-3

Printed on acid-free paper

This Springer imprint is published by Springer Nature
The registered company is Springer Nature Singapore Pte Ltd.
The registered company address is: 152 Beach Road, #21-01/04 Gateway East, Singapore 189721, Singapore

Foreword

A student once asked me after a lecture on climate change impacts to hydrology, 'Sir, What is the problem with increasing temperatures due to global warming. Can't we buy a bigger air conditioner and get back to normal?' The student was correct—one can engineer the effect of temperature rise unless it gets out of hand. The bigger problem is that temperature rise comes with change in rainfall, its intermittency, its distribution in space and in time, and the nature of its extremes that cause floods on one end and droughts on the other. Any change in rainfall will require re-engineering the planet, something that may not even be possible if the change gets too big.

This book by Prof. Srinivasa Raju and Prof. Nagesh Kumar is, to my knowledge, one of the first text-cum-reference books to assess and redesign water resources systems due to our changing climate. It will fill a timely gap to knowledge, given climate change is no longer a topic of debate, but one which countries around the world are learning to adapt to. Part of this adaption requires assessing change to risk for existing water resources infrastructure, put in place to allow us to live in places where water is too much or too little. Part of this adaption is also finding ways of designing new infrastructure that will be needed to combat the new water scarcity or excess a warmer climate will bring. The book draws heavily on the excellent work reported in several PhDs supervized by the authors and their colleagues, along with the considerable literature that has been published on this topic worldwide. The book is meant for those familiar with the principles of water resources systems, their design, management and operation, and for those wishing to learn on how they should be redesigned to cope with the challenges ahead.

The book starts with an explanation of what causes warming, why it is anthropogenic, what changes occur, why they are significant and irrefutable, and what are the implications to hydrology and the design and operation of water resources systems. It then shifts to how one can model future change, the challenges this entails, how climate models should be selected and uncertainty quantified, and how this uncertainty may be reduced through clever combination strategies. Following this, a comprehensive assessment of downscaling approaches is presented, which is needed due to the coarse resolution of climate models. This is followed by chapters detailing the statistical techniques used in assessing model simulations, the hydrologic models needed to simulate changes in flow, and soil moisture from the changed rainfall and evapotranspiration conditions of the future, as well as a number of carefully selected case studies that articulate the range of problems the techniques presented can be used for.

I was especially pleased to see a set of questions at the end of each chapter, providing lecturers examples of questions that could be posed to students, giving an opportunity to assess for themselves what they have learnt and what remains. It is these questions that push students and the rest of us to devise solutions to what is turning out to be one of the biggest challenges humanity has faced till date. Many of these numerical questions enable better understanding of the theory clearly and systematically. While the climate has been changing since eternity, human-induced change is real and significant, with the bigger changes yet unseen and requiring careful assessment and planning. I feel this book is a step in the right direction, as it will provide the knowledge needed to re-engineer the planet and ensure our water resources systems continue to provide the security we have come to expect over the years.

My congratulations to both authors on this excellent accomplishment, and I hope this book forces its readers to ponder not only on the science behind climate change but also the engineering that is required to combat its effects on our way of life and existence.

March 2017

Ashish Sharma
Professor and Future Fellow (ARC)
School of Civil and Environmental Engineering
The University of New South Wales
Sydney, Australia

Preface

Climate change has been emerging as one of the major challenges in the global scenario. Changes in climate may lead to adverse negative impacts on both natural and human systems. Continued emissions of greenhouse gases would further amplify the existing risks and create new complications for people and ecosystems. To analyze the possible impacts of climate change on a river basin, it is required to predict the future climate changes. This ultimately will help in planning and management of water resources in the basin. Effective decision-making to throttle climate change and its risks can be addressed by broad range of analytical and mathematical approaches by predicting the changes. General Circulation/Global Climate Models (GCMs) are one of the most credible tools presently available for modeling climate change. However, accuracy of GCMs, which generally run at coarse grid resolution, decreases with increasingly finer spatial and temporal scales, rendering them unable to represent sub-grid scale features. In other words, GCMs are not able to effectively model sub-grid scale processes which are of prime interest to hydrologists and water resources planners. Downscaling is one of the approaches where GCM outputs are interpolated to the scale of hydrological modeling or local scale requirement.

Over the years, various experts across the world have brought out a number of books on the above subject. Most of the books published so far are rather theoretically based with limited number of examples and case studies. The present book is an amalgamation of available resources and divided into various chapters and information about the chapters are as follows: Chap. 1 provides introduction to climate change and variability, climate feedback, forcing mechanism, atmospheric chemistry, palaeo records, monsoon variability, Holocene, IPCC scenarios, teleconnections, impact of climate change, and organization and utilization of the book. The chapter concludes with revision questions and exercise problems, advanced review questions, references, and suggested further reading. This sort of exercises is provided to all the chapters in the book with exception of Chap. 6 in which case studies are presented. Chapter 2 describes GCMs and their choice, performance indicators for evaluating GCMs, weight estimation, multicriterion decision-making techniques in deterministic and fuzzy scenario, Spearman rank correlation coefficient, and group decision-making. Ensembling methodology of GCMs is also discussed. Chapter 3 describes downscaling techniques. Detailed discussion is presented on statistical downscaling techniques such as multiple regression, artificial neural networks, Statistical Downscaling Model (SDSM), change factor technique, and support vector machine. Brief discussion on nested bias correction is also made. Chapter 4 presents data compression techniques, namely, cluster and fuzzy cluster analysis, Kohonen neural networks, and principal component analysis. Trend detection techniques and optimization techniques, namely, linear and non-linear programming and genetic algorithms, are also discussed. Chapter 5 describes hydrological models, SWMM, HEC-HMS, SWAT, and other modeling techniques. Chapter 6 presents various real-world global case studies in AR3 and AR5 perspective that are related to the theories and techniques explained in the earlier chapters. Even through AR3 is relatively older than AR5, case studies are presented to understand the impact of climate change with temperature anomaly equivalent visualization paths, namely, SRES or RCPs.

Appendix A covers procedures for acquiring data from various sources. Appendices B and C provide representative list of journals and books related to climate. Index is also provided for efficient retrieval of topics.

PowerPoint presentations of selected topics are also provided as an additional study material. Interested individuals can contact publishers for PowerPoint presentations.

The present book can help undergraduate as well as postgraduate programs in the field of hydrology, climate change, and allied fields and can be referred as a text book. It can also be used as a reference book or as supplementary study material for researchers working in this upcoming field. The case studies, PowerPoint presentations, extensive references, limited but informative and illustrative problems, and software information render this book as a valuable source of information for researchers, experts, professionals, teachers, and others who are interested in the field of climate, hydrology, and allied fields.

Special acknowledgements to Dr. Ashish Sharma, Professor and Future Fellow (ARC), School of Civil and Environmental Engineering, The University of New South Wales, Sydney, Australia who readily agreed to write a foreword for the book.

The authors are inspired and motivated by the books, reports and publications of esteemed experts, Dr. Ashish Sharma, Dr. B.C. Bates, Dr. Z.W. Kundzewicz, Dr. T. J. Ross, Dr. S. Wu, Dr. J.P. Palutikof, Dr. D.R. Easterling, Dr. F. Johnson, Dr. R.L. Wilby, Dr. Di Luzio M., Dr. G.S. Rao, Dr. R. Srinivasan, Dr. R. Mehrotra, Dr. Sulochana Gadgil, Dr. R.S. Nanjundiah, Dr. V.V. Srinivas, Dr. S.K. Satheesh, Dr. K.C.Patra, Dr. Danielle Costa Morais, Dr. Adiel Teixeira de Almeida, Dr. Chong-yu Xu, Dr. Lankao, Dr. T.I. Eldho and many others and are greatly benefited from various journal papers, Intergovernmental Panel on Climate Change (IPCC) reports, various climate-related homepages such as IPCC, Climate Prediction Center, Climate Research Unit, etc. It will not be surprising, if the reader finds some of their flavor in this book.

The concerned experts and researchers have generously given permission to utilize their study material. We have included in this book some portions of our own publications and publications of other researchers published in various journals (after obtaining copyright permissions) by giving due reference to the journals at the appropriate places. We sincerely thank the publishers of these journals such as Springer, Elsevier, IWA ASCE, Copernicus, De Gruyter, Inter-Research Science Center (Germany), and Prentice Hall of India for giving us specific permissions to reuse the material. Acknowledgements are also due to Dr. A. Anandhi, Dr. Sonali P., and Dr. T.V. Reshmi Devi, Prof. Ajit Pratap Singh, and Ms. Gayam Akshara for permitting us to utilize material from their works. Special Acknowledgements to Ms. V. Swathi for providing numerical problems in the chapter, hydrological modeling.

Even though efforts were made to quote all the sources in the form of acknowledgements or references, a few may have been missed inadvertently. We sincerely apologize for any such inconvenience caused and assure that these will be duly incorporated in the next edition on noticing the same.

Every effort is made to eliminate typographical calculation and methodological errors but still, some may have been left out. We request the readers to bring it to our notice to rectify them in the next edition. The software information provided does not necessarily indicate that the authors are encouraging to use only those particular software. Similar softwares may also be available which may perform as efficiently as those mentioned or even better. Critical suggestions are welcome for improvement of the contents.

The first author is grateful to Prof. G. Sundar, Director, Dr. A. Vasan, Associate Professor, BITS-Pilani, Hyderabad campus, and Prof. A.K. Sarkar, Director, BITS-Pilani, Pilani campus who provided constant motivation and encouragement for preparing the book. The first author acknowledges the support given by his parents Gopala Rao and Varalakshmi, wife Gayathri Devi, daughter Sai Swetha, and son Sai Satvik. The second author acknowledges the support given by his parents Subrahmanyam and Lakshmi, wife Padma, daughter Sruthi, and son Saketh.

We are also thankful to our colleagues for the encouragement and students for their inquisitive queries.

Last but not the least, authors are grateful to Dr. John Dodson, Series editor, Springer Climate for his valuable suggestions and Ms. Swati Meherishi, Ms. Aparajita Singh, and Ms. Bhavana Purushothaman of Springer for diligently processing the manuscript and for timely publication of the book.

Hyderabad, India
Bangalore, India
April 2017

Komaragiri Srinivasa Raju
Dasika Nagesh Kumar

Contents

About the Authors

Prof. Komaragiri Srinivasa Raju is Professor, Department of Civil Engineering, Birla Institute of Technology and Science, Pilani, Hyderabad Campus, India. He completed M.Tech. from Indian Institute of Technology (IIT) Chennai and Ph.D. from IIT Kharagpur in Water Resources Engineering. He worked with Prof. Lucien Duckstein in the field of Multicriterion Decision-making (MCDM) with reference to Water Resources Planning during his postdoctoral program. His research interests include climate hydrology, water resources systems, MCDM, and soft computing applications in Water Resources Engineering. He has published more than 120 papers in various reputed journals and conferences and has been a reviewer for more than 38 international journals. He has co-authored a book entitled "Multicriterion Analysis in Engineering and Management" along with Prof. D. Nagesh Kumar, published by Prentice Hall of India (PHI), New Delhi. He is presently Associate Editor, ISH Journal of Hydraulic Engineering, Taylor & Francis. He has completed a number of sponsored projects including two collaborative projects with Indian Institute of Science, Bangalore on Climate Modeling and Integrated Urban Flood Management. More information is available at www.bits-pilani.ac.in/hyderabad/ksraju/Profile

Prof. Dasika Nagesh Kumar is working in the Department of Civil Engineering, Indian Institute of Science (IISc), Bangalore, India since May 2002. Earlier, he worked at IIT, Kharagpur (1994–2002) and National Remote Sensing Centre, Hyderabad (1992–1994). He is also the Chairman, Centre for Earth Sciences, and Associate Faculty, Interdisciplinary Centre for Water Research and Divecha Centre for Climate Change, IISc, Bangalore. He obtained Ph.D. (Engg.) from IISc, Bangalore, in 1992. He visited Utah State University, USA in 1999 for 6 months on BOYSCAST fellowship. He worked as Visiting Professor in EMSE, Saint-Etienne, France, in 2012. His research interests include climate hydrology, water resources systems, climate change impact on water resources, artificial neural networks, evolutionary algorithms, fuzzy logic, MCDM, and remote sensing and geographical information system applications in Water Resources Engineering. He has published more than 180 papers in leading international journals and conferences in his research fields. He has co-authored four textbooks, viz., "Multicriterion Analysis in Engineering and Management" published by PHI, New Delhi, "Floods in a Changing

Climate: Hydrologic Modeling" published by Cambridge University Press, U.K., "Hydroclimatic Teleconnection: Indian Perspective", and "Integrated Modelling for Optimal Reservoir Operation for Irrigation", published by Scholars' Press, Germany. He is Associate Editor for ASCE Journal of Hydrologic Engineering. He is the Editor-in-Chief of Open Water Journal, IWA Publishing and in the Editorial Board of Open Hydrology Journal, ISH Journal of Hydraulic Engg., and Journal of Applied Computational Intelligence and Soft Computing. He has been a reviewer for more than 60 international journals including Water Resources Research and Journal of Hydrology. More information is available at civil.iisc.ernet.in/~nagesh/

Introduction

1

Abstract

This chapter provides information about atmospheric activities and impacts of climate change. Causes for climate variability, such as El Niño Southern Oscillation, and its two different phases, La Niña and El Niño for cooling and warming, are discussed. Teleconnections, climate feedback, and forcing mechanisms (radiative and non-radiative, periodic and random, and external and internal) are also parts of the chapter. Direct and indirect effects of aerosols that influence the visibility in the atmosphere are also discussed briefly but critically. Greenhouse gases and consequences of global warming such as variations in rainfall, ice caps and glacier melting, temperature, likelihood increase in frequency of floods and droughts, and acidification due to carbonic acid formation are also explained. In addition, importance of atmospheric chemistry, Palaeo records, monsoon variability, and Holocene is also stressed. Extensive discussion on Intergovernmental Panel on Climate Change (IPCC) scenarios which relate to demographic, economic, technological, and social changes, i.e., Special Report on Emissions Scenarios (SRES) A1, A2, B1, and B2 and Representative Concentration Pathways (RCPs) 2.6, 4.5, 6.0, and 8.5 is also made. Impact of climate change on hydrology, water resources, urbanization, and hydrologic extremes is discussed extensively. In addition, climate change impacts on India are also covered in three aspects: What we know, what could happen, and what can be done? The reader is expected to understand various atmospheric processes/activities, impacts of climate change, and organization and utilization of the book by studying this chapter.

Keywords

Aerosols · Climate change · Forcing · Greenhouse gases · Impact · IPCC · RCP · SRES · Teleconnections

Electronic supplementary material The online version of this chapter (doi:10.1007/978-981-10-6110-3_1) contains supplementary material, which is available to authorized users.

K. Srinivasa Raju and D. Nagesh Kumar, *Impact of Climate Change on Water Resources*, Springer Climate, https://doi.org/10.1007/978-981-10-6110-3_1

1.1 Introduction

Hydrology deals with occurrence as well as distribution of water in the atmosphere, surface, and below surface of the Earth (Definition of Hydrologic 2017). Climatology is the scientific study of climate. It examines variations in atmosphere due to the oceans' circulation, atmospheric gases concentration, and intensity variations of solar radiation both for the long term and short term. A systematic analysis focusing on the spatiotemporal variation of climate in the hydrologic cycle is termed as hydroclimatology. In hydroclimatological studies, complex space and time variations of physical processes by both hydrologic cycle and climate sciences are observed and modeled together, and inferences are drawn for an improved appraisal. Generally, the output of this complex physical process facilitates in determining the impact of frequent extreme events on the primary components of the hydrologic cycle by employing both statistical and deterministic techniques (Global Change Hydrology Program: Hydroclimatology 2017).

The global hydrologic cycle is a logical unifying theme for hydroclimatology and consists of two branches, viz., terrestrial and atmospheric. Terrestrial branch deals with the natural process at or near the land surface (i.e., the common process related to hydrology). The atmospheric branch explains transport of water in the vapor phase, precipitation, and evaporation (Shelton 2009). These two branches are interlinked at the interface of the atmosphere and Earth's surface. Modern hydroclimatology provides a holistic view on how climate changes occur over time with respect to both human and natural actions. It perceives water as a connecting element between hydrology and climatology (Chahine 1992). Continuous increase in global temperatures may result in potential catastrophic events, warranting the need of research on hydrological processes like evapotranspiration and streamflow.

In the present chapter, climate change and variability, climate feedback, forcing mechanism, atmospheric chemistry, Palaeo records, monsoon variability, Holocene, Intergovernmental Panel on Climate Change (IPCC) scenarios, teleconnections, and impact of climate change are discussed (Kumar et al. 1995; Jain and Lall 2001). Glossary of the terms relevant to climate is available at Glossary Relevant to Climate (2017).

1.2 Climate Change and Variability

Average state of the atmosphere represented by temperature and precipitation is termed as climate and it is based on hydrosphere, cryosphere, lithosphere, and biosphere. A continuous increase or decrease in the statistical properties of climate variables is termed as climate change. Climate variability can be caused by climate patterns such as El Niño Southern Oscillation, and its two different phases La Niña and El Niño for cooling and warming, respectively.

Climate change and variability have substantial practical significances associated with climate forcing of the hydrologic cycle both in time and space. There are contemporary studies pertaining to climate change and variability on water resources. Most of these studies focused on coupling hydroclimatic watershed models with global and regional climate models using satellite imagery.

This knowledge will not only increase the awareness of how the hydrological systems may change over the coming century but will also provide information regarding the changes in the present climate and its impact on water resources. Keeping this in view, Intergovernmental Panel on Climate Change (IPCC) was constituted in 1988 by the United Nations Environment Program (UNEP) and World Meteorological Organization (WMO) to conduct climate change research. Till date, IPCC published five assessment reports in the years 1990, 1995, 2001 (IPCC 2001), 2007, (IPCC 2007), and 2014 (IPCC 2014), and prospects of significant changes in the hydrological cycle due to climate change were covered in all the IPCC assessment reports. These reports also address hydrological changes and variability that may cause potential climate hazards. Among the several climate variables, precipitation, temperature, relative humidity, and incident solar radiation are frequently used for climate impact assessment (IPCC 2001). These variables are significant in hydrology and some of their possible impacts are discussed. The following sections elaborate the above discussion.

1.3 Climate Feedback

Climate system is subjected to incoming solar radiation, which is expected to be balanced by the outgoing terrestrial radiation resulting in a situation where Earth continues to neither indefinitely cool down nor heat up. When the climate system responds to the forcing (imposed perturbation/Earth's energy balance variation with space), it results in inequilibrium state.

In a climate system, when a fraction of the output is added to the input of the system, it further modifies the output. The resultant loop is called a feedback, which may accelerate (positive feedback) or decelerate the process (negative feedback) (Hansen et al. 1998). Positive feedback situation is possible when the reaction to primary climate forcing follows the same direction as the initial forcing factor and supplements the climate response to the forcing, which amplifies warming or cooling trends. Example of this nature is ice-albedo feedback due to mountain ranges (Umbgrove 1947). Negative feedback situation is possible when the reaction to primary climate forcing acts in the opposite direction to that of the initial forcing factor and reduces the climate response to the forcing. The various feedback processes occurring in the atmosphere due to forcing cause complex climate changes.

1.4 Forcing Mechanism

Climate change is caused by forcing agents, which may be radiative or non-radiative, periodic or random, and external or internal. It is essential to analyze the climate change for various time steps, e.g., daily, monthly, and yearly.

1.4.1 Radiative and Non-radiative Forcing

Radiative forcing mechanism is the difference between outgoing terrestrial radiation and incoming solar radiation. It is caused due to variations in atmospheric composition, solar radiation, volcanic activity, and Earth's orbit around the Sun. Whereas non-radiative mechanisms do not directly affect the atmosphere energy budget (Shine et al. 1990).

1.4.2 Random and Periodic Forcing

Randomness in climate variation is mainly due to the chaotic and complex climate system behaviors resulting in a large proportion of the unpredictable climate variation (Goodess et al. 1992). In the case of periodic forcing, identifying the forcing mechanisms and their impact on the global climate framework enables to ascertain future climate changes. However, outcome depends on the response of a climate system which cannot be described entirely either as a linear or a non-linear process.

1.4.3 External and Internal Forcing Mechanisms

External forcing mechanisms involve agents acting from the extraterrestrial systems including galactic, orbital (obliquity, orbital shape, and changes in precession) and solar variations (Berger 1978). Internal forcing mechanisms involve agents operating within the climate system (from ocean, atmosphere, and land systems) and are caused by factors such as orogeny, volcanic activity, ocean circulation, land use/land cover changes, and variations in atmospheric composition (including aerosol content and greenhouse gases).

1.4.3.1 Aerosols

Increase in aerosols has two effects on the climate system. The aerosol effect can be direct, due to scattering and absorption of solar radiation by aerosols thereby influencing visibility in the atmosphere (direct radiative forcing). The indirect effect increases the formation of clouds and influences the properties and gas chemistry of aerosols. Aerosol content change in the atmosphere is due to both natural and anthropogenic factors. Some of the challenges to assess their impact on climate change are (a) tiny size of aerosols less than 1 μm (much less than diameter of human hair ~75 μm) and (b) high accurate instrumentation and precise modeling requirements (Satheesh and Srinivasan 2002; Satheesh and Lubin 2003; Satheesh and Krishna Moorthy 2005; Satheesh 2006; Sedlacek and Lee 2007; Khain et al. 2008; Aerosols 2017).

1.4.3.2 Greenhouse Gases

Solar radiation emitted by the surface of Earth is held by greenhouse gas molecules. These are then diffused into the atmosphere in all directions causing warming of

Earth's surface. Carbon dioxide (CO_2), water vapor (H_2O), nitrous oxide (N_2O), chlorofluorocarbons (CFCs), and methane (CH_4) are some of the greenhouse gases that cause warming. Anthropogenic and natural factors are responsible for content change in greenhouse gases in the atmosphere. Natural factors include changes in atmospheric CO_2 and CH_4 concentrations, vegetation, and weathering of rocks, whereas anthropogenic factors include forest clearing, fossil fuel burning, and other industrial processes. CO_2 is one of the important agents responsible for climate warming due to its long persistence in the atmosphere (Houghton et al. 1996). There are many consequences of global warming such as variations in rainfall, ice caps and glacier melting, temperature, likelihood increase in frequency of floods, and droughts and acidification due to carbonic acid formation.

Global Climate Models (GCMs) are developed to evaluate the plausible responses of the climate system to the changes in the behavior of natural and human systems either separately or together. Climate models perform better at continental and large regional scales compared to smaller spatial scales. Coupled Model Intercomparison Project (CMIP) was devised for analyzing the coupled Atmosphere-Ocean General Circulation Models' (AOGCMs) outputs. GCMs from both phases 3 and 5 of CMIP, which are termed as CMIP3 and CMIP5, are used extensively in the last couple of decades worldwide for climate impact assessment studies. Detailed discussion on GCMs is presented in Chap. 2.

1.4.4 Atmospheric Chemistry

Atmospheric chemistry relates to chemistry of the Earth's atmosphere and that of other planets. Atmospheric chemistry has interdisciplinary applications in climatology, meteorology, geology, oceanography, volcanology, physics, and many other areas. This topic is gaining importance due to its scope of identifying and analyzing atmospheric effects such as global warming, greenhouse gases, photochemical smog, ozone depletion, and acid rain. Availability of advanced technology for observation, laboratory infrastructure, high computing facilities, and interdisciplinary nature of research is the reason for its growth (Atmospheric Chemistry 2017).

1.4.5 Palaeo Records

Palaeo/past records are expected to provide information about the past for analyzing changes in the Earth's system. Past records include historical (documented) or prehistorical records, i.e., without documentation/further back in geological time that can consider geomorphology, sedimentology, and ecology (Palaeo Records 2017).

1.4.6 Monsoon Variability

Significant and erratic seasonal spatiotemporal variation of monsoon rainfall affects economy across the world; for example, crop production affects food security and gross domestic product (Gadgil et al. 1999). Gadgil (2003) addressed the challenging problem of relating convection over the surrounding ocean and Pacific to the monsoon variability. This paper also quoted the work of Halley (1686) where differential heating between ocean and land was mentioned as the cause for monsoon rainfall (Webster 1987). This paper also discussed the thought process of Charney (1969), where "monsoon is considered as a manifestation of the seasonal migration of the intertropical convergence zone." This paper observed considerable variation of rainfall over sub-seasonal scales between active and weak spells and quoted the work of Blanford (1886), where the phenomena were described as the fluctuation between spells "during height of rains" and "intervals of droughts". Close correspondence between El Niño and deficit Indian monsoon rainfall is discussed by Sikka (1980) and Pant and Parthasarathy (1981).

1.4.7 Holocene

According to Wikipedia, Holocene is the geological epoch that began after the Pleistocene at approximately 11,700 years Before Present (BP) (Holocene 2017) and marked by variability over century to millennial both at low and high latitudes. Gupta et al. (2006) in their study felt that evidences during historical, archeological, and palaeoclimatic indicate significant dislocation of population and cyclic trend of human migrations. deMenocal (2001) presented impact of late Holocene climate change on population over various time zones: Collapse of Tiwanaku, Mochica, Classic Maya, and Akkadian spanning over 1000, 1500, 1200, and 4200 calendar years BP, respectively. He observed close interaction between human cultural elements and persistent multi-century climate shift. Gupta et al. (2006) also described chronology of events during the Holocene in the Indian subcontinent which spread over seven time zones, anno Domini (AD) 1800 onward, AD 1400–1800, AD 700–1200, 1700 onward, 4000–3500, 7000–4000, and 10,000–7000 calendar years BP, and covered information related to climate in the Indian subcontinent, population response, and agriculture during those time zones.

Srivastava et al. (2003) in their study over central Ganga Plain found significant changes in the morpho-hydrologic conditions during the latest Pleistocene–Holocene. They concluded that hydrologic conditions in the Gangetic plains were influenced largely by climatic changes supported by tectonic activity and observed that increasing agricultural- and human-related activities substantially increased the natural siltation rate of ponds for the past 2 kiloyear (ka). Singhvi and Kale (2009) made extensive and informative studies on paleoclimate studies in India, whereas Singhvi and Krishnan (2014) extensively discussed present and past climate of India including variability of Indian monsoon over time. Saraswat et al. (2016) constructed centennial-scale change in carbonate burial, seawater temperature, upwelling-induced productivity, and evaporation–precipitation from the southeastern Arabian Sea using faunal and geochemical proxies and reported major shift in proxies during the mid-Holocene (6.8–6.2 ka).

1.5 IPCC Climate Scenarios

Various climate scenarios were constructed for assessing the possible impacts of anthropogenic climate alterations and variations of natural climate in future. These are neither predictions nor forecasts of climate conditions (Smith and Hulme 1998; Mearns et al. 2001; Criteria for Selecting Climate Scenarios 2013). The suitability of each scenario is based on its applicability in impact assessment, physical plausibility and consistency with global projections, accessibility, and representation. Various climate scenarios had been explored in impact assessments (Mearns et al. 2001) such as analog scenarios (spatial and temporal) (Bergthórsson et al. 1988) and incremental scenarios (Smith and Hulme 1998). However, some of the limitations of the scenarios based on GCMs are coarse spatial resolution which may not suit regional or local impact assessments, inability to differentiate anthropogenic effect, and different structures for various models.

1.5.1 AR3 Perspective

In 1992, IPCC developed a set of six global emission scenarios (IS92a–f), termed as IS92. These scenarios provide approximate estimates of possible occurrences of greenhouse gases. The IS92 scenarios were further updated in 2000 and the set of updated/new emission scenarios was established in Special Report on Emissions Scenarios (SRES) . Some of the aspects of SRES are (i) improvements of processes of greenhouse gases, (ii) changed geopolitical landscape, and (iii) availability of more driving force data on emissions (Nakicenovic et al. 2000). Four sets of scenarios, A1, A2, B1, and B2 (Table 1.1), were formulated on the basis of four individual story lines which describe about each scenario relating to demographic, economic, technological, and social changes (IPCC Special Report: Emission Scenarios 2000; SRES Emissions Scenarios 2017a, b; Anandhi et al. 2013).

Table 1.1 Information about SRES scenarios (AR3 perspective)

Scenario	Representative characteristics that may emerge in future
A1	Very rapid economic growth and global population that peaks in mid-century and declines thereafter; More efficient and new technologies introduction Sub-classifications: non-fossil energy sources (A1T), fossil intensive (A1F1), balance across all sources (A1B)
A2	Very heterogeneous world with preservation of local identities and self-reliance as underlying theme
B1	Convergent world (similar global population as that of A1) with reduction in material intensity, rapid and significant changes in economic structures, and initiation of clean as well as resource-efficient technologies
B2	Environmental, social, and economic sustainability with increase in global population, more diverse and less rapid technological change, intermediate levels of economic development

1.5.2 AR5 Perspective

Representative Concentration Pathways (RCPs) consider alternative scenarios in aerosol concentrations with greenhouse gas as their origin point. Emission and socioeconomic scenarios were constructed in parallel, namely RCPs 2.6, 4.5, 6.0, and 8.5 (Integrated Assessment Modeling Consortium 2017). RCP 2.6 outlines one pathway where radiative forcing peaks at approximately 3 W/m^2 before 2100 and then decline. RCPs 4.5 and 6.0 outline two intermediate stabilization pathways where forcing is stabilized at approximately 4.5 and 6.0 W/m^2 after 2100. RCP 8.5 outlines one high pathway for which forcing is greater than 8.5 W/m^2 by 2100 and continues to rise for some more time. Temperature anomaly (°C), carbon dioxide equivalent (ppm), and pathway shape for RCPs 2.6, 4.5, 6.0, and 8.5 are 4.9, 1370, rising; 3.0, 850, stabilization without overshoot; 2.4, 650, stabilization without overshoot; and 1.5, 490, peak and decline, respectively.

RCP 8.5, RCP 6.0, and RCP 4.5 are temperature anomalies equivalent to SRES A1F1, SRES B2, and SRES B1 scenarios. This analogousness helps to understand the parallelism between AR3 and AR5 scenarios (Representative Concentration Pathways, Part 3: RCP Technical Summary 2017) .

Moss et al. (2010), Vuuren et al. (2011), and Wayne (2013) provided useful information on RCPs. More information on RCPs is available in Representative Concentration Pathways, Part 1: An Introduction to Scenarios (2017), Representative Concentration Pathways, Part 2: Creating new scenarios (2017), Representative Concentration Pathways, Part 3: RCP Technical Summary (2017), Representative Concentration Pathways description (2017), Scenario Process for AR5 (2017), and Glossary relevant to RCPs (2017).

1.6 Teleconnection Patterns

El Niño (warm) and La Niña (cold) phases together are known as El Niño–Southern Oscillation (ENSO) cycle (Philander 1985, 1990). They describe temperature fluctuations between the atmosphere and ocean in the east-central equatorial Pacific which have large-scale impacts on global weather, ocean processes, and climate. These activities generally last 9–12 months and on an average take place every 2–7 years. El Niño significantly influences marine fisheries, ocean conditions, and weather patterns across significant portions of the globe, whereas La Niña effects tend to be opposite. Kane (1998) related ENSO to Indian Summer Monsoon Rainfall (ISMR). Ramachandran (2007) analyzed ENSO and other related parameters for Indian conditions. Numerous researchers explored ENSO to forecast rainfall over the Indian subcontinent (Sikka 1980; Pant and Parthasarathy 1981; Shukla and Paolino 1983; Parthasarathy et al. 1988; Rao 1997). Relevant information on ENSO is available (Ashok et al. 2004; Gadgil et al. 2003, 2004; Wang et al. 2012; Climate Prediction Center 2017; National Ocean Service 2017).

Other related teleconnection patterns are North Atlantic Oscillation, East Atlantic Pattern, Tropical/Northern Hemisphere, Pacific Transition Pattern, Equatorial Indian Ocean Oscillation, Ocean–Land temperature contrast, etc. (Barnston and Livezey 1987; Barnston et al. 1991; Bell and Basist 1994; Bell and Janowiak 1995; Gadgil et al. 2003, 2004; Bell and Chelliah 2006; Maity and Nagesh Kumar 2006, 2007; Maity et al. 2007; Nanjundiah et al. 2013). In detail, descriptions of these teleconnections are also available in Climate Prediction Center (2017).

1.7 Impact of Climate Change

IPCC reports provide an overview of how (positive/adverse) climate changes may affect natural systems including water resources. Interestingly, water resources are also affected by population size, land use pattern, aging insufficient infrastructure, exploitation of groundwater and reuse of wastewater, dynamically changing economic and social values, accessibility, quality, flood risk mitigation/flood control, energy, hydropower, navigation, etc. Changing patterns may require modified design, relook at/redevelop the operational constraints, infrastructure changes, adaptive management, and extensive research to eliminate knowledge gaps/technology gaps/uncertainties for improved understanding of climate change. Infrastructure deficit if any should be addressed as priority to cater the expected events. Following sections present climate change impact on hydrology, water resources, urbanization, and hydrologic extremes.

1.7.1 Hydrology

- Greenhouse gases concentrations are expected to change the atmosphere radiative balance, causing temperature changes and associated changes in precipitation patterns.
- Amount of precipitation, its spatial distribution, intensity, and its frequency changes affect the runoff.
- Climate change affects discharge into rivers significantly stretching the infrastructure. Increase/decrease in discharge results in under-designed/over-designed reservoirs (IPCC 2008).
- Water quality and temperature in the region are affected by warming of lakes and rivers in the region.
- Climate change effect is visible in hydrological cycle parameters, i.e., precipitation, temperature, evaporation, transpiration, etc.

The fourth assessment report (AR4) of IPCC contains detailed information on the resultant climate change. These reports cover a wide range of topics in hydrology in climate change such as runoff generation, changes in groundwater systems, lakes, floods and droughts, changes in physical and chemical aspects of lakes and rivers, water quality, erosion and sediment transport, and water use.

1.7.2 Water Resources

Numerous studies (Brekke et al. 2009) indicate that climate change is an alarming event (for example, increase in temperature results in increase in evapotranspiration, and decrease in the amount of available water). This necessitates continuous assessment and quantification of impacts of climate change. Bates et al. (2008)in their IPCC report opined that (a) "Observational records and climate projections provide evidence that freshwater resources are most vulnerable and are strongly impacted by climate change in turn have wide-ranging consequences on human societies and ecosystems." (b) "Climate change affects the functioning and operation of existing water infrastructure—including hydropower, structural flood defenses, drainage and irrigation systems—as well as water management practices" (Bates et al. 2008; Milly et al. 2008).

Natural Resources Defense Council (2010) suggested water supply response measures in climate change such as use of climate-smart water management tools, water management at regional level, system reoperation, climate change impact consideration into project design, and incorporation of environmentally sound flood management policies. Wateraid (2017), a water resource, organization, discussed about climate change and its occurrence and highlighted its possible impacts on health, food, land, water, environment, and large-scale changes for temperature rise of 1–5 °C and more than 5 °C and discussed the impacts for Africa and Asia. Islam and Sikka (2010), Gosain et al. (2011), Bhatt and Mall (2015), and India Environment Portal (2017) performed similar studies.

Groundwater, largest available freshwater on Earth, is an integral part of hydrological cycle and is a supporting source for domestic, agriculture, and industrial sectors along with surface water. However, groundwater availability is affected by indiscriminate pumping and unpredictable climate change (Bates et al. 2008; Siebert et al. 2010; Green et al. 2011). For example, drought situation may result in unhealthy and unsustainable pumping of groundwater, which is irreversible. Panwar and Chakrapani (2013) in their extensive studies discussed about India's groundwater utilization and status, behavioral and structural adaptations, promoting groundwater governance, risk zones of groundwater, mapping of climate change, and CO_2 sequestration. Saline water intrusion into groundwater in coastal aquifers due to rise of sea level also affects groundwater quality (Shah 2009). Treidel et al. (2012), Panigrahy et al. (2015), and Refsgaard et al. (2016) discussed impacts on groundwater and resulting uncertainties. Groundwater and surface water management are influenced by human activities and it is a major challenge to the planners.

1.7.3 Urbanization

Lankao (2008) described climate change impact on urbanization. Salient aspects discussed are as follows:

- Climate change may result in change in precipitation levels, mean temperatures, and sea level which impacts energy demand and reduction of the sewers' capacity.
- Floods, droughts, heatwaves, and landslides affect livelihoods of urban population, property, and quality of life.
- Extreme rainfall events escalate flood hazards and landslides and create stress on existing infrastructure and health hazards leading to higher insurance expenses and societal disturbances.
- Social and environmental factors, hazards, public finance, institualization process, flexible climate adaptation, and mitigation strategies can be considered to tackle the effects.

1.7.4 Hydrologic Extremes

Huge amount of damage occurs due to events such as extreme precipitation, temperature, and rise in sea levels (Easterling et al. 2000), resulting in loss of life and property, destruction, and increase in insurance claims. Numerous researchers found that climate change impacts almost all climate variables and trend shows that they impact extreme events also (IPCC 2012). Easterling et al. (2000) envisaged climate extremes, namely more hot summer days and heavy 1-day precipitation, higher maximum and minimum temperatures, increase in heat index, fewer frost days, more intensity multiday precipitation events and heat waves and more intense El Niño events, etc., for observed (twentieth century) and modeling (twentieth-century end) conditions. Tohver et al. (2014), Taye et al. (2015), and Gu et al. (2015) analyzed the impact on hydrological extremes for Pacific northwest region of North America, Blue Nile Basin, and Yangtze River Basin, China, respectively. Related studies on extreme rainfall are available (Haylock and Nicholls 2000).

1.7.5 India: Climate Change Impacts

Impacts of likely temperature increase from 2 to 4 °C in India include extreme heat, rainfall patterns, droughts, groundwater, glacier melt, sea level rise, agricultural and food security, energy and water security, health, and mitigation and conflict (Modified and Adapted from India: Climate Change Impacts 2017) and presented in Table 1.2.

Table 1.2 Climate change impacts on India

Characteristic	What we know	What could happen	What can be done
Extreme heat	India is already experiencing a warming climate	Unusual and unprecedented spells of hot weather are expected to occur far more frequently and cover much larger areas Under 4 °C warming, the west coast and southern India are projected to shift to new, high-temperature climatic regimes with significant impacts on agriculture	With built-up urban areas rapidly becoming "heat-islands", urban planners will need to adopt measures to counteract this effect
Changing rainfall patterns	A decline in monsoon rainfall since the 1950s has already been observed. The frequency of heavy rainfall events has also increased	A 2 °C rise in the world's average temperatures will make India's summer monsoon highly unpredictable At 4 °C warming, an extremely wet monsoon that currently has a chance of occurring only once in 100 years is projected to occur every 10 years by the end of the century An abrupt change in the monsoon could precipitate a major crisis, triggering more frequent droughts as well as greater flooding in large parts of India India's northwest coast to the southeastern coastal region could see higher than average rainfall Dry years are expected to be drier and wet years wetter	Improvements in hydro-meteorological systems for weather forecasting and the installation of flood warning systems can help people move out of harm's way before weather-related disaster strikes Building codes will need to be enforced to ensure that homes and infrastructure can withstand projected changes
Droughts	Evidence indicates that parts of South Asia have become drier since the 1970s with an increase in the number of droughts	Droughts are expected to be more frequent in some areas, especially in northwestern India, Jharkhand, Orissa and Chhattisgarh	Investments in Research and Development on the development of drought-resistant crops can help reduce some of the negative impacts

(continued)

Table 1.2 (continued)

Characteristic	What we know	What could happen	What can be done
	Droughts have major consequences. In 1987 and 2002–2003, droughts affected more than half of India's crop area and led to a huge fall in crop production	Crop yields are expected to fall significantly because of extreme heat by the 2040s	
Groundwater	More than 60% of India's agriculture is rain-fed, making the country highly dependent on groundwater Even without climate change, 15% of India's groundwater resources are overexploited	Although it is difficult to predict future groundwater levels, falling water tables can be expected to reduce further on account of increasing demand for water from a growing population, more affluent lifestyles, as well as from the services sector and industry	The efficient use of groundwater resources will need to be incentivized
Glacier melt	Glaciers in the northwestern Himalayas and in the Karakoram range—where westerly winter winds are the major source of moisture—have remained stable or even advanced On the other hand, most Himalayan glaciers—where a substantial part of the moisture is supplied by the summer monsoon—have been retreating over the past century	At 2.5 °C warming, melting glaciers and the loss of snow cover over the Himalayas are expected to threaten the stability and reliability of northern India's primarily glacier-fed rivers, particularly Indus and Brahmaputra. Ganges will be less dependent on meltwater due to high annual rainfall downstream during the monsoon season Indus and Brahmaputra are expected to see increased flows in spring when the snows melt, with flows reducing subsequently in late spring and summer Alterations in the flows of Indus, Ganges, and Brahmaputra rivers could significantly impact irrigation, affecting the	Major investments in water storage capacity would be needed to benefit from increased river flows in spring and compensate for lower flows later on

(continued)

Table 1.2 (continued)

Characteristic	What we know	What could happen	What can be done
		amount of food that can be produced in their basins as well as the livelihoods of millions of people (209 million in the Indus basin, 478 million in the Ganges basin and 62 million in the Brahmaputra basin in the year 2005)	
Sea level rise	Mumbai has the world's largest population exposed to coastal flooding, with large parts of the city built on reclaimed land, below the high-tide mark. Rapid and unplanned urbanization further increases the risks of sea water intrusion	With India close to the equator, the subcontinent would see much higher rises in sea levels than higher latitudes Sea level rise and storm surges would lead to saltwater intrusion in the coastal areas, impacting agriculture, degrading groundwater quality, contaminating drinking water, and possibly causing a rise in diarrhea cases and cholera outbreaks, as the cholera bacterium survives longer in saline water Kolkata and Mumbai, both the densely populated cities, are particularly vulnerable to the impacts of sea level rise, tropical cyclones, and riverine flooding	Building codes will need to be strictly enforced and urban planning will need to prepare for climate-related disasters Coastal embankments will need to be built where necessary and Coastal Regulation Zone codes enforced strictly
Agriculture and food security	Even without climate change, world food prices are expected to increase due to growing populations and rising incomes, as well as a greater demand for biofuels Rice: While overall rice yields have increased, rising temperatures with	Seasonal water scarcity, rising temperatures, and intrusion of sea water would threaten crop yields, jeopardizing the country's food security If current trends persist, substantial yield reductions in both rice and wheat can be	Crop diversification, more efficient water use, and improved soil management practices, together with the development of drought-resistant crops, can help reduce some of the negative impacts

(continued)

Table 1.2 (continued)

Characteristic	What we know	What could happen	What can be done
	lower rainfall at the end of the growing season have caused a significant loss in India's rice production. Without climate change, average rice yields could have been almost 6% higher (75 million ton in absolute terms) Wheat: Recent studies show that wheat yields peaked in India and Bangladesh around 2001 and have not increased since despite increasing fertilizer applications. Observations show that extremely high temperatures in Northern India—above 34 °C—have had a substantial negative effect on wheat yields, and rising temperatures can only aggravate the situation	expected in the near and medium term Under 2 °C warming by the 2050s, the country may need to import more than twice the amount of food grain than would be required without climate change	
Energy security	Climate-related impacts on water resources can undermine the two dominant forms of power generation in India—hydropower and thermal power generation—both of which depend on adequate water supplies to function effectively To function at full efficiency, thermal power plants need a constant supply of fresh cool water to maintain their cooling systems	The increasing variability and long-term decreases in river flows can pose a major challenge to hydropower plants and increase the risk of physical damage from landslides, flash floods, glacial lake outbursts, and other climate-related natural disasters Decreases in the availability of water and increases in temperature will pose major risk factors to thermal power generation	Projects will need to be planned taking into account climatic risks
Water security	Many parts of India are already experiencing water stress. Even without climate change,	An increase in variability of monsoon rainfall is expected to increase	Improvements in irrigation systems, water harvesting techniques, and more efficient

(continued)

Table 1.2 (continued)

Characteristic	What we know	What could happen	What can be done
	satisfying future demand for water will be a major challenge Urbanization, population growth, economic development, and increasing demand for water from agriculture and industry are likely to aggravate the situation further	water shortages in some areas Studies have found that the threat to water security is very high over Central India, along the mountain ranges of the Western Ghats and in India's northeastern states	agricultural water management can offset some of these risks
Health	Climate change is expected to have major health impacts in India—increasing malnutrition and related health disorders such as child stunting—with the poor likely to be affected most severely. Child stunting is projected to increase by 35% by 2050 compared to a scenario without climate change Malaria, dengue, yellow fever, Chagas disease, schistosomiasis, human African trypanosomiasis and leishmaniasis, and diarrheal infections are a major cause of child mortality that are likely to spread into areas where colder temperatures had previously limited transmission Heat waves are likely to result in a very substantial rise in mortality and death, and injuries from extreme weather events are likely to increase	Health systems will need to be strengthened in identified hotspots	Improvements in hydro-meteorological systems for weather forecasting and the installation of flood warning systems can help people move out of harm's way before weather-related disaster strikes Research and Development on vaccination to face the projected risks is a priority
Migration and conflict	South Asia is a hotspot for the migration of people from	Climate change impacts on agriculture and livelihoods can increase	Regional cooperation on water issues will be needed

(continued)

Table 1.2 (continued)

Characteristic	What we know	What could happen	What can be done
	disaster-affected or degraded areas to other national and international regions The Indus and the Ganges–Brahmaputra–Meghna Basins are major transboundary rivers, and increasing demand for water is already leading to tensions among countries over water sharing	the number of climate refugees	

1.8 Organization of the Book

This book is divided into various chapters and information about the chapters are as follows.

This chapter provides an introduction to climate change and variability, climate feedback, forcing mechanism, atmospheric chemistry, Palaeo records, monsoon variability, Holocene, Intergovernmental Panel on Climate Change (IPCC) scenarios, teleconnections, impact of climate change, and organization and utilization of the book. The chapter concludes with revision questions and exercise problems, advanced review questions, references, and suggested further reading. This sort of exercises is provided to all the chapters in the book with exception of Chap. 6 in which case studies are presented.

Chapter 2 describes GCMs and their choice, performance indicators for evaluating GCMs, weight estimation, multicriterion decision-making techniques in deterministic and fuzzy scenario, Spearman rank correlation coefficient, and group decision-making. Ensembling methodology of GCMs is also discussed.

Chapter 3 describes downscaling techniques. Detailed discussion is presented on statistical downscaling techniques such as multiple regression, artificial neural networks, Statistical Downscaling Model (SDSM), change factor technique, and support vector machine. Brief discussion on nested bias correction is also made.

Chapter 4 presents data compression techniques, namely, cluster and fuzzy cluster analysis, Kohonen neural networks, and principal component analysis. Trend detection techniques and optimization techniques, namely, linear and non-linear programming and genetic algorithms are also discussed.

Chapter 5 describes hydrological models, SWMM, HEC-HMS, SWAT, and other modeling techniques.

Chapter 6 presents various real-world global case studies in AR3 and AR5 perspectives that are related to the theories and techniques explained in the earlier chapters. Even through AR3 is relatively older than AR5, case studies are presented to understand the impact of climate change with temperature anomaly equivalent visualization paths, namely, SRES or RCPs.

Appendix A covers procedures for acquiring data from various sources. Appendices B and C provide representative list of journals and books related to climate. Index is also provided for efficient retrieval of topics.

PowerPoint presentations of selected topics are also provided as an additional study material. Interested individuals can contact publishers for PowerPoint presentations.

1.9 Utilization of the Book

The present book can help undergraduate as well as postgraduate programs in the field of hydrology, climate change, and allied fields, and can be referred as a textbook. It can also be used as a reference book or as supplementary study material for researchers working in this upcoming field. The case studies, PowerPoint presentations, extensive references, limited but informative and illustrative problems, and software information render this book as a valuable source of information for researchers, experts, professionals, teachers, and others who are interested in the field of climate, hydrology, and allied fields.

Next chapter presents the selection of Global Climate Models.

Revision Questions

1.1 What is hydroclimatology? What is its significance?
1.2 Define climate system. What are the various components of global climate system?
1.3 What is climate feedback? How many types of feedbacks are possible? What is their impact?
1.4 How many reports are published by IPCC up to year 2015?
1.5 What is the forcing mechanism?
1.6 What is the difference between (a) radiative forcing, and non-radiative forcing, and (b) random and periodic forcing, and (c) external and internal forcing?
1.7 What is aerosol? What is its impact on atmosphere?
1.8 What are greenhouse gases? What are the factors affecting the changes in the greenhouse gases in atmosphere?
1.9 What is the role of global climatic models?
1.10 What is the purpose of CMIP3 and CMIP5?
1.11 What are the roles of Palaeo records and atmospheric chemistry in climate change?
1.12 What is monsoon variability? How it may affect agriculture?
1.13 What is Holocene?
1.14 What is climate scenario?

1.15 What is story line? How many story lines exist in SRES perspective? What is their practical significance?
1.16 What are scenarios in AR5 perspective?
1.17 What is the purpose of Representative Concentration Pathways? Compare four Representative Concentration Pathways in terms of radioactive forcing, concentration and pathway shape?
1.18 What are La Niña, El Niño, and ENSO?, How ENSO is different as compared to El Niño and La Niña?
1.19 What are the possible teleconnections?
1.20 What is the impact of climate change on (a) surface and groundwater, (b) urbanization, and (c) hydrologic extremes?
1.21 What are the characteristics affecting climate situation in India?

Advanced Review Questions

1.22 What is the difference between CMIP3 and CMIP5?
1.23 What are examples of forcing agents?
1.24 How uncertainties in temperature affect the estimation of greenhouse gases or vice versa?
1.25 What are the guidelines for selection of climate scenario?
1.26 How climate scenarios help India to assess future greenhouse gases and water availability?
1.27 Among four Representative Concentration Pathways, which is suitable for India?
1.28 What are the limitations for using RCPs?
1.29 What is the difference between SRES scenarios and RCPs?
1.30 Do you feel that existing SRES and RCP scenarios are sufficient to assess future climate situations for India? If yes, justify how they are sufficient? If not, provide the reasons and possible improvements? Can you suggest story lines other than the proposed?
1.31 What are the teleconnections impacting Indian monsoon rainfall?
1.32 Is there any activity affected by climate change other than discussed in this chapter?
1.33 Which IPCC reports discussed about hydrologic extremes and water resources?
1.34 Mention names of the two places in India where climate change affected the water resources availability or any other impact preferably one case study in South India and one case study in North India?

References

Aerosols (2017). https://en.wikipedia.org/wiki/Aerosol. Accessed 31 Jan 2017

Anandhi A, Srinivas VV, Nagesh Kumar D (2013) Impact of climate change on hydrometeorological variables in a river basin in India for IPCC SRES scenarios. In: Rao YS, Zhang TC, Ojha CSP, Gurjar BR, Tyagi RD, Kao CM (eds) Climate change modeling, mitigation, and adaptation. American Society of Civil Engineers, pp 327–356

Ashok K, Guan Z, Saji NH, Yamagata T (2004) Individual and combined influences of ENSO and the Indian Ocean Dipole on the Indian summer monsoon. J Clim 17:3141–3155

Atmospheric Chemistry (2017). https://en.wikipedia.org/wiki/Atmospheric_chemistry. Accessed 31 Jan 2017

Barnston AG, Livezey RE (1987) Classification, seasonality and persistence of low-frequency atmospheric circulation patterns. Mon Weather Rev 115:1083–1126

Barnston AG, Livezey RE, Halpert MS (1991) Modulation of Southern Oscillation—Northern hemisphere mid-winter climate relationships by QBO. J Clim 4:203–217

Bates BC, Kundzewicz ZW, Wu S, Palutikof JP (eds) (2008) Climate change and water: technical paper of the intergovernmental panel on climate change. IPCC Secretariat, Geneva, Switzerland. http://ipcc.ch/pdf/technical-papers/climate-change-water-en.pdf. Accessed 31 Jan 2017

Bell GD, Basist AN (1994) Seasonal climate summary: the global climate of December 1992–January 1993: mature ENSO conditions continue in the tropical Pacific, California drought abates. J Clim 7:1581–1605

Bell GD, Janowiak JE (1995) Atmospheric circulation associated with the Midwest floods of 1993. Bull Am Met Soc 5:681–695

Bell GD, Chelliah M (2006) Leading tropical modes associated with interannual and multidecadal fluctuations in North Atlantic hurricane activity. J Clim 19:590–612

Berger AL (1978) Long-term variations of caloric insolation resulting from the Earth's orbital elements. Quatern Res 9(2):139–167

Bergthórsson P, Björnsson H, Dórmundsson O, Gudmundsson B, Helgadóttir A, Jónmundsson JV (1988) The effects of climatic variations on agriculture in Iceland. In: Parry ML, Carter TR, Konijn NT (eds) The impact of climatic climate scenario development variations on agriculture, vol 1, Assessments in cool temperate and cold regions. Kluwer, The Netherlands, pp 381–509

Bhatt D, Mall RK (2015) Surface water resources, climate change and simulation modeling. Aquat Procedia 4:730–738

Blanford HF (1886) Rainfall of India. Mem India Meteorol Dept 2:217–448

Brekke LD, Kiang JE, Olsen JR, Pulwarty RS, Raff DA, Turnipseed DP, Webb RS, White KD (2009) Climate change and water resources management—a federal perspective: U.S. Geological Survey Circular 1331, p 65. http://pubs.usgs.gov/circ/1331/. Accessed 31 Jan 2017

Chahine MT (1992) The hydrological cycle and its influence on climate. Nature 359–379

Charney JG (1969) The intertropical convergence zone and the Hadley circulation of the atmosphere. In: Proceedings of WMO/IUCG symposium on numerical weather prediction, vol III. Japan Meteorological Agency, pp 73–79

Climate Prediction Center (2017). http://www.cpc.ncep.noaa.gov/data/teledoc/telecontents.shtml. Accessed 31 Jan 2017

Criteria for Selecting Climate Scenarios (2013). http://www.ipcc-data.org/guidelines/pages/scen_selection.html. Accessed 31 Jan 2017

Definition of Hydrologic (2017). http://www.thefreedictionary.com/hydrology. Accessed 31 Jan 2017

deMenocal P (2001) Cultural responses to climate change during the Late Holocene. Science 292:667–673

Easterling DR, Meehl GA, Parmesan C, Changnon SA, Karl TR, Mearns LO (2000) Climate extremes: observations, modeling, and impacts. Science 289:2068–2074

Gadgil S, Abrol YP, Rao PRS (1999) On growth and fluctuation of Indian food grain production. Curr Sci 76(4):56–548

Gadgil S (2003) The Indian monsoon and its variability. Annu Rev Earth Planet Sci 31:429–467

Gadgil S, Vinayachandran PN, Francis PA (2003) Droughts of the Indian summer monsoon: role of clouds over the Indian Ocean. Curr Sci 85(12):1713–1719

Gadgil S, Vinayachandran PN, Francis PA, Gadgil S (2004) Extremes of the Indian summer monsoon rainfall, ENSO and equatorial Indian ocean oscillation. Geophys Res Lett 31:L12213

Global Change Hydrology Program: Hydroclimatology (2017). http://water.usgs.gov/osw/programs/globalchange.html. Accessed 31 Jan 2017

Glossary Relevant to Climate (2017). http://glossary.ametsoc.org/wiki/Main_Page. Accessed 31 Jan 2017

Glossary Relevant to RCPs (2017). http://www.ipcc-data.org/guidelines/pages/glossary/glossary_r.html. Accessed 31 Jan 2017

Goodess CM, Palutikof JP, Davies TD (1992) The nature and causes of climate change: assessing the long term future. Belhaven Press, London, p 248

Gosain AK, Rao S, Arora A (2011) Climate change impact assessment of water resources of India. Curr Sci 101:356–371

Green TR, Taniguchi M, Kooi H, Gurdak JJ, Allen DM, Hiscock KM, Treidel H, Aureli A (2011) Beneath the surface of global change: impacts of climate change on groundwater. J Hydrol 405:532–560

Gu H, Yu Z, Wang G, Wang J, Ju Q, Yang C, Fan C (2015) Impact of climate change on hydrological extremes in the Yangtze River Basin, China. Stoch Environ Res Risk Assess 29:693–707

Gupta AK, Anderson DM, Pandey DN, Singhvi AK (2006) Adaptation and human migration, and evidence of agriculture coincident with changes in the Indian summer monsoon during the Holocene. Curr Sci 90(8):1082–1090

Halley E (1686) An historical account of the trade winds and monsoons observable in the seas between and near the tropics with an attempt to assign a physical cause of the said winds. Philos Trans R Soc Lond 16:153–168

Hansen J, Sato M, Lacis A, Ruedy R, Tegen I, Matthews E (1998) Climate forcings in the industrial era. Proc Natl Acad Sci 95:12753–12758

Haylock M, Nicholls N (2000) Trends in extreme rainfall indices for an updated high quality data set for Australia, 1910–1998. Int J Climatol 20:1533–1541

Holocene (2017). https://en.wikipedia.org/wiki/Holocene. Accessed 31 Jan 2017

Houghton JT, Filho MLG, Callander BA, Harris N, Kattenburg A, Maskell K (1996) Climate change 1995: the science of climate change, WGI to the second assessment report of the intergovernmental panel on climate change. Cambridge University Press, Cambridge, p 584

India: Climate Change Impacts (2017). http://www.worldbank.org/en/news/feature/2013/06/19/india-climate-change-impacts. Accessed 31 Jan 2017

India Environment Portal (2017) Climate change impacts on water resources in India. http://www.indiaenvironmentportal.org.in/files/india-climate-5-water-DEFRA.pdf. Accessed 31 Jan 2017

Integrated Assessment Modeling Consortium (2017). http://www.globalchange.umd.edu/iamc/. Accessed 31 Jan 2017

IPCC Special Report: Emission Scenarios (2000). https://www.ipcc.ch/pdf/special-reports/spm/sres-en.pdf. Accessed 31 Jan 2017

IPCC (2001) The scientific basis: third assessment report of the intergovernmental panel on climate change. Cambridge University Press, Cambridge

IPCC (2007) The physical science basis: fourth assessment report of the intergovernmental panel on climate change. Cambridge University Press, Cambridge

IPCC (2008) Special report on managing the risks of extreme events and disasters to advance climate change adaptation. http://www.ipcc-wg2.gov/SREX/images/uploads/SREX-All_FINAL.pdf. Accessed 31 Jan 2017

IPCC (2012) Managing the risks of extreme events and disasters to advance climate change adaptation. A special report of working groups I and II of the intergovernmental panel on climate change. Cambridge University Press, Cambridge, p 582

IPCC (2014) The physical science basis: working group 1: contribution to the IPCC fifth assessment report. Cambridge University Press, Cambridge

Islam A, Sikka AK (2010) Climate change and water resources in india: impact assessment and adaptation strategies. In: Jha MK (ed) Natural and anthropogenic disasters: vulnerability, preparedness and mitigation. Springer, The Netherlands and Capital Publishing Company, New Delhi, pp 386–412

Jain S, Lall U (2001) Floods in a changing climate: does the past represent the future? Water Resour Res 37(12):3193–3205

Kane RP (1998) ENSO relationship to the rainfall of Sri Lanka. Int J Climatol 18(8):859–872

Khain AP, Benmoshe N, Pokrovsky A (2008) Factors determining the impact of aerosols on surface precipitation from clouds: attempt of classification. J Atmos Sci 65:1721–1748

Kumar KK, Soman MK, Rupakumar K (1995) Seasonal forecasting of Indian summer monsoon rainfall: a review. Weather 150:449–467

Lankao PR (2008) Urban areas and climate change: review of current issues and trends. http://www.ral.ucar.edu/staff/prlankao/GRHS_2011_IssuesPaperfinal.pdf. Accessed 31 Jan 2017

Maity R, Nagesh Kumar D (2006) Bayesian dynamic modeling for monthly Indian summer monsoon rainfall using El Niño Southern Oscillation (ENSO) and Equatorial Indian Ocean Oscillation (EQUINOO). J Geophys Res 111:D07104

Maity R, Nagesh Kumar D (2007) Hydroclimatic teleconnection between global sea surface temperature and rainfall over India at subdivisional monthly scale. Hydrol Process 21 (14):1802–1813

Maity R, Nagesh Kumar D, Nanjundiah RS (2007) Review of hydroclimatic teleconnection between hydrologic variables and large-scale atmospheric circulation indices with Indian perspective. ISH J Hydraul Eng 13(1):77–92

Mearns LO, Hulme M, Carter TR, Leemans R, Lal M, Whetton P (2001) Climate scenario development. In: Houghton JT, Ding Y, Griggs DJ, Noguer M, van der Linden PJ, Xiaosu D, Maskell K (eds) Climate change 2001: the scientific basis. Cambridge University Press, Cambridge, pp 739–768

Milly PCD, Bettencourt J, Falkenmark M, Hirsch RM, Kundezewicz ZW, Lettenmaier DP, Stouffer RJ (2008) Stationarity is dead—whither water management? Science 319:573–574

Moss RH, Edmonds JA, Hibbard KA, Manning MR, Rose SK, Vuuren DPV, Carter TR, Emori S, Kainuma M, Kram T, Meehl GA, Mitchell JFB, Nakicenovic N, Riahi K, Smith SJ, Stouffer RJ, Thomson AM, Weyant JP, Wilbanks TJ (2010) The next generation of scenarios for climate change research and assessment. Nature 463:747–756

Nakicenovic N, Alcamo J, Davis G, de Vries B, Fenhann J, Gaffin S, Gregory K, Grübler A, Jung TY, Kram T, La Rovere EL, Michaelis L, Mori S, Morita T, Pepper W, Pitcher H, Price L, Raihi K, Roehrl A, Rogner HH, Sankovski A, Schlesinger M, Shukla P, Smith S, Swart R, van Rooijen S, Victor N, Dadi Z (2000) IPCC special report on emissions scenarios. Cambridge University Press, Cambridge, p 599

Nanjundiah RS, Francis PA, Ved M, Gadgil S (2013) Predicting the extremes of Indian summer monsoon rainfall with coupled ocean—atmosphere models. Curr Sci 104(10):1380–1393

National Ocean Service (2017). http://oceanservice.noaa.gov/facts/ninonina.html. Accessed 31 Jan 2017

Natural Resources Defense Council (2010) Climate change and water resource management: adaptation strategies for protecting people and the environment. https://www.nrdc.org/water/files/waterandclimate.pdf. Accessed 31 Jan 2017

Palaeo Records (2017). http://climatica.org.uk/climate-science-information/palaeo-records. Accessed 31 Jan 2017

Panigrahy BP, Singh PK, Tiwari AK, Kumar B (2015) Impact of climate change on groundwater resources. Int Res J Environ Sci 4(3):86–92

Pant GB, Parthasarathy B (1981) Some aspects of an association between the southern oscillation and Indian summer monsoon. Arch Meteorol Geophys Bioclimatol Ser B 29:245–251

Panwar S, Chakrapani GJ (2013) Climate change and its influence on groundwater resources. Curr Sci 105(1):37–46

Parthasarathy B, Diaz HF, Eischeid JK (1988) Prediction of all India summer monsoon rainfall with regional and large-scale parameters. J Geophys Res 93(5):5341–5350

Philander SG (1985) El Niño and La Niña. J Atmos Sci 42:2652–2662

Philander SG (1990) El Niño, La Niña, and the Southern Oscillation. Academic Press, London, p 289

Ramachandran R (2007) Cloud of mystery: the atmosphere over the Indian Ocean perhaps holds the key to seasonal rainfall, and scientists are just beginning to understand its role. http://www.frontline.in/static/html/fl2421/stories/20071102506009600.htm. Accessed 31 Jan 2017

Rao GN (1997) Interannual variations of monsoon rainfall in Godavari river basin-connections with the southern oscillation. J Clim 11:768–771

Refsgaard JC, Sonnenborg TO, Butts MB, Christensen JH, Christensen S, Drews M, Jensen KH, Jørgensen F, Jørgensen LF, Larsen MAD, Rasmussen SH, Seaby LP, Seifert D, Vilhelmsen TN (2016) Climate change impacts on groundwater hydrology—where are the main uncertainties and can they be reduced? Hydrol Sci J. doi:10.1080/02626667.2015.1131899

Representative Concentration Pathways, Part 1: An Introduction to Scenarios (2017). http://www.skepticalscience.com/rcp.php?t=1. Accessed 31 Jan 2017

Representative Concentration Pathways, Part 2: Creating New Scenarios (2017). http://www.skepticalscience.com/rcp.php?t=2. Accessed 31 Jan 2017

Representative Concentration Pathways, Part 3: RCP Technical Summary (2017). http://www.skepticalscience.com/rcp.php?t=3. Accessed 31 Jan 2017

Representative Concentration Pathways Description (2017). http://sedac.ipcc-data.org/ddc/ar5_scenario_process/RCPs.html. Accessed 31 Jan 2017

Saraswat R, Naik DK, Nigam R, Gaur AS (2016) Timing, cause and consequences of mid-Holocene climate transition in the Arabian Sea. Quatern Res 86(2):162–169

Satheesh SK, Srinivasan J (2002) Enhanced aerosol loading over Arabian Sea during pre-monsoon season: natural or anthropogenic? Geophys Res Lett doi:10.1029/2002GL015687

Satheesh SK, Lubin D (2003) Short wave versus long wave radiative forcing due to aerosol over Indian Ocean: role of sea-surface winds. Geophys Res Lett 30(13): Art no 1695

Satheesh SK, Krishna Moorthy K (2005) Radiative effects of natural aerosols: a review. Atmos Environ 39(11):2089–2110

Satheesh SK (2006) Pollution, aerosols, and the climate, leader page articles. The Hindu, Published on 04 Sept 2006

Scenario Process for AR5 (2017). http://sedac.ipcc-data.org/ddc/ar5_scenario_process/scenario_overview.html. Accessed 31 Jan 2017

Sedlacek A, Lee J (2007) Photothermal interferometric aerosol absorption spectrometry. Aerosol Sci Technol 41(12):1089–1101

Shah T (2009) Climate change and groundwater: India's opportunities for mitigation and adaptation. Environ Res Lett 4(3):13

Shelton ML (2009) Hydroclimatology: perspectives and applications. Cambridge University Press, Cambridge

Shine KP, Derwent RG, Wuebbles DJ, Morcette JJ (1990) Radiative forcing of climate. In: Houghton JT, Jenkins GJ, Ephraums JJ (eds) Climate change: the IPCC scientific assessment, intergovernmental panel on climate change (IPCC). Cambridge University Press, Cambridge, pp 41–68

Shukla J, Paolino DA (1983) The southern oscillation and long-range forecasting of the summer monsoon rainfall over India. Mon Weather Rev 111:1830–1837

Siebert S, Burke J, Faures JM, Frenken K, Hoogeveen J, Döll P, Portmann FT (2010) Groundwater use for irrigation—a global inventory. Hydrol Earth Syst Sci 14:1863–1880

Sikka DR (1980) Some aspects of the large-scale fluctuations of summer monsoon rainfall over India in relation to fluctuations in the planetary and regional scale circulation parameters. Proc Indian Acad Sci—Earth Planet Sci 89(2):179–195

Singhvi AK, Kale VS (2009) Paleoclimate studies in India: last iceage to the present. Indian National Science Academy ICRPWCRP-SCOPE report series 4, New Delhi

Singhvi AK, Krishnan R (2014) Past and the present climate of India. In: Kale VS (ed) Landscapes and landforms of India, world geomorphological landscapes. Springer, pp 15–23

Smith JB, Hulme M (1998) Climate change scenarios in Chapter 3. In: Feenstra J, Burton I, Smith JB, Tol RSJ (eds) Handbook on methods of climate change impacts assessment and adaptation strategies, October, UNEP/IES, Amsterdam

SRES Emissions Scenarios (2017a). http://sedac.ipcc-data.org/ddc/sres/index.html. Accessed 31 Jan 2017

SRES Emissions Scenarios (2017b). http://www.ipcc.ch/ipccreports/sres/emission/index.php?idp=3. Accessed 31 Jan 2017

Srivastava P, Singh IB, Sharma S, Shukla UK, Singhvi AK (2003) Late Pleistocene–Holocene hydrologic changes in the interfluve areas of the central Ganga Plain, India. Geomorphology 54279–54292

Taye MT, Willems P, Block P (2015) Implications of climate change on hydrological extremes in the Blue Nile basin: a review. J Hydrol: Reg Stud 4:280–293

Tohver IM, Hamlet AF, Lee SY (2014) Impacts of 21st-century climate change on hydrologic extremes in the Pacific Northwest Region of North America. J Am Water Resour Assoc 50 (6):1461–1476

Treidel H, Martin-Bordes JL, Gurdak J (eds) (2012) Climate change effects on groundwater resources: a global synthesis of findings and recommendations. CRC Press/Balkema

Umbgrove JHF (1947) The pulse of the Earth. Springer, Netherlands, p 385

Vuuren DPV, Edmonds J, Kainuma M, Riahi K, Thomson A, Hibbard K, Hurtt GC, Kram T, Krey V, Lamarque JF, Masui T, Meinshausen M, Nakicenovic N, Smith SJ, Rose SK (2011) The representative concentration pathways: an overview. Clim Change 109:5–31

Wang C, Deser C, Yu J, DiNezio P, Clement A (2012) El Niño and Southern Oscillation (ENSO): a review. http://www.aoml.noaa.gov/phod/docs/ENSO_Revision.pdf. Accessed 31 Jan 2017

Wateraid (2017) Climate Change and Water Resources. http://www.wateraid.org/. Accessed 31 Jan 2017

Wayne G (2013) The beginner's guide to representative concentration pathways. Skeptical Sci

Webster PJ (1987) The elementary monsoon. In: Fein JS, Stephens PL (eds) Monsoons. Wiley, New York, pp 3–32

Suggested Further Reading

Bai X (2003) The process and mechanism of urban environmental change: an evolutionary view. Int J Environ Pollut 19(5):528–541

Betsill MM, Bulkeley H (2007) Looking back and thinking ahead: a decade of cities and climate change research. Local Environ 12(5):447–456

Klein Tank AMG, Peterson TC, Quadir DA, Dorji S, Zou X, Tang H, Santhosh K, Joshi UR, Jaswal AK, Kolli RK, Sikder AB, Deshpande NR, Revadekar JV, Yeleuova K, Vandasheva S, Faleyeva M, Gomboluudev P, Budhathoki KP, Hussain A, Afzaal M, Chandrapala L, Anvar H, Amanmurad D, Asanova VS, Jones PD, New MG, Spektorman T (2006) Changes in daily temperature and precipitation extremes in central and South Asia. J Geophys Res 111:D16105

Knutti R, Sedlacek J (2013) Robustness and uncertainties in the new CMIP5 climate model projections. Nat Clim Change 3:369–373

Konikow LF (2011) Contribution of global groundwater depletion since 1900 to sea-level rise. Geophys Res Lett 38:L17401

Manton MJ, Della-Marta PM, Haylock MR, Hennessy KJ, Nicholls N, Chambers LE, Collins DA, Daw G, Finet A, Gunawan D, Inape K, Isobe H, Kestin TS, Lefale P, Leyu CH, Lwin T, Maitrepierre L, Ouprasitwong N, Page CM, Pahalad J, Plummer N, Salinger MJ, Suppiah R, Tran VJ, Trewin B, Tibig I, Yee D (2001) Trend in extreme daily rainfall and temperature in Southeast Asia and South Pacific: 1961–1998. Int J Climatol 21:269–284

McGranahan G, Balk D, Anderson B (2007) The rising tide: assessing the risks of climate change and human settlements in low-elevation coastal zones. Environ Urban 19(1):17–37

Muller M (2007) Adapting to climate change: water management for urban resilience. Environ Urban 9(1):99–113

Preethi B, Kripalani RH (2010) Indian summer monsoon rainfall variability in global coupled ocean-atmospheric models. Clim Dyn 35:1521–1539

Rasmusson EM, Carpenter TH (1983) The relationship between the eastern Pacific sea surface temperature and rainfall over India and Sri Lanka. Mon Weather Rev 111:84–354

Sherif MM, Singh VP (1999) Effect of climate change on sea water intrusion in coastal aquifers. Hydrol Process 13:1277–1287

WWAP (2009) The United Nations World Water Development Report 3: water in a changing world, World water assessment programme. Paris, UNESCO Publishing, UNESCO, p 349

Selection of Global Climate Models

2

Abstract

This chapter describes Global Climate Models (GCMs), limitations and uncertainties associated with the formulation of GCMs due to the effect of aerosols which are differently parameterized in GCMs, initial and boundary conditions for each GCM, parameters and model structure of GCMs, randomness, future greenhouse gas emissions, and scenarios leading to significant variability across model simulations of future climate. This chapter discussed the necessity of performance indicators for evaluating GCMs and explained mathematical description of these indicators. It also emphasized on normalization approach, weight computing techniques such as entropy and rating, ranking approaches, namely, compromise programming, cooperative game theory, TOPSIS, weighted average, PROMETHEE, and fuzzy TOPSIS. Spearman rank correlation which measures consistency in ranking pattern and group decision-making that aggregates individual rankings obtained by different techniques to form a single group preference is also part of this chapter. Ensembling methodology of GCMs is also discussed. Reader is expected to understand various uncertainties associated, role of decision-making techniques for ranking of GCMs by studying this chapter.

Keywords

Correlation · GCMs · Group decision · Normalization · Performance indicators
Ranking · Uncertainty · Weight

Electronic supplementary material The online version of this chapter (doi:10.1007/978-981-10-6110-3_2) contains supplementary material, which is available to authorized users.

K. Srinivasa Raju and D. Nagesh Kumar, *Impact of Climate Change on Water Resources*, Springer Climate, https://doi.org/10.1007/978-981-10-6110-3_2

2.1 Introduction

Multicriterion decision-making (MCDM) techniques are capable of selecting the best global climate model (GCM). Selected GCMs can be explored for applications, e.g., downscaling and adaptation studies (Bogardi and Nachtnebel 1994). The present chapter discussed about GCMs, normalization, and weight estimation techniques, MCDM techniques, Spearman rank correlation coefficient, group decision-making, and methodology of ensembling of GCMs. The discussed techniques are demonstrated with numerical examples in climate modeling situations. In the present book, methods and techniques are used interchangeably.

2.2 Global Climate Models

The Earth climate system is the result of interaction between various components such as atmosphere, snow, ice, land surface, ocean, and other water bodies, and living beings like humans and animals. Human-induced factors such as deforestation and burning of fossil fuels also lead to change in atmospheric composition in addition to various external factors. Due to this there will be climate change, which affects the variability of the parameters that continue to exist for a long period. For example, nitrous oxide, carbon dioxide, methane, ozone, hydrofluorocarbons, and sulfur hexafluoride are the greenhouse gases that are increasing and affecting global temperature over the past half century and are expected to follow the similar trend in future. These greenhouse gases affect the absorption, scattering, and emission of radiation in the Earth surface and atmosphere. These aspects necessitated to study the relationship between greenhouse gases and the global climate. This is possible through climate modeling or simply climate models. Various relevant climate models are described by Thompson and Perry (1997), Goudie and Cuff (2001), and Kendal and Henderson-Sellers (2013):

- Energy balance models are one dimensional in nature, which relate latitude and sea surface-level temperature variation.
- Radiative-convective models are one dimensional in nature. They analyze vertical temperature, explicit profile modeling of radiative process and convective adjustment.
- General circulation or global climate models (GCMs) are sophisticated numerical tools, which are three dimensional in nature. They simulate Earth's climate with different climate variables, initial and boundary conditions, and structure. GCMs are increasingly being employed to solve or to assess regional/local issues (What is a GCM 2013). Wilby et al. (2009) described GCMs as numerical solutions of a partial differential equation(s). GCMs are formulated on the principles of movement of energy, momentum of a particle, and conservation of mass.

- Coupled atmosphere–ocean global climate models combine the interactions of the atmospheric GCMs and oceanic GCMs.

Xu (1999) mentioned that GCMs are found to be capable of projecting average precipitation, temperature, etc., over future decades or centuries. He and numerous researchers, however, cautioned about the limitations of GCMs such as

- Accuracy of GCMs, which generally run at coarse grid resolution ($\sim 3^\circ \times 3^\circ$), decreases with increasingly finer spatial and temporal scales, rendering them unable to represent sub-grid-scale features. In other words, GCMs are not able to effectively model sub-grid-scale processes which are of prime interest to hydrologists and water resources planners.
- Accuracy of GCMs decreases from free tropospheric to surface variables, whereas surface variables have significant application in water balance computations.

The uncertainties associated with the formulation of GCMs arise due to the effect of aerosols which are differently parameterized in GCMs, initial and boundary conditions for each GCM, parameter and model structure of GCMs, randomness, future greenhouse gas emissions, and Representative Concentration Pathways (RCPs) leading to significant variability across model simulations of future climate (Raje and Mujumdar 2010). These uncertainties accumulate from various levels such as GCM to downscaling level and may propagate to the local levels, which may affect the adaptation studies that would be used as the basis for implementation. In brief, the uncertainty begins at selection of suitable GCMs, selection of downscaling technique(s), and selection of suitable hydrologic model(s). Numerous authors suggested uncertainty minimization by employing ensembling of relevant models, which may provide increased confidence while projecting climate change. For example, instead of single GCM output, output of multiple GCMs can be used as the basis to feed inputs to downscaling techniques. Similarly outputs from some of the downscaling techniques such as Statistical Downscaling Modeling (SDSM) (Wilby et al. 2002), Multiple Linear Regression (MLR), and Artificial Neural Networks (ANN) can be ensembled (or these outputs can be studied) and can be passed on to hydrologic models. Similar experimentation can be performed in hydrologic modeling where few hydrologic models can be coupled or even hybridized (Semenov and Stratonovitch 2010).

2.3 Performance Indicators for Evaluating GCMs

Pierce et al. (2009) in their studies raised the issues such as "What effect does picking different global models have on the regional climate study? If different global models give different downscaled results what strategy should be used for selecting the global models? Are there overall strategies that can be used to guide the choice of models?" The above queries necessitated to evaluate the available GCMs for accuracy and their adaptability (Legates and McCabe 1999). Hence, the GCMs are to be evaluated to assess their performance by simulating the historic observations. This enables to choose GCMs of higher performance so that the relevant output obtained from the suitable/best GCMs can be used for further analysis (Mujumdar and Nagesh Kumar 2012).

A performance indicator is a measure of any GCM to determine how well it simulates the observed data. Simple, effective, and meaningful metrics are required to evaluate the GCMs across space and time and to evolve a subset of models that can be employed for hydrologic modeling applications. These indicators may provide the basis to assess the confidence level of outputs of GCMs (Helsel and Hirsch 2002; Gleckler et al. 2008; Johnson and Sharma 2009; Macadam et al. 2010; Wilks 2011; Sonali and Nagesh Kumar 2013; Ojha et al. 2014).

Different researchers used various performance indicators, such as, Sum of Squares of Deviation (SSD), Mean Square Deviation (MSD), Root Mean Square Deviation (RMSD), Normalized Root Mean Square Deviation (NRMSD), Absolute Normalized Mean Bias Deviation (ANMBD), Average Absolute Relative Deviation (AARD), Pearson Correlation Coefficient (CC), Nash–Sutcliffe Efficiency (NSE), and Skill Score (SS). Among all, SSD, MSD, RMSD, NRMSD, ANMBD, AARD are of deviation/error category. The mathematical descriptions of these indicators are as follows:

(a) SSD is addition of the squared difference between the observed values and the GCM-simulated values. x_i and y_i are observed and simulated values respectively. T is number of datasets.

$$SSD = \sum_{i=1}^{T} (x_i - y_i)^2 \tag{2.1}$$

(b) MSD is the average of squares of deviation.

$$MSD = \frac{1}{T}\sum_{i=1}^{T} (x_i - y_i)^2 \tag{2.2}$$

(c) RMSD is square root of mean square of deviation.

$$RMSD = \sqrt{\frac{1}{T}\sum_{i=1}^{T} (x_i - y_i)^2} \tag{2.3}$$

(d) NRMSD is the ratio of RMSD and mean of observed values. Less value of NRMSD is preferred.

$$NRMSD = \frac{\sqrt{\frac{1}{T}\sum_{i=1}^{T}(x_i - y_i)^2}}{\bar{x}} \tag{2.4}$$

(e) ANMBD is ratio of the mean of the differences between the observed and the GCM-simulated values to the mean of observed values. Less value of ANMBD is preferred.

$$ANMBD = \left| \frac{\frac{1}{T}\left(\sum_{i=1}^{T}(y_i - x_i)\right)}{\bar{x}} \right| \tag{2.5}$$

(f) AARD is the mean of the absolute values of relative deviation. Less value of AARD is preferred.

$$AARD = \frac{1}{T}\sum_{i=1}^{T}|ARD_i|; \quad \text{where } ARD_i = \frac{(y_i - x_i)}{x_i} \tag{2.6}$$

(g) CC relates strength of the linear relationship between the observed and the GCM-simulated values. Here, $\bar{x}, \bar{y}$ are average of observed and simulated values, whereas σ_{obs} and σ_{sim} are the standard deviations. CC value near to 1.0 indicates good model performance. In all chapters wherever applicable, the word "correlation coefficient" is used as a generalization.

$$CC = \frac{\sum_{i=1}^{T}(x_i - \bar{x})(y_i - \bar{y})}{(T-1)\sigma_{obs}\sigma_{sim}} \tag{2.7}$$

(h) NSE is defined (Nash and Sutcliffe 1970) as:

$$\text{NSE} = 1 - \frac{\sum_{i=1}^{T}(x_i - y_i)^2}{\sum_{i=1}^{T}(x_i - \bar{x})^2} \tag{2.8}$$

NSE ranges from $-\infty$ to 1. If a model simulates the observed conditions perfectly, NSE value will be 1 (Nash–Sutcliffe Efficiency 2017).

(i) SS (Maximo et al. 2008) measures the similarity between the Probability Density Functions (PDFs) of the observed and simulated values across the entire PDF and expressed as

$$SS = \frac{1}{T}\sum_{i=1}^{nb} \min(f_{m,} f_o) \quad (2.9)$$

where nb is number of bins used to calculate the PDF for a given region. f_m, f_o are the frequencies of values in the given bin from the chosen GCM and of the observed values. Skill score varies between zero and one.

Numerical Problem 2.1 Global climate model-simulated temperature data (in °K) in a given region in India along with the observed/historic data are presented in Table 2.1. Compute the performance of the GCM for its simulating capability with that of historic data in terms of SSD, MSD, RMSD, CC, NRMSD, ANMBD, AARD, NSE, and SS.

Table 2.1 Historic/observed data and simulated data by GCM

Datasets	Historic/observed data (°K)	GCM-simulated data (°K)
1	243	244
2	244	248
3	245	251
4	246	258
5	247	248
6	248	264
7	248	253
8	249	252
9	249	253
10	250	264
11	250	256
12	251	256
13	253	265
14	254	245
15	255	258
16	256	267
17	257	259
18	258	261
19	259	260
20	260	260
21	262	266
22	263	270
23	265	266
24	269	264
25	270	273
26	271	270
27	273	272

Solution:
Notation:

x_i	Observed temperature value (°K)
y_i	Simulated value (°K)
$\bar{x}$	Mean of the observed values (°K)
$\bar{y}$	Mean of simulated values (°K)
T	Number of observations recorded
σ_{obs}	Standard deviation of the observed value set (°K)
σ_{sim}	Standard deviation of simulated value set (°K)

Estimated parameters

$$\bar{x} = 255.37\,°\mathrm{K} \qquad \sum_{i=1}^{T} (x_i - y_i)^2 = 1232\,°\mathrm{K}^2$$

$$\bar{y} = 259.37\,°\mathrm{K} \qquad \sum_{i=1}^{T} (x_i - \bar{x})^2 = 2016.29\,°\mathrm{K}^2$$

$$\sum_{i=1}^{T} (y_i - \bar{y})^2 = 1750.29\,°\mathrm{K}^2 \qquad \sum_{i=1}^{T} (x_i - \bar{x})(y_i - \bar{y}) = 1483.30\,°\mathrm{K}^2$$

Standard deviation of a set of observations is calculated by the following formula:

$$\sigma = \sqrt{\frac{\sum_{i=1}^{T} (x_i - \bar{x})^2}{(T-1)}}$$

$$\text{Standard deviation of observed data} = \sigma_{obs} = \sqrt{\frac{2016.29}{26}} = 8.8062\,°\mathrm{K}$$

$$\text{Standard deviation of simulated data} = \sigma_{sim} = \sqrt{\frac{1750.29}{26}} = 8.2048\,°\mathrm{K}$$

Computation of performance indicators (Refer Table 2.2)

Table 2.2 Computation of relevant parameters

Dataset	x_i	y_i	$(x_i - y_i)$	$(x_i - y_i)^2$	$(x_i - \bar{x})$	$(x_i - \bar{x})^2$	$(y_i - \bar{y})$	$(y_i - \bar{y})^2$	$(x_i - \bar{x}) * (y_i - \bar{y})$	$\left\lvert \frac{(y_i - x_i)}{x_i} \right\rvert$
1	243	244	−1	1	−12.3704	153.0261	−15.3704	236.2492	190.138	0.0041
2	244	248	−4	16	−11.3704	129.2853	−11.3704	129.286	129.286	0.0164
3	245	251	−6	36	−10.3704	107.5446	−8.3704	70.0636	86.8044	0.0245
4	246	258	−12	144	−9.3704	87.8038	−1.3704	1.8779	12.8412	0.0488
5	247	248	−1	1	−8.3704	70.0631	−11.3704	129.286	95.1748	0.0040
6	248	264	−16	256	−7.3704	54.3224	4.6296	21.4332	−34.122	0.0645
7	248	253	−5	25	−7.3704	54.3224	−6.3704	40.582	46.9524	0.0202
8	249	252	−3	9	−6.3704	40.5816	−7.3704	54.3228	46.9524	0.0120
9	249	253	−4	16	−6.3704	40.5816	−6.3704	40.582	40.582	0.0161
10	250	264	−14	196	−5.3704	28.8409	4.6296	21.4332	−24.8628	0.0560
11	250	256	−6	36	−5.3704	28.8409	−3.3704	11.3596	18.1004	0.0240
12	251	256	−5	25	−4.3704	19.1001	−3.3704	11.3596	14.73	0.0199
13	253	265	−12	144	−2.3704	5.6187	5.6296	31.6924	−13.3444	0.0474
14	254	245	9	81	−1.3704	1.8779	−14.3704	206.5084	19.6932	0.0354
15	255	258	−3	9	−0.3704	0.1372	−1.3704	1.8779	0.5075	0.0118
16	256	267	−11	121	0.6296	0.3964	7.6296	58.2108	4.8035	0.0430
17	257	259	−2	4	1.6296	2.6557	−0.3704	0.1371	−0.6036	0.0078
18	258	261	−3	9	2.6296	6.9150	1.6296	2.6555	4.2851	0.0116
19	259	260	−1	1	3.6296	13.1742	0.6296	0.3963	2.2851	0.0039
20	260	260	0	0	4.6296	21.4335	0.6296	0.3963	2.9147	0.0000
21	262	266	−4	16	6.6296	43.9520	6.6296	43.9516	43.9516	0.0153
22	263	270	−7	49	7.6296	58.2112	10.6296	112.9884	81.0996	0.0266
23	265	266	−1	1	9.6296	92.7298	6.6296	43.9516	63.8404	0.0038

(continued)

Table 2.2 (continued)

Dataset	x_i	y_i	$(x_i - y_i)$	$(x_i - y_i)^2$	$(x_i - \bar{x})$	$(x_i - \bar{x})^2$	$(y_i - \bar{y})$	$(y_i - \bar{y})^2$	$(x_i - \bar{x}) * (y_i - \bar{y})$	$\left\lvert\frac{(y_i - x_i)}{x_i}\right\rvert$
24	269	264	5	25	13.6296	185.7668	4.6296	21.4332	63.0996	0.0186
25	270	273	−3	9	14.6296	214.0261	13.6296	185.766	199.3956	0.0111
26	271	270	1	1	15.6296	244.2853	10.6296	112.9884	166.1364	0.0037
27	273	272	1	1	17.6296	310.8038	12.6296	159.5068	222.6548	0.0037
Total	6895	7003	−108	1232		2016.2964		1750.296	1483.296	0.5541

$$SSD = \sum_{i=1}^{T} (x_i - y_i)^2 = 1232\,^\circ\mathrm{K}^2$$

$$MSD = \frac{1}{T}\sum_{i=1}^{T} (x_i - y_i)^2 = \frac{1232}{27} = 45.63\,^\circ\mathrm{K}^2$$

$$RMSD = \sqrt{\frac{1}{T}\sum_{i=1}^{T} (x_i - y_i)^2} = \sqrt{45.63} = 6.7549\,^\circ\mathrm{K}$$

$$\mathrm{CC} = \frac{\sum_{i=1}^{T} (x_i - \bar{x})(y_i - \bar{y})}{(T-1)\sigma_{obs}\sigma_{sim}} = \frac{1483.29}{26 * 8.8062 * 8.2048} = 0.7896$$

$$NRMSD = \frac{\sqrt{\frac{1}{T}\sum_{i=1}^{T} (x_i - y_i)^2}}{\bar{x}} = \frac{6.7549}{255.37} = 0.02645$$

$$ANMBD = \left|\frac{\frac{1}{T}\sum_{i=1}^{T} (y_i - x_i)}{\bar{x}}\right| = \left|\frac{108}{27 * 255.37}\right| = 0.0156$$

$$AARD = \frac{1}{T}\sum_{i=1}^{T} |ARD_i| = \frac{1}{T}\sum_{i=1}^{T} \left|\frac{(y_i - x_i)}{x_i}\right| = \frac{0.5541}{27} = 0.02052$$

$$\mathrm{NSE} = 1 - \frac{\sum_{i=1}^{T} (x_i - y_i)^2}{\sum_{i=1}^{T} (x_i - \bar{x})^2} = 1 - \frac{1232}{2016.29} = 1 - 0.6110 = 0.3890$$

Computation of Skill Score (SS)

- The maximum and minimum temperatures of the complete dataset (observed and simulated as shown in Table 2.1) is found out. Here, in this case the maximum value is 273 °K, whereas the minimum value is 243 °K.
- Now an appropriate bin width is to be chosen. Here it is chosen as 5 °K.
- Hence, the number of bins (nb) is calculated as: (maximum − minimum)/bin width = (273 − 243)/5 = 6.

Then, the values are segregated into the bins, to find the frequencies f_o and f_m as follows (Table 2.3):

Table 2.3 Computation of skill score

Bin	f_o (frequency of observed datasets in the chosen bin)	f_m (frequency of simulated datasets in the chosen bin)	Minimum of f_o and f_m
243–248	7	4	4
249–253	6	4	4
254–258	5	4	4
259–263	4	4	4
264–268	1	7	1
269–273	4	4	4

$$\text{Sum of the minimum of } f_o \,\&\, f_m = 4+4+4+4+1+4 = 21$$

$$\text{Skill Score} = \frac{1}{T}\sum_{i=1}^{nb} \min(f_{m,} f_o) = \frac{21}{27} = 0.7777$$

Table 2.4 presents summary of performance indicators

Table 2.4 Summary of computed performance indicators

Indicator	Value	Remarks
Sum of squares of deviation (SSD)	1232 $(°K)^2$	Lower value is preferable. Near to zero is ideal
Mean square deviation (MSD)	45.63 $(°K)^2$	Lower value is preferable. Near to zero is ideal
Root mean square deviation (RMSD)	6.7549 (°K)	Lower value is preferable. Near to zero is ideal
Pearson correlation coefficient (CC)	0.7896 (no unit)	Higher value is preferable. Near to one is ideal
Normalized root mean square deviation (NRMSD)	0.02645 (no unit)	Lower value is preferable. Near to zero is ideal
Absolute normalized mean bias deviation (ANMBD)	0.0156 (no unit)	Lower value is preferable. Near to zero is ideal
Average absolute relative deviation (AARD)	0.02052 (no unit)	Lower value is preferable. Near to zero is ideal
Nash–Sutcliffe efficiency (NSE)	0.3890 (no unit)	Higher value is preferable. Near to one is ideal
Skill score (SS)	0.7777 (no unit)	Higher value is preferable. Near to one is ideal

2.4 Ranking of Global Climate Models

Procedural steps for selection of the best GCM are presented in Fig. 2.1 (Raju and Nagesh Kumar 2014a; Duckstein et al. 1989).

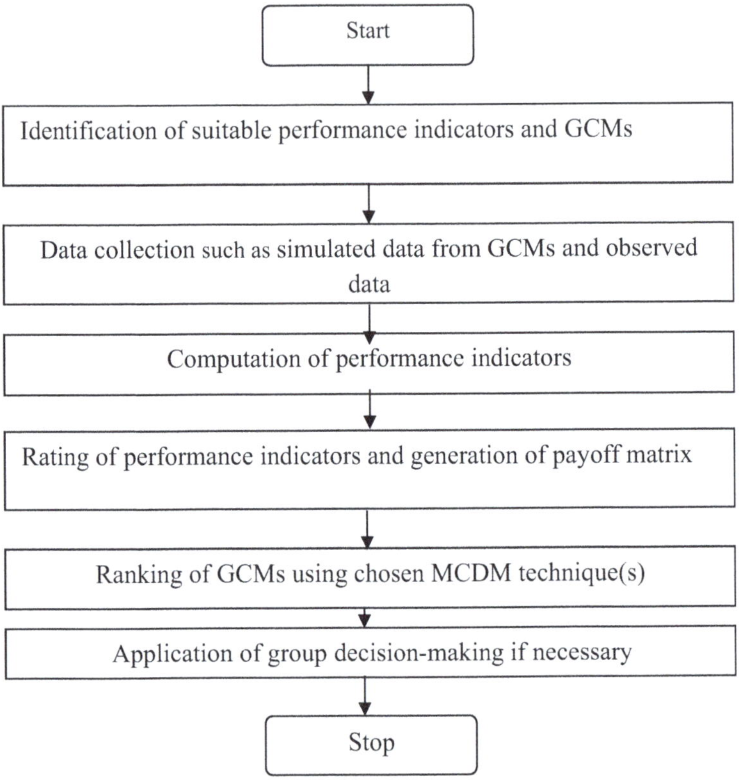

Fig. 2.1 Flowchart of procedural steps for selection of the best GCM

2.4.1 Normalization Techniques

Normalization facilitates the conversion of different non-commensurable indicators to the same space. In the present study, simple normalization technique (denoted as type 3) is presented (Table 2.5). More details of normalization techniques are available from Pomerol and Romero (2000) and Raju and Nagesh Kumar (2014a).

Table 2.5 Description of chosen normalization technique (Type 3)

Description	Mathematical representation
Normalized value k_{aj}	$k_{aj} = \frac{K_j(a)}{\sum_{a=1}^{T} K_j(a)}$ where $K_j(a)$ is the value of indicator j for GCM a; T represents the total number of GCMs

2.4.2 Weight Computing Techniques

A number of techniques are available for determination of weights. In the present chapter, only two techniques are described, namely, entropy and rating.

2.4.2.1 Entropy Technique

The methodology is explained in Table 2.6 (Pomerol and Romero 2000; Raju and Nagesh Kumar 2014a).

Table 2.6 Methodology of entropy technique

Step	Description	Mathematical expression/remark
1	Normalize the payoff matrix if required	k_{aj}
2	Entropy for each indicator j	$En_j = -\frac{1}{\ln(T)}\sum_{a=1}^{T} k_{aj}\ln(k_{aj})$ for $j = 1, \ldots, J$ a is index for GCMs; ($j = 1,2, \ldots, J$) where J is number of indicators; T represents total number of GCMs
3	Degree of diversification	$Dd_j = 1 - En_j$
4	Normalized weight of indicators	$r_j = \frac{Dd_j}{\sum_{j=1}^{J} Dd_j}$

Numerical Problem 2.2 Eleven GCMs in Coupled Model Intercomparison Project (CMIP3), namely, BCCR-BCCM2.0, INGV-ECHAM4, GFDL2.0, GFDL2.1, GISS, IPSL-CM4, MIROC3, MRI-CGCM2, NCAR-PCMI, UKMO-HADCM3, and UKMO-HADGEM1 are analyzed for the variable, precipitation. Five indicators, namely, CC, NRMSD, ANMBD, AARD, SS are the performance indicators. Payoff matrix (11 GCMs vs. 5 indicators] is presented in Table 2.7. Apply entropy technique for determination of weights. Normalization technique 3 can be explored (Raju and Nagesh Kumar 2014b).

Table 2.7 Values of performance indicators obtained for the 11 GCMs

GCM	CC	NRMSD	ANMBD	AARD	SS
BCCR	0.7751	0.7960	0.2744	1.7127	0.7717
ECHAM	0.7866	0.7573	0.1619	1.8639	0.6833
GFDL2.0	0.7868	0.8286	0.4157	0.8080	0.8150
GFDL2.1	0.7395	0.7871	0.1551	1.2731	0.8350
GISS	0.8275	0.8221	0.4786	0.7539	0.7783
IPSL	0.4740	1.2539	0.7082	1.0124	0.6583
MIROC3	0.8416	0.6224	0.0613	1.3811	0.8567
CGCM2	0.7708	0.9386	0.4985	0.6556	0.7550
PCMI	0.3553	1.1779	0.4899	1.6149	0.6283
HADCM3	0.8018	0.8793	0.5092	0.8002	0.8100
HADGEM1	0.8064	0.9422	0.5686	0.7010	0.7883

Table 2.8 Transformed values of performance indicators obtained for the 11 GCMs

GCM	CC	NRMSD[a]	ANMBD[a]	AARD[a]	SS
BCCR	0.7751	−0.7960	−0.2744	−1.7127	0.7717
ECHAM	0.7866	−0.7573	−0.1619	−1.8639	0.6833
GFDL2.0	0.7868	−0.8286	−0.4157	−0.8080	0.8150
GFDL2.1	0.7395	−0.7871	−0.1551	−1.2731	0.8350
GISS	0.8275	−0.8221	−0.4786	−0.7539	0.7783
IPSL	0.4740	−1.2539	−0.7082	−1.0124	0.6583
MIROC3	0.8416	−0.6224	−0.0613	−1.3811	0.8567
CGCM2	0.7708	−0.9386	−0.4985	−0.6556	0.7550
PCMI	0.3553	−1.1779	−0.4899	−1.6149	0.6283
HADCM3	0.8018	−0.8793	−0.5092	−0.8002	0.8100
HADGEM1	0.8064	−0.9422	−0.5686	−0.7010	0.7883

[a]Minimum NRMSD, ANMBD, AARD are desirable. Negative sign is incorporated before values of indicators to represent in maximization perspective, i.e., (−min) = max

Table 2.9 Entropy values, degree of diversification, and weight of indicators

Characteristic of indicator j	CC	NRMSD	ANMBD	AARD	SS
Entropy value En_j (Step 2, Table 2.6)	0.9896	0.9922	0.9416	0.9719	0.9982
Degree of diversification Dd_j (Step 3, Table 2.6)	0.0104	0.0078	0.0584	0.0281	0.0018
Weight r_j (Step 4, Table 2.6)	0.0976	0.0732	0.5484	0.2639	0.0169

Solution:

Applied normalization technique: 3 (Sect. 2.4.1, Table 2.5); Table 2.8 presents values of transformed payoff matrix. Table 2.9 presents entropy values, degree of diversification, and weight of indicators.

2.4.2.2 Rating Technique

Rating technique facilitates the rating of indicators on a numeral scale. However, there is likely chance of subjectivity while rating the indicators by individual experts and chosen numeral scales (Raju and Nagesh Kumar 2014a).

2.4.3 Multicriterion Decision-Making Techniques in Deterministic Scenario

Number of MCDM techniques can be applied to rank GCMs. However, in the present chapter only few techniques are discussed. Researchers are suggested to refer to Raju and Nagesh Kumar (2014a) for more details about various MCDM techniques.

2.4.3.1 Compromise Programming (CP):

It is established on distance measure *Lp* metric (Raju and Nagesh Kumar 2014a). The methodology is explained in Table 2.10.

Table 2.10 Methodology of compromise programming

Step	Description	Mathematical expression/remark
1	Normalize the payoff matrix if required	Choose suitable normalization technique (Sect. 2.4.1)
2	Ideal value for each indicator j among available GCMs	k_j^* $j = 1,2, \ldots, J$ where J is the number of indicators
3	Lp metric value for each GCM a	$L_{pa} = \left[\sum_{j=1}^{J} r_j^p \left\|k_j^* - k_j(a)\right\|^p\right]^{\frac{1}{p}}$ $k_j(a)$ = Value of indicator j for GCM a; r_j = Weight assigned to the indicator j; p = Parameter (1, 2, … ∞)
4	Rank the GCMs built on the L_{pa} values.	Lower L_{pa} indicates suitable GCM

Numerical Problem 2.3 Compute L_p metric values of GCMs and corresponding ranking pattern for the payoff matrix presented in Table 2.11 using compromise programming technique taking parameter $p = 2$. All the 36 GCMs of CMIP5 (Coupled Model Intercomparison Project 5), namely, ACCESS1.0, ACCESS1.3, BCC-CSM1.1, BCC-CSM1.1-m, BNU-ESM, CCSM4, CESM1-BGC, CESM1-CAM5, CESM1-FAST CHEM, CESM1-WACCM, CNRM-CM5, CSIRO-Mk3.6, CanESM2, FGOALS-s2, FIO-ESM, GFDL-CM3, GFDL-ESM2G, GFDL-ESM2M, GISS-E2-H, GISS-E2-R-CC, GISS-E2-R, HadCM3, HadGEM2-AO, INM-CM4, IPSL-CM5A-LR, IPSL-CM5A-MR, IPSL-CM5B-LR, MIROC4h, MIROC5, MIROC-ESM-CHEM, MIROC-ESM, MPI-ESM-LR, MPI-ESM-MR, MPI-ESM-P, MRI-CGCM3, and NorESM1-M are evaluated on three indicators, namely, SS, CC, and NRMSD on the climate variable maximum temperature. Weight of indicators for SS, CC, and NRMSD are 0.0483, 0.0435, and 0.9083 respectively (Raju et al. 2017).

Solution:

Sample calculation for ACCESS 1.0

Values of SS, CC, NRMSD = 0.8280, 0.9269, −0.1664

Ideal values of SS, CC, NRMSD = 0.9378, 0.9875, −0.1104 (Step 2, Table 2.10)

Weights of SS, CC, NRMSD = 0.0483, 0.0435, 0.9083

L_p metric value for ACCESS 1.0 from ideal solution (Step 3, Table 2.10) is:

Table 2.11 Indicator values for 36 GCMs (inputs), L_p Metric value, and rank (outputs)

GCM (1)	SS (2)	CC (3)	NRMSD[a] (4)	L_p metric (5)	Rank (6)
ACCESS1.0	0.8280	0.9269	−0.1664	0.0512	6
ACCESS1.3	0.8036	0.8699	−0.2907	0.1640	30
BCC-CSM1.1	0.9230	0.9110	−0.2933	0.1662	31
BCC-CSM1.1-m	0.9093	0.9267	−0.1835	0.0665	10
BNU-ESM	0.8166	0.9731	−0.2621	0.1380	25
CCSM4	0.8326	0.9738	−0.2025	0.0838	18
CESM1-BGC	0.8519	0.9755	−0.1753	0.0591	9
CESM1-CAM5	0.8715	0.9647	−0.1893	0.0717	12
CESM1-FAST CHEM	0.8405	0.9730	−0.2004	0.0819	17
CESM1-WACCM	0.7304	0.9250	−0.2713	0.1466	27
CNRM-CM5	0.9218	0.9772	−0.1104	0.0009	1
CSIRO-Mk3.6	0.8160	0.9081	−0.2474	0.1246	24
CanESM2	0.8007	0.9360	−0.1910	0.0736	13
FGOALS-s2	0.9378	0.9875	−0.1364	0.0237	3
FIO-ESM	0.8660	0.9216	−0.2874	0.1609	29
GFDL-CM3	0.8264	0.9210	−0.3898	0.2539	34
GFDL-ESM2G	0.7614	0.9428	−0.4552	0.3133	35
GFDL-ESM2M	0.8895	0.9425	−0.3555	0.2226	33
GISS-E2-H	0.6387	0.8448	−0.1698	0.0562	7
GISS-E2-R-CC	0.7172	0.8151	−0.1994	0.0819	16
GISS-E2-R	0.6283	0.8036	−0.1976	0.0810	15
HadCM3	0.9078	0.5585	−0.6980	0.5340	36
HadGEM2-AO	0.8928	0.9455	−0.1952	0.0771	14
INM-CM4	0.9050	0.9054	−0.2813	0.1553	28
IPSL-CM5A-LR	0.7442	0.9132	−0.1736	0.0582	8
IPSL-CM5A-MR	0.7374	0.9242	−0.3548	0.2222	32
IPSL-CM5B-LR	0.7801	0.8562	−0.2184	0.0985	19
MIROC4h	0.8852	0.9704	−0.1880	0.0706	11
MIROC5	0.9052	0.9031	−0.1625	0.0475	5
MIROC-ESM-CHEM	0.9002	0.9789	−0.1313	0.0191	2
MIROC-ESM	0.8968	0.9749	−0.1417	0.0285	4
MPI-ESM-LR	0.8245	0.9573	−0.2286	0.1075	20
MPI-ESM-MR	0.8812	0.9356	−0.2310	0.1096	21
MPI-ESM-P	0.8350	0.9634	−0.2400	0.1179	22
MRI-CGCM3	0.7919	0.8674	−0.2460	0.1235	23
NorESM1-M	0.8199	0.9822	−0.2623	0.1381	26
Maximum value	0.9378	0.9875	−0.1104		

[a]Minimum NRMSD is desirable. Negative sign is incorporated before values of indicator to represent in maximization perspective, i.e., (−min) = max

$$\left[\sqrt{\begin{array}{l}[0.0483(0.9378-0.8280)]^2+[0.0435(0.9875-0.9269)]^2+\\ [0.9083(-0.1104+0.1664)]^2\end{array}}\right]=0.0512$$

Similarly, L_p metric value for other GCMs are computed. Suitable GCM is the one, which is having minimum L_p metric value from ideal solution (Step 4, Table 2.10). Columns 5 and 6 of Table 2.11 present the L_p metric values of GCMs and corresponding ranking pattern.

- Rank of 1 to the lowest L_p metric value and last rank to the highest L_p metric are to be given. The lower the rank, the better is the GCM, i.e., the GCM with rank 1 is the best and the GCM with rank 2 is the next best, and so on. The GCM with the highest rank is the least suitable for the case.
- L_p metric value is varying between 0.0009 (first rank) and 0.5340 (least preferred) among 36 ranks.
- CNRM-CM5, MIROC-ESM-CHEM, and FGOALS-s2 are occupying the first three positions with L_p metric values of 0.0009, 0.0191, and 0.0237, respectively, and can be explored further for downscaling and adaptation studies (Table 2.11).
- GFDL-ESM2G and HadCM3 with L_p metric values of 0.3133 and 0.5340 occupied 35th and 36th positions (Table 2.11), which are the least suitable for the chosen data (Raju et al. 2017).

2.4.3.2 Cooperative Game Theory (CGT):

It is established on distance measure, i.e., as "far" as possible to "anti-ideal" solution (Gershon and Duckstein 1983; Raju and Nagesh Kumar 2014a). The methodology is explained in Table 2.12.

Table 2.12 Methodology of CGT

Step	Description	Mathematical expression/remark
1	Normalize the payoff matrix if required	Choose suitable normalisation technique (Sect. 2.4.1)
2	Anti-ideal value for each indicator j among available GCMs	k_j^{**} $j = 1, 2, \ldots, J$ where J is the number of indicators
3	Geometric distance for GCM a	$D_a = \prod_{j=1}^{J} \left\lvert k_j(a) - k_j^{**} \right\rvert^{r_j}$ $k_j(a)$ = Value of indicator j for GCM a; r_j = Weight assigned to the indicator
4	Rank the GCMs built on the D_a values	Higher D_a indicates suitable GCM

Numerical Problem 2.4 Solve Numerical Problem 2.3 using CGT. Use payoff matrix data in Table 2.11.

Solution:

Sample calculation for ACCESS 1.0

Values of SS, CC, NRMSD = 0.8280, 0.9269, −0.1664
Anti-ideal values of SS, CC, NRMSD = 0.6283, 0.5585, −0.698 (Step 2, Table 2.12)
Weights of SS, CC, NRMSD = 0.0483, 0.0435, 0.9083
Geometric distance value D_a for ACCESS 1.0 from anti-ideal solution (Step 3, Table 2.12) is:

$$\prod_{j=1}^{3}\left|(0.8280-0.6283)^{0.0483}(0.9269-0.5585)^{0.0435}(-0.1664-(-0.698))^{0.9083}\right|$$

$$D_a \text{ for ACCESS 1.0} = 0.9251 * 0.9574 * 0.5633 = 0.4989$$

Similarly, computation of D_a values for other GCMs are made. Suitable GCM is the one, which is having maximum value of D_a (Step 4, Table 2.12). Table 2.13 presents the D_a values of GCMs and corresponding ranking pattern.

Table 2.13 D_a value and corresponding ranking pattern: CGT

GCM	D_a value	Rank	GCM	D_a value	Rank
ACCESS1.0	0.4990	6	GISS-E2-H	0.4254	22
ACCESS1.3	0.3865	30	GISS-E2-R-CC	0.4457	18
BCC-CSM1.1	0.3961	28	GISS-E2-R	0.0000	35
BCC-CSM1.1-m	0.4924	8	HadCM3	0.0000	36
BNU-ESM	0.4176	25	HadGEM2-AO	0.4819	11
CCSM4	0.4711	15	INM-CM4	0.4053	26
CESM1-BGC	0.4968	7	IPSL-CM5A-LR	0.4793	12
CESM1-CAM5	0.4861	10	IPSL-CM5A-MR	0.3256	32
CESM1-FAST CHEM	0.4737	14	IPSL-CM5B-LR	0.4443	19
CESM1-WACCM	0.3956	29	MIROC4h	0.4888	9
CNRM-CM5	0.5599	1	MIROC5	0.5088	5
CSIRO-Mk3.6	0.4272	21	MIROC-ESM-CHEM	0.5399	2
CanESM2	0.4751	13	MIROC-ESM	0.5303	4
FGOALS-s2	0.5393	3	MPI-ESM-LR	0.4468	17
FIO-ESM	0.3977	27	MPI-ESM-MR	0.4491	16
GFDL-CM3	0.3038	33	MPI-ESM-P	0.4383	20
GFDL-ESM2G	0.2406	34	MRI-CGCM3	0.4232	23
GFDL-ESM2M	0.3397	31	NorESM1-M	0.4182	24

- Rank of 1 to the highest D_a and last rank to the lowest D_a value are to be given. The lower the rank, the better is the GCM, i.e., the GCM with rank 1 is the best and the GCM with rank 2 is the second best, and so on. The GCM with the highest rank is the least suitable for the case.
- D_a value is varying between 0.0 (last rank) and 0.5599 (first rank) among 36 ranks.
- CNRM-CM5, MIROC-ESM-CHEM, and FGOALS-s2 with D_a values of 0.5599, 0.5399, and 0.5393 are occupying the first three positions, respectively, (Table 2.13) and can be explored further for downscaling and adaptation studies.
- GISS-E2-R and HadCM3 with D_a values of 0.0 occupied 35th and 36th positions respectively.

2.4.3.3 Technique for Order Preference by Similarity to an Ideal Solution (TOPSIS)

It is established on the distance measure between ideal and anti-ideal solutions (Opricovic and Tzeng 2004; Raju and Nagesh Kumar 2014a, 2015a). The methodology is explained in Table 2.14.

Table 2.14 Methodology of TOPSIS

Step	Description	Mathematical expression/remark
1	Normalize the payoff matrix if required	Choose suitable normalisation technique (Sect. 2.4.1)
2	Ideal value and anti-ideal value for each indicator j among available GCMs	k_j^*, k_j^{**} $j = 1, 2, \ldots, J$ where J is number of indicators
3	Separation measure DS_a^+ of GCM a from the ideal solution	$DS_a^+ = \sqrt{\sum_{j=1}^{J} r_j(k_j(a) - k_j^*)^2}$
4	Separation measure DS_a^- of GCM a from the anti-ideal solution	$DS_a^- = \sqrt{\sum_{j=1}^{J} r_j(k_j(a) - k_j^{**})^2}$
5	Relative closeness CR_a	$CR_a = \frac{DS_a^-}{(DS_a^- + DS_a^+)}$
6	Rank the GCMs built on the CR_a values	Higher CR_a indicates suitable GCM

Numerical Problem 2.5 Eleven GCMs in Coupled Model Intercomparison Project (CMIP3), namely, BCCR-BCCM2.0, INGV-ECHAM4, GFDL2.0, GFDL2.1, GISS, IPSL-CM4, MIROC3, MRI-CGCM2, NCAR-PCMI, UKMO-HADCM3, and UKMO-HADGEM1 are evaluated for climate variables, precipitation (PR), and temperature at 3 levels, i.e., 500, 700, 850 mb (and referred from now as T500, T700, T850) on performance indicator SS. Equal weight of 0.25 is considered for each skill score indicator, SPR, ST500, ST700, and ST850. Determine ranking of GCMs using TOPSIS. Relevant data is presented in Table 2.15. Use abbreviations

Table 2.15 Skill score values for the chosen 11 GCMs

GCM	SPR	ST500	ST700	ST850
BCCR	0.7717	0.5533	0.2583	0.3300
ECHAM	0.6833	0.3483	0.2800	0.3917
GFDL2.0	0.8150	0.5533	0.4000	0.4183
GFDL2.1	0.8350	0.5533	0.2633	0.3783
GISS	0.7783	0.2833	0.2033	0.3150
IPSL	0.6583	0.2833	0.2017	0.4167
MIROC3	0.8567	0.3583	0.3867	0.3800
CGCM2	0.7550	0.4517	0.2367	0.5250
PCMI	0.6283	0.3217	0.1783	0.3367
HADCM3	0.8100	0.3483	0.4100	0.4100
HADGEM1	0.7883	0.2850	0.2067	0.3983

BCCR, ECHAM, GFDL2.0, GFDL2.1, GISS, IPSL, MIROC3, CGCM2, PCMI, HADCM3, and HADGEM1 for GCMs for easy computation (Raju and Nagesh Kumar 2015a). Assume ideal values of each indicator as 1 and anti-ideal values of each indicator as 0.

Solution:

Sample calculation for BCCR-BCCM 2.0

Values of SPR, ST500, ST700, ST850: 0.7717, 0.5533, 0.2583, 0.3300
Ideal values of SPR, ST500, ST700, ST850 = 1 each (Step 2, Table 2.14)
Anti-ideal values of SPR, ST500, ST700, ST850 = 0 each (Step 2, Table 2.14)
Weight of SPR, ST500, ST700, ST850 = 0.25 each
Separation measure of BCCR from ideal solution, i.e., DS^+ for BCCR is (Step 3, Table 2.14):

$$\left[0.25 * \sqrt{\begin{matrix}(0.7717-1.00)^2+(0.5533-1.00)^2+(0.2583-1.000)^2+\\(0.3300-1.000)^2\end{matrix}}\right]$$
$$= 0.2796$$

Separation measure of BCCR from anti-ideal solution, i.e., DS^- for BCCR is (Step 4, Table 2.14):

$$0.25 * \left[\sqrt{\begin{matrix}(0.7717-0.00)^2+(0.5533-0.00)^2+(0.2583-0.000)^2+\\(0.3300-0.000)^2\end{matrix}}\right]$$
$$= 0.2595$$

Relative closeness of BCCR with reference to anti-ideal solution, i.e., CR_a for BCCR is (Step 5, Table 2.14): $\frac{DS^-_{BCCR}}{(DS^-_{BCCR}+DS^+_{BCCR})} = \frac{0.2595}{(0.2595+0.2796)} = 0.4814$

Table 2.16 DS_a^+, DS_a^-, CR_a values and corresponding ranking pattern: TOPSIS

GCM	DS_a^+	DS_a^-	CR_a	Rank
BCCR	0.2796	0.2595	0.4814	6
ECHAM	0.2972	0.2264	0.4324	8
GFDL2.0	0.2414	0.2856	0.5420	1
GFDL2.1	0.2688	0.2757	0.5063	2
GISS	0.3228	0.2273	0.4132	9
IPSL	0.3170	0.2133	0.4022	10
MIROC3	0.2730	0.2688	0.4961	3
CGCM2	0.2703	0.2629	0.4931	5
PCMI	0.3272	0.2005	0.3800	11
HADCM3	0.2689	0.2638	0.4952	4
HADGEM1	0.3110	0.2377	0.4332	7

Similarly, CR_a values for other GCMs are computed. Suitable GCM is the one with maximum relative closeness CR_a value from anti-ideal solution (Step 6, Table 2.14). Table 2.16 presents the DS_a^+, DS_a^-, CR_a value of GCMs and corresponding ranking pattern.

- GFDL2.0, GFDL2.1, MIROC3 with CR_a values of 0.5420, 0.5063, and 0.4961, respectively, occupied the first three positions (Table 2.16).
- IPSL, PCMI with CR_a values of 0.4022 and 0.3800 occupied the last the two positions (Table 2.16).

2.4.3.4 Weighted Average Technique

It is utility-related technique (Raju and Nagesh Kumar 2014a). The methodology is explained in Table 2.17.

Table 2.17 Methodology of weighted average technique

Step	Description	Mathematical expression/remark
1	Normalize the payoff matrix if required	Choose suitable normalization technique (Sect. 2.4.1)
2	Utility of GCM a	$V_a = \left[\sum_{j=1}^{J} r_j k_j\right]$ k_j = Value of indicator j for GCM a; r_j = Weight assigned to the indicator j
3	Rank the GCMs built on the V_a values	Higher V_a indicates suitable GCM

Numerical Problem 2.6 Solve Numerical Problem 2.3 using weighted average technique. Use payoff matrix data in Table 2.11.

Solution:

Sample calculation for ACCESS 1.0

Values of SS, CC, NRMSD = 0.8280, 0.9269, −0.1664
Weight of SS, CC, NRMSD = 0.0483, 0.0435, 0.9083
Weighted Average value V for ACCESS 1.0 (Step 2, Table 2.17) is:

$$\begin{aligned} \text{V for ACCESS 1.0} &= 0.8280 \times 0.0483 + 0.9269 \times 0.0435 + (-0.1664) \\ &\quad \times 0.9083 \\ &= -0.0708 \end{aligned}$$

Similarly, weighted average values for other GCMs are computed. Best GCM is the one with maximum weighed average value V_a (Step 3, Table 2.17). Table 2.18 presents the V_a values of GCMs and corresponding ranking pattern.

- Utility value of GCMs are varying between −0.0132 (first rank) and −0.5658 (least preferred) among 36 ranks.
- CNRM-CM5, MIROC-ESM-CHEM, and FGOALS-s2 with utility values of −0.0132, −0.0332, and −0.0356, respectively, are occupying the first three positions.
- GFDL-ESM2G and HadCM3 with utility values of −0.3356 and −0.5658 occupied 35th and 36th positions respectively.

Table 2.18 V_a value and corresponding ranking pattern: weighted average technique

GCM	V_a	Rank	GCM	V_a	Rank
ACCESS1.0	−0.0708	6	GISS-E2-H	−0.0866	11
ACCESS1.3	−0.1873	31	GISS-E2-R-CC	−0.1110	17
BCC-CSM1.1	−0.1821	30	GISS-E2-R	−0.1141	18
BCC-CSM1.1-m	−0.0824	9	HadCM3	−0.5658	36
BNU-ESM	−0.1562	26	HadGEM2-AO	−0.0930	13
CCSM4	−0.1013	16	INM-CM4	−0.1724	28
CESM1-BGC	−0.0756	7	IPSL-CM5A-LR	−0.0820	8
CESM1-CAM5	−0.0878	12	IPSL-CM5A-MR	−0.2464	33
CESM1-FAST CHEM	−0.0991	15	IPSL-CM5B-LR	−0.1234	19
CESM1-WACCM	−0.1709	27	MIROC4h	−0.0857	10
CNRM-CM5	−0.0132	1	MIROC5	−0.0645	5
CSIRO-Mk3.6	−0.1458	23	MIROC-ESM-CHEM	−0.0332	2
CanESM2	−0.0941	14	MIROC-ESM	−0.0429	4
FGOALS-s2	−0.0356	3	MPI-ESM-LR	−0.1261	20
FIO-ESM	−0.1791	29	MPI-ESM-MR	−0.1265	21
GFDL-CM3	−0.2740	34	MPI-ESM-P	−0.1357	22
GFDL-ESM2G	−0.3356	35	MRI-CGCM3	−0.1474	24
GFDL-ESM2M	−0.2389	32	NorESM1-M	−0.1559	25

Types of generalized indicator functions			Preference function values for various types of indicator functions
1	Usual	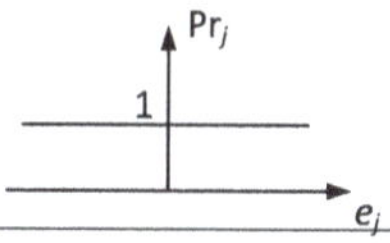	$\Pr_j = \begin{bmatrix} 0 & if & e_j \le 0 \\ 1 & if & e_j \succ 0 \end{bmatrix}$
2	Quasi	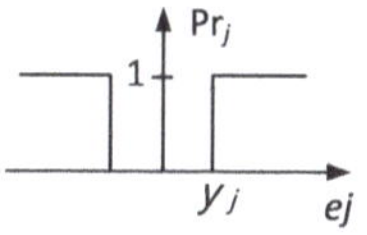	$\Pr_j = \begin{bmatrix} 0 & if & e_j \le y_j \\ 1 & if & e_j \succ y_j \end{bmatrix}$
3	Linear preference and no indifference area	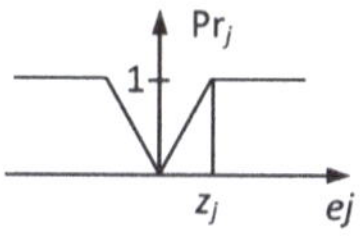	$\Pr_j = \begin{bmatrix} \frac{e_j}{z_j} & if & e_j \le z_j \\ 1 & if & e_j \succ z_j \end{bmatrix}$
4	Level	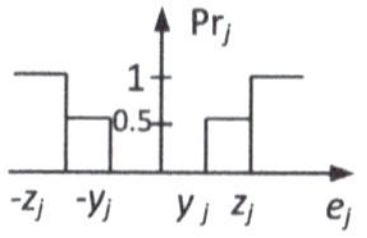	$\Pr_j = \begin{bmatrix} 0 & if & e_j \le y_j \\ 0.5 & if & y_j \prec e_j \le z_j \\ 1 & if & e_j \succ z_j \end{bmatrix}$
5	Linear preference and indifference area	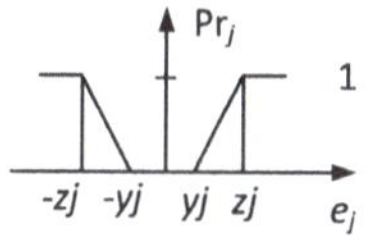	$\Pr_j = \begin{bmatrix} 0 & if & e_j \le y_j \\ \frac{(e_j - y_j)}{(z_j - y_j)} & if & y_j \prec e_j \le z_j \\ 1 & if & e_j \succ z_j \end{bmatrix}$
6	Gaussian	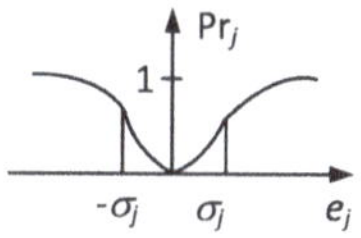	$\Pr_j = [1 - e^{\frac{-e_j^2}{2\sigma_j^2}}]$

Fig. 2.2 Types of various indicator functions and relevant preference function values in PROMETHEE-2

2.4.3.5 Preference Ranking Organization Method of Enrichment Evaluation (PROMETHEE-2)

It is built-in preference function concept (Pomerol and Romero 2000; Raju and Nagesh Kumar 2014a, b; Brans et al. 1986). Preference function $\Pr_j(a, b)$ depends on the pairwise difference e_j between the evaluations $k_j(a)$ and $k_j(b)$ of GCMs a and b for indicator j, chosen indicator function and corresponding parameters such as indifference and preference thresholds y_j and z_j. Six types of indicator functions are available (Fig. 2.2). The methodology is explained in Table 2.19.

Table 2.19 Methodology of PROMETHEE-2

Step	Description	Mathematical expression/remark
1	Multi indicator preference index (MIPI)	$\pi(a,b) = \frac{\sum_{j=1}^{J} r_j \Pr_j(a,b)}{\sum_{j=1}^{J} r_j}$
2	Outranking index of GCM a in the T GCMs [T is number of GCMs]	$\phi^+(a) = \frac{\sum_A \pi(a,b)}{(T-1)}$
3	Outranked index of GCM a in the T GCMs	$\phi^-(a) = \frac{\sum_A \pi(b,a)}{(T-1)}$
4	Net ranking of GCM a in the T GCMs	$\phi(a) = \phi^+(a) - \phi^-(a)$
5	Rank the GCMs built on the $\phi(a)$ values	Higher $\phi(a)$ indicates suitable GCM

Numerical Problem 2.7 Eleven GCMs in Coupled Model Intercomparison Project (CMIP3), namely, BCCR-BCCM2.0, INGV-ECHAM4, GFDL2.0, GFDL2.1, GISS, IPSL-CM4, MIROC3, MRI-CGCM2, NCAR-PCMI, UKMO-HADCM3, and UKMO-HADGEM1, are evaluated on indicators CC, NRMSD, ANMBD, AARD, and SS for climate variable precipitation. Payoff matrix (GCMs vs. indicators) is presented in Table 2.20. Weights of the indicators obtained by entropy technique are 0.0976, 0.0729, 0.5481, 0.2640, and 0.0174 respectively. Using PROMETHEE-2 technique, determine the ranking of GCMs using the above weights. Assume usual indicator function for all indicators. Use abbreviations BCCR, ECHAM, GFDL2.0, GFDL2.1, GISS, IPSL, MIROC3, CGCM2, PCMI, HADCM3, and HADGEM1 for GCMs for easy computations (Raju and Nagesh Kumar 2014b).

Table 2.20 Values of performance indicators obtained for the 11 GCMs

GCM	CC	NRMSD[a]	ANMBD[a]	AARD[a]	SS
BCCR	0.7751	0.7960	0.2744	1.7127	0.7717
ECHAM	0.7866	0.7573	0.1619	1.8639	0.6833
GFDL2.0	0.7868	0.8286	0.4157	0.8080	0.8150
GFDL2.1	0.7395	0.7871	0.1551	1.2731	0.8350
GISS	0.8275	0.8221	0.4786	0.7539	0.7783
IPSL	0.4740	1.2539	0.7082	1.0124	0.6583
MIROC3	0.8416	0.6224	0.0613	1.3811	0.8567
CGCM2	0.7708	0.9386	0.4985	0.6556	0.7550
PCMI	0.3553	1.1779	0.4899	1.6149	0.6283
HADCM3	0.8018	0.8793	0.5092	0.8002	0.8100
HADGEM1	0.8064	0.9422	0.5686	0.7010	0.7883

Solution:

Pairwise difference between values of GCMs and preference functions for indicator CC:

Pairwise difference between values of GCMs for each indicator (five in this case) are to be computed. For example, for indicator CC, the pairwise difference of values in Table 2.21, between BCCR and GFDL2.0 is 0.7751 − 0.7868 = −0.0117 (Table 2.22a) and so the corresponding value of preference function under usual indicator function is 0 (as −0.0117 < 0). Vice versa, pairwise difference between GFDL2.0 and BCCR for CC is 0.0117 (Table 2.22a) and corresponding value of preference function is 1 (as 0.0117 > 0) as in the case of usual indicator function (Table 2.23a), elements of preference function matrix are either 0 or 1. Pairwise preference function values are computed for each indicator in the similar format. Table 2.22a–e presents pairwise difference matrix for CC, NRMSD, ANMBD, AARD, SS whereas Table 2.23a–e present preference function values for CC, NRMSD, ANMBD, AARD, and SS.

Table 2.21 Transformed values of performance indicators obtained for the 11 GCMs

GCM	CC	NRMSD[a]	ANMBD[a]	AARD[a]	SS
BCCR	0.7751	−0.7960	−0.2744	−1.7127	0.7717
ECHAM	0.7866	−0.7573	−0.1619	−1.8639	0.6833
GFDL2.0	0.7868	−0.8286	−0.4157	−0.8080	0.8150
GFDL2.1	0.7395	−0.7871	−0.1551	−1.2731	0.8350
GISS	0.8275	−0.8221	−0.4786	−0.7539	0.7783
IPSL	0.4740	−1.2539	−0.7082	−1.0124	0.6583
MIROC3	0.8416	−0.6224	−0.0613	−1.3811	0.8567
CGCM2	0.7708	−0.9386	−0.4985	−0.6556	0.7550
PCMI	0.3553	−1.1779	−0.4899	−1.6149	0.6283
HADCM3	0.8018	−0.8793	−0.5092	−0.8002	0.8100
HADGEM1	0.8064	−0.9422	−0.5686	−0.7010	0.7883

[a]Minimum NRMSD, ANMBD, AARD are desirable. Negative sign is incorporated before values of indicators to represent in maximization perspective, i.e., (−min) = max

Multi Indicator Preference Index, $\pi(ECHAM, BCCR)$ for pairwise GCMs (ECHAM, BCCR) is computed as follows (Step 1, Table 2.19):

Preference function values for ECHAM and BCCR for indicators CC, NRMSD, ANMBD, AARD, SS are 1, 1, 1, 0, 0. Corresponding weights of indicators are 0.0976, 0.0729, 0.5481, 0.2640, and 0.0174 respectively.

Multi Indicator Preference Index for pair of GCMs (ECHAM and BCCR)

$$\frac{[0.0976 \times 1 + 0.0729 \times 1 + 0.5481 \times 1 + 0.2640 \times 0 + 0.0174 \times 0]}{[0.0976 + 0.0729 + 0.5481 + 0.2640 + 0.0174]} = \frac{0.7186}{1}$$
$$= 0.7186$$

Computations are repeated for all possible pairs for all indicators resulting in Table 2.24.

Table 2.22 Pairwise difference matrix

GCM	BCCR	ECHAM	GFDL 2.0	GFDL 2.1	GISS	IPSL	MIRO C3	CGCM2	PCMI	HADCM3	HAD GEM1
(a) Pairwise difference matrix for indicator CC											
BCCR	0	−0.0115	−0.0117	−0.0356	0.0524	−0.3011	0.0665	0.0043	0.4198	−0.0267	0.0313
ECHAM	0.0115	0	−0.0002	0.0471	−0.0409	0.3126	−0.055	0.0158	0.4313	−0.0152	−0.0198
GFDL2.0	0.0117	0.0002	0	0.0473	−0.0407	0.3128	−0.0548	0.016	0.4315	−0.015	−0.0196
GFDL2.1	−0.0356	−0.0471	−0.0473	0	−0.088	0.2655	−0.1021	−0.0313	0.3842	−0.0623	−0.0669
GISS	0.0524	0.0409	0.0407	0.088	0	0.3535	−0.0141	0.0567	0.4722	0.0257	0.0211
IPSL	−0.3011	−0.3126	−0.3128	−0.2655	−0.3535	0	−0.3676	−0.2968	0.1187	−0.3278	−0.3324
MIROC3	0.0665	0.055	0.0548	0.1021	0.0141	0.3676	0	0.0708	0.4863	0.0398	0.0352
CGCM2	−0.0043	−0.0158	−0.016	0.0313	−0.0567	0.2968	−0.0708	0	0.4155	−0.031	−0.0356
PCMI	−0.4198	−0.4313	−0.4315	−0.3842	−0.4722	−0.1187	−0.4863	−0.4155	0	−0.4465	−0.4511
HADCM3	0.0267	0.0152	0.015	0.0623	−0.0257	0.3278	−0.0398	0.031	0.4465	0	−0.0046
HADGEM1	0.0313	0.0198	0.0196	0.0669	−0.0211	0.3324	−0.0352	0.0356	0.4511	0.0046	0
(b) Pairwise difference matrix for indicator NRMSD											
BCCR	0	−0.0387	0.0326	−0.0089	0.0261	0.4579	0.1736	0.1426	0.3819	0.0833	0.1462
ECHAM	0.0387	0	0.0713	0.0298	0.0648	0.4966	−0.1349	0.1813	0.4206	0.122	0.1849
GFDL2.0	−0.0326	−0.0713	0	−0.0415	−0.0065	0.4253	−0.2062	0.11	0.3493	0.0507	0.1136
GFDL2.1	0.0089	−0.0298	0.0415	0	0.035	0.4668	−0.1647	0.1515	0.3908	0.0922	0.1551
GISS	−0.0261	−0.0648	0.0065	−0.035	0	0.4318	−0.1997	0.1165	0.3558	0.0572	0.1201
IPSL	−0.4579	−0.4966	−0.4253	−0.4668	−0.4318	0	−0.6315	−0.3153	−0.076	−0.3746	−0.3117
MIROC3	0.1736	0.1349	0.2062	0.1647	0.1997	0.6315	0	0.3162	0.5555	0.2569	0.3198
CGCM2	−0.1426	−0.1813	−0.11	−0.1515	−0.1165	0.3153	−0.3162	0	0.2393	−0.0593	0.0036
PCMI	−0.3819	−0.4206	−0.3493	−0.3908	−0.3558	0.076	−0.5555	−0.2393	0	−0.2986	−0.2357
HADCM3	−0.0833	−0.122	−0.0507	−0.0922	−0.0572	0.3746	−0.2569	0.0593	0.2986	0	0.0629
HADGEM1	−0.1462	−0.1849	−0.1136	−0.1551	−0.1201	0.3117	−0.3198	−0.0036	0.2357	−0.0629	0

(continued)

Table 2.22 (continued)

GCM	BCCR	ECHAM	GFDL 2.0	GFDL 2.1	GISS	IPSL	MIRO C3	CGCM2	PCMI	HADCM3	HAD GEM1
(c) Pairwise difference matrix for indicator ANMBD											
BCCR	0	−0.1125	0.1413	−0.1193	0.2042	0.4338	−0.2131	0.2241	0.2155	0.2348	0.2942
ECHAM	0.1125	0	0.2538	−0.0068	0.3167	0.5463	−0.1006	0.3366	0.328	0.3473	0.4067
GFDL2.0	−0.1413	−0.2538	0	−0.2606	0.0629	0.2925	−0.3544	0.0828	0.0742	0.0935	0.1529
GFDL2.1	0.1193	0.0068	0.2606	0	0.3235	0.5531	−0.0938	0.3434	0.3348	0.3541	0.4135
GISS	−0.2042	−0.3167	−0.0629	−0.3235	0	0.2296	−0.4173	0.0199	0.0113	0.0306	0.09
IPSL	−0.4338	−0.5463	−0.2925	−0.5531	−0.2296	0	−0.6469	−0.2097	−0.2183	−0.199	−0.1396
MIROC3	0.2131	0.1006	0.3544	0.0938	0.4173	0.6469	0	0.4372	0.4286	0.4479	0.5073
CGCM2	−0.2241	−0.3366	−0.0828	−0.3434	−0.0199	0.2097	−0.4372	0	−0.0086	0.0107	0.0701
PCMI	−0.2155	−0.328	−0.0742	−0.3348	−0.0113	0.2183	−0.4286	0.0086	0	0.0193	0.0787
HADCM3	−0.2348	−0.3473	−0.0935	−0.3541	−0.0306	0.199	−0.4479	−0.0107	−0.0193	0	0.0594
HADGEM1	−0.2942	−0.4067	−0.1529	−0.4135	−0.09	0.1396	−0.5073	−0.0701	−0.0787	−0.0594	0
(d) Pairwise difference matrix for indicator AARD											
BCCR	0	0.1512	−0.9047	−0.4396	−0.9588	−0.7003	−0.3316	−1.0571	−0.0978	−0.9125	−1.0117
ECHAM	−0.1512	0	−1.0559	−0.5908	−1.11	−0.8515	−0.4828	−1.2083	−0.249	−1.0637	−1.1629
GFDL2.0	0.9047	1.0559	0	0.4651	−0.0541	0.2044	0.5731	−0.1524	0.8069	−0.0078	−0.107
GFDL2.1	0.4396	0.5908	−0.4651	0	−0.5192	−0.2607	0.108	−0.6175	0.3418	−0.4729	−0.5721
GISS	0.9588	1.11	0.0541	0.5192	0	0.2585	0.6272	−0.0983	0.861	0.0463	−0.0529
IPSL	0.7003	0.8515	−0.2044	0.2607	−0.2585	0	0.3687	−0.3568	0.6025	−0.2122	−0.3114
MIROC3	0.3316	0.4828	−0.5731	−0.108	−0.6272	−0.3687	0	−0.7255	0.2338	−0.5809	−0.6801
CGCM2	1.0571	1.2083	0.1524	0.6175	0.0983	0.3568	0.7255	0	0.9593	0.1446	0.0454
PCMI	0.0978	0.249	−0.8069	−0.3418	−0.861	−0.6025	−0.2338	−0.9593	0	−0.8147	−0.9139
HADCM3	0.9125	1.0637	0.0078	0.4729	−0.0463	0.2122	0.5809	−0.1446	0.8147	0	−0.0992
HADGEM1	1.0117	1.1629	0.107	0.5721	0.0529	0.3114	0.6801	−0.0454	0.9139	0.0992	0

(continued)

Table 2.22 (continued)

GCM	BCCR	ECHAM	GFDL 2.0	GFDL 2.1	GISS	IPSL	MIRO C3	CGCM2	PCMI	HADCM3	HAD GEM1
(e) Pairwise difference matrix for indicator SS											
BCCR	0	0.0884	−0.0433	−0.0633	−0.0066	0.1134	−0.085	0.0167	0.1434	−0.0383	−0.0166
ECHAM	−0.0884	0	−0.1317	−0.1517	−0.095	0.025	−0.1734	−0.0717	0.055	−0.1267	−0.105
GFDL2.0	0.0433	0.1317	0	−0.02	0.0367	0.1567	−0.0417	0.06	0.1867	0.005	0.0267
GFDL2.1	0.0633	0.1517	0.02	0	0.0567	0.1767	−0.0217	0.08	0.2067	0.025	0.0467
GISS	0.0066	0.095	−0.0367	−0.0567	0	0.12	−0.0784	0.0233	0.15	−0.0317	−0.01
IPSL	−0.1134	−0.025	−0.1567	−0.1767	−0.12	0	−0.1984	−0.0967	0.03	−0.1517	−0.13
MIROC3	0.085	0.1734	0.0417	0.0217	0.0784	0.1984	0	0.1017	0.2284	0.0467	0.0684
CGCM2	−0.0167	0.0717	−0.06	−0.08	−0.0233	0.0967	−0.1017	0	0.1267	−0.055	−0.0333
PCMI	−0.1434	−0.055	−0.1867	−0.2067	−0.15	−0.03	−0.2284	−0.1267	0	−0.1817	−0.16
HADCM3	0.0383	0.1267	−0.005	−0.025	0.0317	0.1517	−0.0467	0.055	0.1817	0	0.0217
HADGEM1	0.0166	0.105	−0.0267	−0.0467	0.01	0.13	−0.0684	0.0333	0.16	−0.0217	0

Table 2.23 Preference function matrix

GCM	BCCR	ECHAM	GFDL 2.0	GFDL 2.1	GISS	IPSL	MIRO C3	CGCM2	PCMI	HADCM3	HAD GEM1
(a) Preference function matrix for indicator CC											
BCCR	0	0	0	1	0	1	0	1	1	0	0
ECHAM	1	0	0	1	0	1	0	1	1	0	0
GFDL2.0	1	1	0	1	0	1	0	1	1	0	0
GFDL2.1	0	0	0	0	0	1	0	0	1	0	0
GISS	1	1	1	1	0	1	0	1	1	1	1
IPSL	0	0	0	0	0	0	0	0	1	0	0
MIROC3	1	1	1	1	1	1	0	1	1	1	1
CGCM2	0	0	0	1	0	1	0	0	1	0	0
PCMI	0	0	0	0	0	0	0	0	0	0	0
HADCM3	1	1	1	1	0	1	0	1	1	0	0
HADGEM1	1	1	1	1	0	1	0	1	1	1	0
(b) Preference function matrix for indicator NRMSD											
BCCR	0	0	1	0	1	1	0	1	1	1	1
ECHAM	1	0	1	1	1	1	0	1	1	1	1
GFDL2.0	0	0	0	0	0	1	0	1	1	1	1
GFDL2.1	1	0	1	0	1	1	0	1	1	1	1
GISS	0	0	1	0	0	1	0	1	1	1	1
IPSL	0	0	0	0	0	0	0	0	0	0	0
MIROC3	1	1	1	1	1	1	0	1	1	1	1
CGCM2	0	0	0	0	0	1	0	0	1	0	1
PCMI	0	0	0	0	0	1	0	0	0	0	0
HADCM3	0	0	0	0	0	1	0	1	1	0	1
HADGEM1	0	0	0	0	0	1	0	0	1	0	0

(continued)

Table 2.23 (continued)

GCM	BCCR	ECHAM	GFDL 2.0	GFDL 2.1	GISS	IPSL	MIRO C3	CGCM2	PCMI	HADCM3	HAD GEM1
(c) Preference function matrix for indicator ANMBD											
BCCR	0	0	1	0	1	1	0	1	1	1	1
ECHAM	1	0	1	0	1	1	0	1	1	1	1
GFDL2.0	0	0	0	0	1	1	0	1	1	1	1
GFDL2.1	1	1	1	0	1	1	0	1	1	1	1
GISS	0	0	0	0	0	1	0	1	1	1	1
IPSL	0	0	0	0	0	0	0	0	0	0	0
MIROC3	1	1	1	1	1	1	0	1	1	1	1
CGCM2	0	0	0	0	0	1	0	0	0	1	1
PCMI	0	0	0	0	0	1	0	1	0	1	1
HADCM3	0	0	0	0	0	1	0	0	0	0	1
HADGEM1	0	0	0	0	0	1	0	0	0	0	0
(d) Preference function matrix for indicator AARD											
BCCR	0	1	0	0	0	0	0	0	0	0	0
ECHAM	0	0	0	0	0	0	0	0	0	0	0
GFDL2.0	1	1	0	1	0	1	1	0	1	0	0
GFDL2.1	1	1	0	0	0	0	1	0	1	0	0
GISS	1	1	1	1	0	1	1	0	1	1	0
IPSL	1	1	0	1	0	0	1	0	1	0	0
MIROC3	1	1	0	0	0	0	0	0	1	0	0
CGCM2	1	1	1	1	1	1	1	0	1	1	1
PCMI	1	1	0	0	0	0	0	0	0	0	0
HADCM3	1	1	1	1	0	1	1	0	1	0	0
HADGEM1	1	1	1	1	1	1	1	0	1	1	0

(continued)

Table 2.23 (continued)

GCM	BCCR	ECHAM	GFDL 2.0	GFDL 2.1	GISS	IPSL	MIRO C3	CGCM2	PCMI	HADCM3	HAD GEM1
(e) Preference function matrix for indicator SS											
BCCR	0	1	0	0	0	1	0	1	1	0	0
ECHAM	0	0	0	0	0	1	0	0	1	0	0
GFDL2.0	1	1	0	0	1	1	0	1	1	1	1
GFDL2.1	1	1	1	0	1	1	0	1	1	1	1
GISS	1	1	0	0	0	1	0	1	1	0	0
IPSL	0	0	0	0	0	0	0	0	1	0	0
MIROC3	1	1	1	1	1	1	0	1	1	1	1
CGCM2	0	1	0	0	0	1	0	0	1	0	0
PCMI	0	0	0	0	0	0	0	0	0	0	0
HADCM3	1	1	0	0	1	1	0	1	1	0	1
HADGEM1	1	1	0	0	1	1	0	1	1	0	0

Table 2.24 Multi indicator preference index

GCM	BCCR	ECHAM	GFDL 2.0	GFDL 2.1	GISS	IPSL	MIRO C3	CGCM2	PCMI	HADCM3	HAD GEM1
BCCR	0	0.2814	0.621	0.0976	0.621	0.736	0	0.736	0.736	0.621	0.621
ECHAM	0.7186	0	0.621	0.1705	0.621	0.736	0	0.7186	0.736	0.621	0.621
GFDL2.0	0.379	0.379	0	0.3616	0.5655	1	0.264	0.736	1	0.6384	0.6384
GFDL2.1	0.9024	0.8295	0.6384	0	0.6384	0.736	0.264	0.6384	1	0.6384	0.6384
GISS	0.379	0.379	0.4345	0.3616	0	1	0.264	0.736	1	0.9826	0.7186
IPSL	0.264	0.264	0	0.264	0	0	0.264	0	0.379	0	0
MIROC3	1	1	0.736	0.736	0.736	0.736	0	0.736	1	0.736	0.736
CGCM2	0.264	0.2814	0.264	0.3616	0.264	1	0.264	0	0.4519	0.8121	0.885
PCMI	0.264	0.264	0	0	0	0.621	0	0.5481	0	0.5481	0.5481
HADCM3	0.379	0.379	0.3616	0.3616	0.0174	1	0.264	0.1879	0.4519	0	0.6384
HADGEM1	0.379	0.379	0.3616	0.3616	0.2814	1	0.264	0.115	0.4519	0.3616	0

Computation of ϕ^+ (Step 2, Table 2.19)

$$\phi^+ \text{ for BCCR} = \frac{[0 + 0.2814 + 0.6210 + 0.0976 + 0.6210 + 0.7360 + 0 + 0.7360 + 0.7360 + 0.6210 + 0.6210]}{10}$$
$$= 0.5071$$

Computation of ϕ^- (Step 3, Table 2.19)

$$\phi^- \text{ for BCCR} = \frac{[0 + 0.7186 + 0.3790 + 0.9024 + 0.3790 + 0.2640 + 1 + 0.2640 + 0.2640 + 0.3790 + 0.3790]}{10}$$
$$= 0.4929$$

Computation of net ϕ (Step 4, Table 2.19)

$$\text{Net } \phi \text{ for BCCR} = \phi^+ \text{ for BCCR} - \phi^- \text{for BCCR} = 0.5071 - 0.4929 = 0.0142$$

ϕ^+, ϕ^-, net ϕ values for GCMs are computed and presented in Table 2.25. Suitable GCM is the one, which is having the highest net ϕ value (Step 5, Table 2.19). Table 2.25 presents ϕ^+, ϕ^-, net ϕ, and corresponding ranking pattern of all GCMs.

Table 2.25 Values of ϕ^+, ϕ^-, net ϕ and ranks of GCMs

GCM	ϕ^+	ϕ^-	Net ϕ	Rank
BCCR	0.5071	0.4929	0.0142	6
ECHAM	0.5564	0.4436	0.1127	5
GFDL2.0	0.5962	0.4038	0.1924	4
GFDL2.1	0.6924	0.3076	0.3848	2
GISS	0.6255	0.3745	0.2511	3
IPSL	0.1435	0.8565	−0.7130	11
MIROC3	0.8152	0.1848	0.6304	1
CGCM2	0.4848	0.5152	−0.0304	7
PCMI	0.2793	0.7207	−0.4413	10
HADCM3	0.4041	0.5959	−0.1918	8
HADGEM1	0.3955	0.6045	−0.2090	9

- MIROC3 and GFDL2.1 with ϕ values of 0.6304, 0.3848 occupied the first two positions.
- IPSL with ϕ value of −0.7130 occupied last position (Raju and Nagesh Kumar 2014b).

2.4.4 Multicriterion Decision-Making Technique in Fuzzy Scenario

Imprecision in indicator values that may arise due to interpolation, averaging procedures, and approximations can be tackled through fuzzy logic. Extension of TOPSIS in fuzzy named as Fuzzy TOPSIS is presented as an initiation. Table 2.26 presents the methodology (Opricovic and Tzeng 2004; Raju and Nagesh Kumar 2014a, 2015b).

Table 2.26 Methodology of Fuzzy TOPSIS

Step	Description	Mathematical expression/remark
1	Input payoff matrix and specification of membership function; for triangular membership function $\tilde{Y}_{ij}(p_{ij}, q_{ij}, s_{ij})$ where p, q, s are lower, middle, and upper values	Payoff matrix will change depending on the chosen membership function 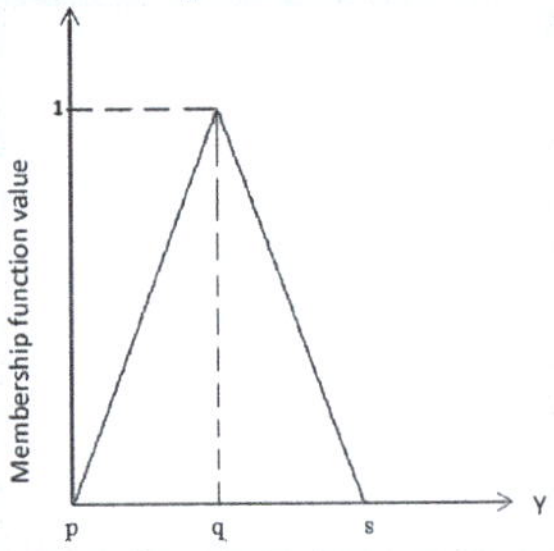 Typical triangular membership function
2	Ideal value and anti-ideal value for each indicator j among available GCMs	$\tilde{Y}_j^*$ (ideal), $\tilde{Y}_j^{**}$ (anti-ideal) represent with elements (p_j^*, q_j^*, s_j^*), $(p_j^{**}, q_j^{**}, s_j^{**})$; $j = 1, 2, \ldots, J$ where J is the number of indicators
3	Separation measure of each GCM a from the ideal solution	$DS_a^+ = \sum_{j=1}^{J} d(\tilde{Y}_{aj}, \tilde{Y}_j^*) = \sqrt{\frac{[(p_{aj}-p_j^*)^2+(q_{aj}-q_j^*)^2+(s_{ia}-s_j^*)^2]}{3}}$
4	Separation measure of each GCM a from the anti-ideal solution	$DS_a^- = \sum_{j=1}^{J} d(\tilde{Y}_{aj}, \tilde{Y}_j^{**}) = \sqrt{\frac{[(p_{aj}-p_j^{**})^2+(q_{aj}-q_j^{**})^2+(s_{aj}-s_j^{**})^2]}{3}}$
5	Relative closeness CR_a	$CR_a = \frac{DS_a^-}{(DS_a^- + DS_a^+)}$
6	Rank the GCMs built on the CR_a values	Higher CR_a indicates suitable GCM

Numerical Problem 2.8 Eleven GCMs in CMIP3 scenario mentioned in Table 2.27 are analyzed for the variable, precipitation. CC, NRMSD, and SS are the performance indicators. Payoff matrix (11 GCMs vs. 5 indicators] is presented in Table 2.27. Rank the GCMs using fuzzy TOPSIS. Assume equal weights for indicators. Ideal values of CC, NRMSD, SS as (1, 1, 1) each whereas anti-ideal values of CC, NRMSD, SS as (0, 0, 0) each (Raju and Nagesh Kumar 2015b).

Table 2.27 Indicators obtained for 11 GCMs

Model	CC			NRMSD			SS		
	p_{ij}	q_{ij}	s_{ij}	p_{ij}	q_{ij}	s_{ij}	p_{ij}	q_{ij}	s_{ij}
UKMO-HAD GEM1	0.649	0.806	0.964	0.390	0.466	0.578	0.714	0.788	0.863
GISS	0.670	0.828	0.985	0.436	0.534	0.687	0.704	0.778	0.853
GFDL2.0	0.629	0.787	0.945	0.433	0.529	0.680	0.741	0.815	0.889
BCCR-BCCM 2.0	0.617	0.775	0.933	0.448	0.551	0.717	0.697	0.772	0.846
IPSL-CM4	0.316	0.474	0.632	0.305	0.350	0.410	0.584	0.658	0.733
UKMO-HADCM3	0.644	0.802	0.960	0.413	0.499	0.631	0.736	0.810	0.884
GFDL2.1	0.582	0.740	0.897	0.452	0.557	0.727	0.761	0.835	0.909
INGV-ECHAM 4	0.629	0.787	0.944	0.466	0.579	0.765	0.609	0.683	0.758
MIROC3	0.684	0.842	0.999	0.544	0.705	1.000	0.782	0.857	0.931
MRI-CGCM2	0.613	0.771	0.929	0.391	0.467	0.581	0.681	0.755	0.829
NCAR- PCMI	0.198	0.355	0.513	0.322	0.372	0.441	0.554	0.628	0.703

Solution:

Sample calculation for GISS

Values of CC, NRMSD, SS in triangular membership function: (0.670, 0.828, 0.985), (0.436, 0.534, 0.687), (0.704, 0.778, 0.853)
Ideal values of CC, NRMSD, SS = (1, 1, 1) each
Anti-ideal values of CC, NRMSD, SS = (0, 0, 0) each

(i) Separation measure of GISS from ideal solution (Step 3, Table 2.26):

$$DS^{I}_{GISS} = \sqrt{\frac{[(p_{aj} - p_j^*)^2 + (q_{aj} - q_j^*)^2 + (s_{aj} - s_j^*)^2]}{3}}$$

$$= \sqrt{\frac{[(0.670 - 1)^2 + (0.828 - 1)^2 + (0.985 - 1)^2]}{3}} \text{ for correlation coefficient}$$

$$+ \sqrt{\frac{[(0.436 - 1)^2 + (0.534 - 1)^2 + (0.687 - 1)^2]}{3}} \text{ for normalized root mean square deviation}$$

$$+ \sqrt{\frac{[(0.704 - 1)^2 + (0.778 - 1)^2 + (0.853 - 1)^2]}{3}} \text{ for skill score}$$

$$= 0.2150 + 0.4594 + 0.2299 = 0.9043$$

(ii) Separation measure of GISS from anti-ideal solution (Step 4, Table 2.26):

$$
\begin{aligned}
DS_{GISS} &= \sqrt{\frac{\left[(p_{aj}-p_j^{**})^2+(q_{aj}-q_j^{**})^2+(s_{aj}-s_j^{**})^2\right]}{3}}\\
&= \sqrt{\frac{\left[(0.670-0)^2+(0.828-0)^2+(0.985-0)^2\right]}{3}}\text{for correlation coefficient}\\
&\quad+\sqrt{\frac{\left[(0.436-0)^2+(0.534-0)^2+(0.687-0)^2\right]}{3}}\text{for normalized root mean square deviation}\\
&\quad+\sqrt{\frac{\left[(0.704-0)^2+(0.778-0)^2+(0.853-0)^2\right]}{3}}\text{for skill score}\\
&= 0.8376+0.5619+0.7807=2.1802
\end{aligned}
$$

(iii) Relative closeness of GISS with reference to anti-ideal measure (Step 5, Table 2.26):

$$
CR_{GISS}=\frac{DS^{-}_{GISS}}{(DS^{-}_{GISS}+DS^{+}_{GISS})}=\frac{2.1802}{(2.1802+0.9043)}=0.7068
$$

Table 2.28 presents DS_a^+ , DS_a^-, CR_a, and ranking pattern for 11 global climate models.

Table 2.28 Ranking pattern of global climate models

Model	DS_a^+	DS_a^-	CR_a	Rank
UKMO-HADGEM1	0.9804	2.0914	0.6808	7
GISS	0.9043	2.1802	0.7068	3
GFDL2.0	0.9076	2.1714	0.7052	4
BCCR-BCCM 2.0	0.9378	2.1424	0.6955	6
IPSL-CM 4	1.5351	1.5100	0.4959	10
UKMO-HADCM3	0.9296	2.1466	0.6978	5
GFDL2.1	0.9024	2.1776	0.7070	2
INGV-ECHAM 4	0.9869	2.0989	0.6802	8
MIROC3	0.6732	2.4833	0.7867	1
MRI-CGCM2	1.0413	2.0251	0.6604	9
NCAR-PCMI	1.6576	1.3906	0.4562	11

- MIROC3 occupied first position with DS_a^+ , DS_a^-, CR_a values, 0.6732, 2.4833, 0.7867, respectively, followed by GFDL 2.1 with DS_a^+ , DS_a^-, CR_a values, 0.9024, 2.1776, 0.7070 respectively. Third position is occupied by GISS with, DS_a^+ , DS_a^-, CR_a values, 0.9043, 2.1802, 0.7068 respectively.
- Relative closeness is almost same with slight difference of 0.0002 for GFDL 2.1 and GISS for both second and third positions (Raju and Nagesh Kumar 2015b).

2.4.5 Spearman Rank Correlation Coefficient

Spearman rank correlation coefficient (R) measures the correlation (Gibbons 1971) between ranks (Table 2.29).

Table 2.29 Methodology of Spearman rank correlation coefficient

Step	Description	Mathematical expression/remark
1	Spearman rank correlation coefficient	$R = 1 - \frac{6\sum_{a=1}^{T} e_a^2}{T(T^2-1)}$ e_a is difference between ranks for the same GCM a; T is number of GCMs; R value varies between −1 and 1

Numerical Problem 2.9 Ranking of 11 GCMs obtained by compromise programming and TOPSIS are presented in Table 2.30. Compute Spearman rank correlation coefficient (R).

Table 2.30 Ranking pattern obtained by compromise programming and TOPSIS

GCM	Compromise programming	TOPSIS
1	6	8
2	5	7
3	4	4
4	2	3
5	3	2
6	11	11
7	1	1
8	7	9
9	10	10
10	8	5
11	9	6

Solution:

e_a values are −2, −2, 0, −1, 1, 0, 0, −2, 0, 3, 3; e_a^2 values are 4, 4, 0, 1, 1, 0, 0, 4, 0, 9, 9

$\sum e_a^2$ value is 32 (Step 1, Table 2.29)

$$R = 1 - \frac{6 * 32}{11(11^2 - 1)} = 0.8545$$

Spearman rank correlation coefficient value is 0.8545 (Step 1, Table 2.29).

2.4.6 Group Decision-Making

Group decision-making is a procedure in which ranking pattern with reference to individual ranking techniques are integrated to form a single group preference. Table 2.31 presents the methodology (Morais and Almeida 2012; Raju et al. 2017):

Table 2.31 Computation of strength, weakness, and net strength of each GCM

Step	Description	Mathematical expression/remark
1	Division of the ranks	The descending order rankings are divided into upper and lower portions: X = T/2 for even number of GCMs and T/2 + 1 for odd number of GCMs and Y = X + 1 where T is the number of the GCMs. The GCMs with rankings from 1 to X constitute the upper portion
2	Strength of each GCM a	$ST_a = \sum_{k=1}^{m}\sum_{z}^{X}(X - z + 1)q_{az}^{k} \quad \forall a, k \quad \forall z = 1, \ldots x$ where q_{az}^{k} = 1 if GCM a is in the position z for the ranking technique k and 0 otherwise. a corresponds to the GCMs in the upper portion; z is the position in upper portion ranging from the first position to the Xth position (z = 1st, … xth) and k represents a ranking technique (k = 1, 2, … m)
3	Weakness of the GCM a	$WE_a = \sum_{k=1}^{m}\sum_{z=y}^{T}(z - Y + 1)q_{az}^{k} \quad \forall a, k \quad \forall z = y, \ldots T$ where, q_{az}^{k} = 1 if GCM a is in the position z for the MCDM technique k and 0 otherwise. a corresponds to the GCMs in the lower portion; z is the position in lower portion ranging from the first position to the lower portion (Yth) up to the last ranking in the lower portion
4	Net strength of GCM a	$NS_a = ST_a - WE_a$
5	Rank the GCMs built on the NS_a values	Higher NS_a indicates suitable GCM

Numerical Problem 2.10 Four MCDM techniques, namely, CP, TOPSIS, WA, and PROMETHEE ranked 20 GCMs, G1 to G20 (Table 2.32). Compute the group ranking of GCMs on strength and weakness perspective (Raju et al. 2017).

Table 2.32 Ranking pattern by CP, TOPSIS, WA, PROMETHEE

GCMs	CP	TOPSIS	WA	PROMETHEE
G1	9	9	9	4
G2	7	7	7	9
G3	12	12	12	8
G4	2	2	2	5

(continued)

Table 2.32 (continued)

GCMs	CP	TOPSIS	WA	PROMETHEE
G5	6	6	5	14
G6	13	13	13	15
G7	5	5	6	12
G8	1	1	1	6
G9	8	8	8	1
G10	11	11	11	13
G11	16	16	16	16
G12	9	9	9	4
G13	10	10	10	2
G14	15	15	15	7
G15	4	4	4	3
G16	11	11	11	13
G17	9	9	9	4
G18	14	14	14	11
G19	14	14	14	11
G20	3	3	3	10

Solution:

Table 2.33 presents GCMs in the descending order of ranking. Here X = T/2 where T is number of GCMs; accordingly X is fixed as 10 (However, X can be fixed any other value considering the intuition of decision maker). In this regard, upper portion consists of GCMs having ranks from 1 to 10.

Table 2.33 Ranking of GCMs in the descending order

Rank	CP	TOPSIS	WA	PROMETHEE
1	G8	G8	G8	G9
2	G4	G4	G4	G13
3	G20	G20	G20	G15
4	G15	G15	G15	G1, G12, G17
5	G7	G7	G5	G4
6	G5	G5	G7	G8
7	G2	G2	G2	G14
8	G9	G9	G9	G3
9	G1, G12, G17	G1, G12, G17	G1, G12, G17	G2
10	G13	G13	G13	G20
11	G10,G16	G10, G16	G10,G16	G18, G19
12	G3	G3	G3	G7
13	G6	G6	G6	G10, G16
14	G18, G19	G18, G19	G18, G19	G5
15	G14	G14	G14	G6
16	G11	G11	G11	G11

Strength of a GCM can be stated as the sum of the positional count of a GCM (Step 2, Table 2.31). For example, strength of GCM G5 is explained as follows (Table 2.34): G5 occupied fifth rank (z = 5) for WA which means that $q_{az}^{k} = 1$ and (X − z + 1) is (10 − 5 + 1) = 6; Sixth rank (z = 6) for CP and TOPSIS; $q_{az}^{k} = 1$ and (X − z + 1) is (10 − 6 + 1) = 5: Note that G5 has occupied fifth and sixth ranks in upper portion. According to Step 2, Table 2.31, strength of G5 is computed as: 6 ∗ 1 + 5 ∗ 2 = 16 (Table 2.34). Similar procedure is repeated for lower portion (weakness perspective; Step 3, Table 2.31). G5 occupied fourteenth rank (z = 14) for PROMETHEE which means that $q_{az}^{k} = 1$ and (z − Y + 1) is (14 − 11 + 1) = 4. Accordingly, weakness of G5 is 4 ∗ 1 = 4 (Table 2.35).

Table 2.34 Computation of strength of GCM G5(X = 10)

Rank/z	q_{az}^{k}				(X − z + 1)	Strength (ST$_a$) = (X − z + 1)* q_{az}^{k}
	CP	TOPSIS	WA	PROMETHEE		
1	0	0	0	0	(10 − 1 + 1) = 10	0
2	0	0	0	0	(10 − 2 + 1) = 9	0
3	0	0	0	0	(10 − 3 + 1) = 8	0
4	0	0	0	0	(10 − 4 + 1) = 7	0
5	0	0	1	0	(10 − 5 + 1) = 6	6 ∗ 1 = 6
6	1	1	0	0	(10 − 6 + 1) = 5	5 ∗ 2 = 10
7	0	0	0	0	(10 − 7 + 1) = 4	0
8	0	0	0	0	(10 − 8 + 1) = 3	0
9	0	0	0	0	(10 − 9 + 1) = 2	0
10	0	0	0	0	(10 − 10 + 1) = 1	0
Total strength of G5						16

Table 2.35 Computation of weakness of GCM G5 (X = 10; Y = X + 1 = 11)

Rank/j	q_{az}^{k}				(z − Y + 1)	Weakness (WE$_a$) = (z − Y + 1)* q_{az}^{k}
	CP	TOPSIS	WA	PROMETHEE		
11	0	0	0	0	(11 − 11 + 1) = 1	0
12	0	0	0	0	(12 − 11 + 1) = 2	0
13	0	0	0	0	(13 − 11 + 1) = 3	0
14	0	0	0	1	(14 − 11 + 1) = 4	4
15	0	0	0	0	(15 − 11 + 1) = 5	0
16	0	0	0	0	(16 − 11 + 1) = 6	0
Total weakness of G5						4

Net strength of G5 is 16 − 4 = 12 (Step 4, Table 2.31). Similarly, strength, weakness, and net strength of other GCMs are computed and presented in Table 2.36. G8, G4, G15, G20 occupied first four positions with net strengths of 35, 33, 29, 25. However, the last position (fifteenth rank) is occupied by G11 with a net strength of −24. GCM with high net strength is desirable. Accordingly, all the GCMs were ranked.

Table 2.36 Strength, weakness and net strength of GCMs

GCMs	Strength (ST_a)	sum	Weakness (WE_a)	sum	$NS_a = ST_a - WE_a$	Rank
G1	10(0) + 9(0) + 8 (0) + 7(1) + 6(0) + 5 (0) + 4(0) + 3(0) + 2 (1 + 1 + 1) + 1(0)	13	1(0) + 2(0) + 3(0) + 4 (0) + 5(0) + 6(0) + 7 (0) + 8(0) + 9(0) +10(0)	0	13	8
G2	10(0) + 9(0) + 8 (0) + 7(0) + 6(0) + 5 (0) + 4(1 + 1 + 1) + 3 (0) + 2(1) + 1(0)	14	1(0) + 2(0) + 3(0) + 4 (0) + 5(0) + 6(0) + 7 (0) + 8(0) + 9(0) + 10 (0)	0	14	7
G3	10(0) + 9(0) + 8 (0) + 7(0) + 6(0) + 5 (0) + 4(0) + 3(1) + 2 (0) + 1(0)	3	1(0) + 2(1 + 1 + 1) + 3 (0) + 4(0) + 5(0) + 6 (0) + 7(0) + 8(0) + 9 (0) + 10(0)	6	−3	10
G4	10(0) + 9 (1 + 1 + 1) + 8(0) + 7 (0) + 6(1) + 5(0) + 4 (0) + 3(0) + 2(0) + 1 (0)	33	1(0) + 2(0) + 3(0) + 4 (0) + 5(0) + 6(0) + 7 (0) + 8(0) + 9(0) + 10 (0)	0	33	2
G5	10(0) + 9(0) + 8 (0) + 7(0) + 6(1) + 5 (1 + 1) + 4(0) + 3 (0) + 2(0) + 1(0)	16	1(0) + 2(0) + 3(0) + 4 (1) + 5(0) + 6(0) + 7 (0) + 8(0) + 9(0) + 10 (0)	4	12	9
G6	10(0) + 9(0) + 8 (0) + 7(0) + 6(0) + 5 (0) + 4(0) + 3(0) + 2 (0) + 1(0)	0	1(0) + 2(0) + 3 (1 + 1 + 1) + 4(0) + 5 (1) + 6(0) + 7(0) + 8 (0) + 9(0) + 10(0)	14	−14	14
G7	10(0) + 9(0) + 8 (0) + 7(0) + 6(2) + 5 (1) + 4(0) + 3(0) + 2 (0) + 1(0)	17	1(0) + 2(1) + 3(0) + 4 (0) + 5(0) + 6(0) + 7 (0) + 8(0) + 9(0) + 10 (0)	2	15	6
G8	10(1 + 1 + 1) + 9 (0) + 8(0) + 7(0) + 6 (0) + 5(1) + 4(0) + 3 (0) + 2(0) + 1(0)	35	1(0) + 2(0) + 3(0) + 4 (0) + 5(0) + 6(0) + 7 (0) + 8(0) + 9(0) + 10 (0)	0	35	1
G9	10(1) + 9(0) + 8 (0) + 7(0) + 6(0) + 5 (0) + 4(0) + 3 (1 + 1 + 1) + 2(0) + 1 (0)	19	1(0) + 2(0) + 3(0) + 4 (0) + 5(0) + 6(0) + 7 (0) + 8(0) + 9(0) + 10 (0)	0	19	5
G10	10(0) + 9(0) + 8 (0) + 7(0) + 6(0) + 5 (0) + 4(0) + 3(0) + 2 (0) + 1(0)	0	1(1) + 2(0) + 3(1) + 4 (0) + 5(0) + 6(0) + 7 (0) + 8(0) + 9(0) + 10 (0)	6	−6	11
G11	10(0) + 9(0) + 8 (0) + 7(0) + 6(0) + 5 (0) + 4(0) + 3(0) + 2 (0) + 1(0)	0	1(0) + 2(0) + 3(0) + 4 (0) + 5(0) + 6 (1 + 1 + 1 + 1) + 7 (0) + 8(0) + 9(0) + 10 (0)	24	−24	15

(continued)

Table 2.36 (continued)

GCMs	Strength (ST_a)	sum	Weakness (WE_a)	sum	$NS_a = ST_a - WE_a$	Rank
G12	10(0) + 9(0) + 8 (0) + 7(1) + 6(0) + 5 (0) + 4(0) + 3 (1 + 1 + 1) + 2(0) + 1 (0)	13	1(0) + 2(0) + 3(0) + 4 (0) + 5(0) + 6(0) + 7 (0) + 8(0) + 9(0) + 10 (0)	0	13	8
G13	10(0) + 9(1) + 8 (0) + 7(0) + 6(0) + 5 (0) + 4(0) + 3(0) + 2 (0) + 1(1 + 1 + 1)	12	1(0) + 2(0) + 3(0) + 4 (0) + 5(0) + 6(0) + 7 (0) + 8(0) + 9(0) + 10 (0)	0	12	9
G14	10(0) + 9(0) + 8 (0) + 7(0) + 6(0) + 5 (0) + 4(1) + 3(0) + 2 (0) + 1(0)	4	1(0) + 2(0) + 3(0) + 4 (0) + 5(1 + 1 + 1) + 6 (0) + 7(0) + 8(0) + 9 (0) + 10(0)	15	−11	12
G15	10(0) + 9(0) + 8 (0) + 7(1 + 1 + 1) + 6 (0) + 5(0) + 4(0) + 3 (0) + 2(0) + 1(0)	29	1(0) + 2(0) + 3(0) + 4 (0) + 5(0) + 6(0) + 7 (0) + 8(0) + 9(0) + 10 (0)	0	29	3
G16	10(0) + 9(0) + 8 (0) + 7(0) + 6(0) + 5 (0) + 4(0) + 3(0) + 2 (0) + 1(0)	0	1(3) + 2(0) + 3(1) + 4 (0) + 5(0) + 6(0) + 7 (0) + 8(0) + 9(0) + 10 (0)	6	−6	11
G17	10(0) + 9(0) + 8 (0) + 7(1) + 6(0) + 5 (0) + 4(0) + 3(0) + 2 (1 + 1 + 1) + 1(0)	13	1(0) + 2(0) + 3(0) + 4 (0) + 5(0) + 6(0) + 7 (0) + 8(0) + 9(0) + 10 (0)	0	13	8
G18	10(0) + 9(0) + 8 (0) + 7(0) + 6(0) + 5 (0) + 4(0) + 3(0) + 2 (0) + 1(0)	0	1(1) + 2(0) + 3(0) + 4 (1 + 1 + 1) + 5(0) + 6 (0) + 7(0) + 8(0) + 9 (0) + 10(0)	13	−13	13
G19	10(0) + 9(0) + 8 (0) + 7(0) + 6(0) + 5 (0) + 4(0) + 3(0) + 2 (0) + 1(0)	0	1(1) + 2(0) + 3(0) + 4 (1 + 1 + 1) + 5(0) + 6 (0) + 7(0) + 8(0) + 9 (0) + 10(0)	13	−13	13
G20	10(0) + 9(0) + 8 (1 + 1 + 1) + 7(0) + 6 (0) + 5(0) + 4(0) + 3 (0) + 2(0) + 1(1)	25	1(0) + 2(0) + 3(0) + 4 (0) + 5(0) + 6(0) + 7 (0) + 8(0) + 9 (0) + 10 (0)	0	25	4

*3(1) means product of 3 and 1

Numerous authors worked on various aspects of MCDM, entropy technique, group decision-making, and Spearman rank correlation with relevance to GCM selection and weight of performance indicators (Anandhi et al. 2011; Johnson et al. 2011; Taylor et al. 2012; Fu et al. 2013; Su et al. 2013; Perkins et al. 2013; Raju and Nagesh Kumar 2014a, b, 2015a, b, 2016; Raju et al. 2017; Hughes et al. 2014).

2.4.7 Ensemble of GCMs

Forecasting future climate projections will be helpful for efficient planning in order to mitigate and adapt to changing the climate. GCMs are widely used for this purpose but the common practice is to employ output of a single GCM or single scenario which ultimately results in various uncertainties. Policy decisions formulated on these results cannot be agreed upon because they reflect only a partial assessment of the risk involved.

Uncertainties in these projections have to be assessed to provide higher quality and more quantitative climate change information. In order to address the underlying uncertainties in climate modeling, a number of GCMs and emission scenarios are employed and termed as Multi Model Ensemble (MME). The process of integrating and ensembling of models can be done by taking simple arithmetical average or by following a weighting procedure developed on the performance of the GCMs simulating historic climate data. The models considered in the ensemble process should be reliable, i.e., they should represent the present-day climate factually and involves comparing GCM simulations with observed climate by considering performance measures. The best performing models can be employed for formulating MME.

The present chapter dealt with a description of climate models, performance evaluation, MCDM methodology. Forthcoming chapter discusses various downscaling techniques.

Software (Information as on 30.12.2016)

Researchers can write their computer programs in any of the programming environment after ascertaining the structure of the algorithm discussed in various chapters of the book. However, the following information is provided for better understanding of the representative tools that may be employed.

PROMETHEE: Visual PROMETHEE 1.4: http://www.promethee-gaia.net/software.html.

Spearman Rank Correlation Coefficient: SPSS (Statistical Package for Social Sciences) (http://www-03.ibm.com/software/products/en/spss-statistics).

Revision Questions and Exercise Problems

2.1 What are different types of available climate models?
2.2 What is one-dimensional radiative-convective (RC) model?

2.3 What are global climate models or general circulation models and their purpose?
2.4 What is the opinion of researcher Xu on the GCMs?
2.5 What is a coupled atmosphere–ocean GCM?
2.6 What are the various components of GCMs?
2.7 What are the various uncertainties involved while handling GCMs and related aspects?
2.8 What are the expansions of RCP and SDSM?
2.9 What is the purpose of SDSM?
2.10 What is the purpose of performance indicators? What are the ideal requirements to be a performance indicator?
2.11 What are the available performance indicators?
2.12 Differentiate skill score, correlation coefficient, normalized root mean square deviations, and Nash–Sutcliffe efficiency.
2.13 Name three deviation/error-related performance measures.
2.14 What are the limitations of GCMs? How can these be tackled for effective application of GCMs?
2.15 What are issues raised by researchers Pierce and his team on climate models?
2.16 What are greenhouse gases (GHG) and how changes in greenhouse gases are attributed to natural and anthropogenic factors?
2.17 Mention four researchers who contributed extensively to evaluation of GCMs.
2.18 Precipitation data simulated from a GCM is 2, 3, 6, 7, 8, 10, 11, 12.3, 16.3, 17.2, 18.3, 18.7, and 19.1 whereas observed data is 11.2, 15.8, 13.2, 17.2, 19.3, 8.2, 6.7, 17.3, 16.2, 9.3, 12.1, 13.2, and 23.1. Compute the sum of squares of deviation, mean square deviation, root mean square deviation, Pearson correlation coefficient, normalized root mean square deviation, absolute normalized mean bias deviation, average absolute relative deviation, skill score, and Nash–Sutcliffe efficiency. Use four bins while computing skill score. Discuss the outcome in detail.
2.19 Relative humidity simulated from a GCM is 0.34, 0.56, 0.32, 0.23, 0.14, 0.10, and 0.23 whereas observed data is 0.23, 0.45, 0.42, 0.76, 0.33, 0.12, and 0.32. Compute skill score, normalized root mean square deviations and correlation coefficient.
2.20 What are the procedural steps for selection of best GCM?
2.21 Differentiate between compromise programming and cooperative game theory. How are they efficient in ranking GCMs?
2.22 How can strength and weakness of each GCM be computed in group decision-making?
2.23 How do the weights of indicators affect the ranking of GCMs? Is it necessary to have different weights for different indicators?
2.24 Solve Numerical Problem 2.2 using entropy technique. Consider data from Table 2.7 and consider only CC, ANMBD, AARD, NRMSD for analysis.

2.25 Solve Numerical Problem 2.3 using compromise programming. Use data in Table 2.11. Assume equal weights for all the indicators.

2.26 Solve Numerical Problem 2.3 using cooperative game theory. Use data in Table 2.11. Assume equal weights for all the indicators.

2.27 Solve Numerical Problem 2.5 using TOPSIS. Use data in Table 2.15. Assume weights of 0.2, 0.2, 0.5, 0.1 for the indicators respectively.

2.28 Solve Numerical Problem 2.3 using weighted average technique. Use data in Table 2.11. Assume equal weights for all the indicators.

2.29 Solve Numerical Problem 2.7 using PROMETHEE. Use data in Table 2.20. Assume equal weights for all the indicators. Analyze the numerical problem assuming (a) usual indicator (b) quasi-indicator with indifference value as 0.2.

2.30 Solve Numerical Problem 2.8 using fuzzy TOPSIS. Use data in Table 2.27. Assume weights of indicators CC, NRMSD, SS as (0.1, 0.1, 0.1), (0.2, 0.2, 0.2), (0.3, 0.3, 0.3).

2.31 Nine GCMs in CMIP3 environment as mentioned in Table 2.37 are analyzed for C1 and C2. Payoff matrix is trapezoidal membership function (Nine GCMs vs. two indicators) is presented in Table 2.37. Rank the GCMs using fuzzy TOPSIS. Assume equal weights for indicators. Take ideal values of C1, C2 as (1, 1, 1, 1) each whereas anti-ideal values of C1, C2 as (0, 0, 0, 0) each.

Table 2.37 Data matrix of 9 GCMs and random assignment of clusters

GCMs	C_1	C_2
(1)	(2)	(3)
A1	(0.2, 0.3, 0.4, 0.5)	(0.4, 0.55, 0.66, 0.88)
A2	(0.5, 0.6, 0.7, 0.8)	(0.3, 0.5, 0.6, 0.8)
A3	(0.3, 0.6, 0.9, 1.0)	(0.2, 0.4, 0.6, 0.8)
A4	(0.6, 0.7, 0.8, 0.9)	(0.1, 0.2, 0.4, 0.8)
A5	(0.5, 0.7, 0.8, 0.9)	(0.4, 0.6, 0.8, 1)
A6	(0.2, 0.4, 0.6, 0.8)	(0.22, 0.44, 0.66, 0.88)
A7	(0.2, 0.3, 0.5, 0.8)	(0.44, 0.8, 0.9, 1)
A8	(0.2, 0.4, 0.8, 1.0)	(0.11, 0.44, 0.8, 0.9)
A9	(0.1, 0.4, 0.7, 0.9)	(0.3, 0.5, 0.9, 1)

Hint: Distance between two trapezoidal fuzzy numbers is $\sqrt{\frac{[(p_{aj}-p_j)^2+(q_{aj}-q_j)^2+(s_{ia}-s_j)^2+(t_{ia}-t_j)^2]}{4}}$ Where (p, q, s, t) are elements of Trapezoidal fuzzy number.

2.32 Solve Numerical Problem 2.9 using Spearman rank correlation coefficient. Use data in Table 2.30. Consider the first six GCMs for the analysis.

2.33 Solve Numerical Problem 2.10 using group decision-making. Consider ranking of CP and PROMETHEE only for group decision-making calculations. Use data in Table 2.32. Consider first ten ranks of GCMs for the analysis. Compute strength, weakness, and net strength of each GCM.

Advanced Review Questions

2.34 Why the word global is affixed before climate while naming global climate models?
2.35 Why the word circulation is suffixed after general while naming general circulation models?
2.36 Name various laws on which GCMs are developed?
2.37 If given data is imprecise, it is expected that computed performance measures may not be accurate. How can this be encountered?
2.38 In your opinion, can any other performance measures be initiated or developed? Discuss in detail the limitations of the existing performance measures mentioned in the present chapter and possible improvements?
2.39 List the GCMs that are available under CMIP3 and CMIP5.
2.40 Mention two case studies where researchers used any of the performance measures mentioned in the present chapter.
2.41 What are the similarities between CP and TOPSIS? Is there any way PROMETHEE-2 and CP can be related?
2.42 How may ranking pattern differ for different p values in compromize programming? Discuss mathematically?
2.43 How do ideal and anti-ideal values affect the outcome? Is it necessary to normalize the indicator values?
2.44 How is Spearman rank correlation coefficient related to MCDM techniques? Explain same in the present context?
2.45 Is there any relation between group decision-making and Spearman rank correlation coefficient?
2.46 Is it possible to correlate output of entropy with rating techniques with reference to weight estimation?
2.47 How may group decision-making affect if MCDM techniques are not of equal importance? How can this be considered in the group decision-making analysis?

References

Anandhi A, Frei A, Pradhanang SM, Zion MS, Pierson DC, Schneiderman EM (2011) AR4 climate model performance in simulating snow water equivalent over Catskill Mountain Watersheds, New York, USA. Hydrol Process 25:3302–3311

Bogardi JJ, Nachtnebel HP (eds) (1994) Multicriteria decision analysis in water resources management. In: International hydrological programme, UNESCO, Paris

Brans JP, Vincke P, Mareschal B (1986) How to select and how to rank projects: the PROMETHEE method. Eur J Oper Res 24:228–238

Duckstein L, Tecle A, Nachnebel HP, Hobbs BF (1989) Multicriterion analysis of hydropower operation. J Energy Eng 115(3):132–153

Fu G, Zhaofei L, Charles SP, Xu Z, Zhijun Y (2013) Score-based method for assessing the performance of GCMs: a case study of Southeastern Australia. J Geophys Res Atmos 118:4154–4167

Gershon M, Duckstein L (1983) Multiobjective approaches to river Basin planning. J Water Res Plann Manage 109(1):13–28. ASCE

Gibbons JD (1971) Nonparametric statistical inference. McGraw-Hill, New York

Gleckler PJ, Taylor KE, Doutriaux C (2008) Performance metrics for climate models. J Geophys Res 113:D06104

Goudie A, Cuff DJ (2001) Encyclopedia of global change: environmental change and human society. Oxford University Press

Helsel DR, Hirsch RM (2002) Statistical methods in water resources, U.S. geological survey techniques of water resources investigations, Book 4, Chapter A3

Hughes DA, Mantel S, Mohobane T (2014) An assessment of the skill of downscaled GCM outputs in simulating historical patterns of rainfall variability in South Africa. Hydrol Res 45 (1):134–147

Johnson FM, Sharma A (2009) Measurement of GCM skill in predicting variables relevant for hydro climatological assessments. J Clim 22:4373–4382

Johnson F, Westra S, Sharma A, Pitman AJ (2011) An assessment of GCM skill in simulating persistence across multiple time scales. J Clim 24:3609–3623

Kendal MG, Henderson-Sellers A (2013) A climate modelling primer. Wiley

Legates DR, McCabe GJ (1999) Evaluating the use of goodness-of-fit measures in hydrologic and hydroclimatic model validation. Water Resour Res 35(1):233–241

Macadam I, Pitman AJ, Whetton PH, Abramowitz G (2010) Ranking climate models by performance using actual values and anomalies: implications for climate change impact assessments. Geophys Res Lett 37:L16704

Maximo CC, McAvaney BJ, Pitman AJ, Perkins SE (2008) Ranking the AR4 climate models over the Murray-Darling Basin using simulated maximum temperature minimum temperature and precipitation. Int J Climatol 28:1097–1112

Morais DC, Almeida AT (2012) Group decision making on water resources based on analysis of individual rankings. Omega 40:42–52

Mujumdar PP, Nagesh Kumar D (2012) Floods in a changing climate: hydrologic modeling. International hydrology series. Cambridge University Press

Nash JE, Sutcliffe JV (1970) River flow forecasting through conceptual models part I—a discussion of principles. J Hydrol 10(3):282–290

Nash Sutcliffe Efficiency (2017). http://www.rforge.net/doc/packages/hydroGOF/NSE.html. Accessed 31 Jan 2017

Ojha R, Kumar DN, Sharma A, Mehrotra R (2014) Assessing GCM convergence for the Indian region using the variable convergence score. J Hydrol Eng 19(6):1237–1246

Opricovic S, Tzeng GH (2004) Compromise solution by MCDM methods: a comparative analysis of VIKOR and TOPSIS. Eur J Oper Res 156:445–455

Perkins SE, Pitman AJ, Sissonb SA (2013) Systematic differences in future 20 year temperature extremes in AR4 model projections over Australia as a function of model skill. Int J Climatol 33:1153–1167

Pierce DW, Barnett TP, Santer BD, Gleckler PJ (2009) Selecting global climate models for regional climate change studies. Proc Natl Acad Sci USA 106:8441–8446

Pomerol JC, Romero SB (2000) Multicriterion decision in management: principles and practice. Kluwer Academic, Netherlands

Raje D, Mujumdar PP (2010) Constraining uncertainty in regional hydrologic impacts of climate change: nonstationarity in downscaling. Water Resour Res 46:W07543

Raju KS, Nagesh Kumar D (2014a) Multicriterion analysis in engineering and management. Prentice Hall of India, New Delhi
Raju KS, Nagesh Kumar D (2014b) Ranking of global climatic models for India using multicriterion analysis. Clim Res 60:103–117
Raju KS, Nagesh Kumar D (2015a) Ranking general circulation models for India using TOPSIS. J Water Clim Change 6(2):288–299
Raju KS, Nagesh Kumar D (2015b) Fuzzy approach to rank global climate models. In: Volume 415 of the series advances in intelligent systems and computing. Springer, pp. 53–61
Raju KS, Nagesh Kumar D (2016) Selection of global climate models for India using cluster analysis. J Water Clim Change 7(4):764–774
Raju KS, Sonali P, Nagesh Kumar D (2017) Ranking of CMIP5-based global climate models for India using compromise programming. Theor Appl Climatol 128(3):563–574. doi:10.1007/s00704-015-1721-6
Semenov M, Stratonovitch P (2010) Use of multi-model ensembles from global climate models for assessment of climate change impacts. Clim Res 41:1–14
Sonali P, Nagesh Kumar D (2013) Review of trend detection methods and their application to detect temperature changes in India. J Hydrol 476:212–227
Su F, Duan X, Chen D, Hao Z, Cuo L (2013) Evaluation of the global climate models in the CMIP5 over the Tibetan Plateau. J Clim 26:3187–3208
Taylor KE, Stouffer RJ, Meehl GA (2012) An overview of CMIP5 and the experiment design. Bull Am Meteorol Soc 93:485–498
Thompson RD, Perry AH (1997) Applied climatology: principles and practice. Psychology Press
What is a GCM (2013). http://www.ipcc-data.org/guidelines/pages/gcm_guide.html. Accessed 31 Jan 2017
Wilby RL, Dawson CW, Barrow EM (2002) SDSM—a decision support tool for the assessment of regional climate change impacts. Environ Model Softw 17:147–159
Wilby RL, Troni J, Biot Y, Tedd L, Hewitson BC, Smith DM, Sutton RT (2009) Review of climate risk information for adaptation and development planning. Int J Climatol 29:1193–1215
Wilks DS (2011) Statistical methods in the atmospheric sciences. International geophysics series. Academic Press, San Diego
Xu CY (1999) Climate change and hydrologic models: a review of existing gaps and recent research developments. Water Resour Manage 13:369–382

Suggested Further Reading

Knutti R, Abramowitz G, Collins M, Eyring V, Gleckler PJ, Hewitson B, Mearns L (2010) Good practice guidance paper on assessing and combining multi model climate projections. In: Stocker TF, Qin D, Plattner GK, Tignor M, Midgley PM (eds) Meeting report of the intergovernmental panel on climate change expert meeting on assessing and combining multi model climate projections. IPCC Working Group I Technical Support Unit, University of Bern, Bern, Switzerland
Meehl GA, Stocker TF, Collins WD, Friedlingstein P, Gaye AT, Gregory JM, Kitoh A, Knutti R, Murphy JM, Noda A, Raper SCB, Watterson IG, Weaver AJ, Zhao ZC (2007) Global climate projections. In: Solomon S, Qin D, Manning M, Chen Z, Marquis M, Averyt KB, Tignor M, Miller HL (eds) Climate change 2007: the physical science basis. Contribution of working group I to the fourth assessment report of the intergovernmental panel on climate change. Cambridge University Press, Cambridge, pp 747–846
Murphy J (2004) Quantification of modeling uncertainties in a large ensemble of climate change simulations. Nature 430:768–772

Perkins SE, Pitman AJ, Holbrook NJ, McAveney J (2007) Evaluation of the AR4 climate models' simulated daily maximum temperature, minimum temperature and precipitation over australia using probability density functions. J Clim 20:4356–4376

Randall DA, Wood, RA, Bony S, Colman R, Fichefet T, Fyfe J, Kattsov V, Pitman A, Shukla J, Srinivasan J, Stouffer RJ, Sumi A, Taylor KE (2007) Climate models and their evaluation. In: Solomon S, Qin D, Manning M, Chen Z, Marquis M, Averyt KB, Tignor M, Miller HL (eds) Climate change 2007: the physical science basis. Contribution of working group I to the fourth assessment report of the IPCC. Cambridge University Press, Cambridge, UK

Downscaling Techniques in Climate Modeling

3

Abstract

Describes downscaling techniques where GCM outputs are interpolated to the scale of hydrological modeling or local scale requirement. Statistical downscaling techniques that facilitate statistical relationships that metamorphose large-scale atmospheric variables/predictors simulated by GCMs to local scale variables/predictand are discussed in detail. Techniques include Linear and Non-linear regression, Artificial Neural Networks, Statistical Downscaling Model (SDSM), Change Factor, Least-Square, and Standard Support Vector Machines. Detailed discussion about Artificial Neural Networks that includes information about preprocessing, weights, epoch, activation function, training, learning rate, momentum factor, weight updation procedures, and challenges are also presented. SDSM, combination of regression and conditional weather generator techniques, Change Factor, and Support Vector Machine are also briefed. Nested Bias Correction technique which addresses bias across prespecified multiple timescales is also part of this chapter. Reader is expected to understand various statistical downscaling techniques by studying this chapter.

Keywords

Artificial neural networks · Change factor · Downscaling · Nested bias correction · SDSM · Support vector machine

Electronic supplementary material The online version of this chapter (doi:10.1007/978-981-10-6110-3_3) contains supplementary material, which is available to authorized users.

K. Srinivasa Raju and D. Nagesh Kumar, *Impact of Climate Change on Water Resources*, Springer Climate, https://doi.org/10.1007/978-981-10-6110-3_3

3.1 Introduction

Accuracy of Global Climate Models (GCMs), developed for coarse grid resolution decreases with increase in finer spatial and temporal scales, rendering them unable to replicate sub-grid scale features. However, features at sub-grid scale are important to hydrologists and water resources planners. Downscaling is one of the approaches where GCM outputs are interpolated to the scale of hydrological modeling or local scale requirement (Wang et al. 2004; Mujumdar and Nagesh Kumar 2012). Two categories of downscaling exist, dynamic and statistical. In the dynamic downscaling, Regional Climate Model (RCM) is integrated into GCM for obtaining output at the location of user's choice and this process is computationally burdensome. Statistical downscaling uses mainly statistical and related relationships for getting the output at the location of user's choice and it is easier and flexible compared to dynamic downscaling. In the present chapter, statistical downscaling techniques are discussed in detail. Detailed description of downscaling techniques is available in Fowler et al. (2007) and Sylwia and Emilie (2014).

3.2 Statistical Downscaling

The statistical downscaling techniques facilitate statistical relationships that metamorphose large-scale atmospheric variables/predictors simulated by GCM to local scale variables/predictand. The hydrologic variable to be predicted is called predictand and the climatic variables used as input to the model are called predictors (Karl et al. 1990; Wigley et al. 1990). Weather generators (Wilby et al. 2004; Wilby and Dawson 2007), weather typing/weather classification schemes (Wilby et al. 2004), and transfer functions (Wilby and Dawson 2007) are under this category. Techniques such as linear/non-linear regression, Artificial Neural Networks, Change Factor, Least-Square, and Standard Support Vector Machines (Tripathi et al. 2006) are under transfer function category which is the main focus of the present chapter.

3.2.1 Multiple Regression

Regression is a statistical technique to investigate the relation between a dependent variable y and number of independent variables x. There are many types of regression techniques that can be explored. A regression model which relates dependent variable and multiple independent variables is termed as multiple regression model. Mathematical expression for multiple linear regression is (Wilby et al. 2004; Milton and Arnold 2007; Anandhi et al. 2008):

$$y = b_0 + b_1x_1 + b_2x_2 + \ldots + b_nx_n + e \tag{3.1}$$

where $b_1, b_2, \ldots .b_n$ are coefficients of data $x_1, x_2, \ldots .x_n$; b_0 is the crisp constant and e is the crisp error. Other forms of regression are multiple non-linear regression where relationship between input and outputs is non-linear of any order.

Different types of regression available are as follows:

(a) Simple Linear Regression

$$y = b_0 + b_1 x_1 \tag{3.2}$$

$$b_1 = \frac{n\sum xy - [(\sum x)(\sum y)]}{n\sum x^2 - (\sum x)^2};\ b_0 = \bar{y} - b_1\bar{x} \tag{3.3}$$

$$S_{xx} = \frac{\left[n\sum x^2 - (\sum x)^2\right]}{n};\ S_{yy} = \frac{\left[n\sum y^2 - (\sum y)^2\right]}{n};\ S_{xy} = \frac{[n\sum xy - (\sum x)(\sum y)]}{n} \tag{3.4}$$

Estimated correlation between X and Y is:

$$R = \frac{S_{xy}}{\sqrt{S_{xx}S_{yy}}} \tag{3.5}$$

Here $\bar{x}, \bar{y}$ are mean of x and y observations; n is number of observations. S represents standard deviation.

(b) Polynomial regression

$$y = b_0 + b_1 x + b_2 x^2 \tag{3.6}$$

Three equations are required to obtain three unknowns b_0, b_1, b_2

$$nb_0 + b_1 \sum_{i=1}^{n} x_i + b_2 \sum_{i=1}^{n} x_i^2 = \sum_{i=1}^{n} y_i \tag{3.7}$$

$$b_0 \sum_{i=1}^{n} x_i + b_1 \sum_{i=1}^{n} x_i^2 + b_2 \sum_{i=1}^{n} x_i^3 = \sum_{i=1}^{n} x_i y_i \tag{3.8}$$

$$b_0 \sum_{i=1}^{n} x_i^2 + b_1 \sum_{i=1}^{n} x_i^3 + b_2 \sum_{i=1}^{n} x_i^4 = \sum_{i=1}^{n} x_i^2 y_i \tag{3.9}$$

(c) Multiple regression model linear in nature: $y = b_0 + b_1x_1 + b_2x_2$ (with two variables). Three equations are required to obtain three unknowns b_0, b_1, and b_2.

$$nb_0 + b_1\sum_{i=1}^{n} x_{1i} + b_2\sum_{i=1}^{n} x_{2i} = \sum_{i=1}^{n} y_i \tag{3.10}$$

$$b_0\sum_{i=1}^{n} x_{1i} + b_1\sum_{i=1}^{n} x_{1i}^2 + b_2\sum_{i=1}^{n} x_{1i}x_{2i} = \sum_{i=1}^{n} x_{i1}y_i \tag{3.11}$$

$$b_0\sum_{i=1}^{n} x_{2i} + b_1\sum_{i=1}^{n} x_{2i}x_{1i} + b_2\sum_{i=1}^{n} x_{2i}^2 = \sum_{i=1}^{n} x_{2i}y_i \tag{3.12}$$

Detailed information on regression is available from Milton and Arnold (2007).

Numerical Problem 3.1 Maximum temperature (in °Celsius) data generated by India Meteorological Department (IMD) is dependent on number of parameters that are available from National Centers for Environmental Prediction (NCEP). Notable among them are P5_U (Zonal Velocity Component at 500 Hpa Height in m/s) and P5_V (Meridoinal Velocity Component at 500 Hpa Height in m/s). Data is presented in Table 3.1. Establish (a) linear relationship between predictand (temperature) and predictor (Zonal Velocity Component at 500 Hpa Height in m/s) (b) polynomial relationship between predictand (temperature) and predictor (Zonal Velocity Component at 500 Hpa Height in m/s) (c) multiple linear relationship between predictand (temperature) and predictors (Zonal Velocity Component at 500 Hpa Height in m/s and Meridoinal Velocity Component at 500 Hpa Height in m/s).

Solution

(a) Linear relationship between predictand (temperature) and predictor (Zonal Velocity Component at 500 Hpa Height in m/s) in the form of $y = b_0 + b_1x$ (Table 3.2).

$$\sum x = 7.9808; \sum y = 792.2191; \sum x^2 = 3.4202;$$
$$\sum y^2 = 25341.90;$$
$$\sum xy = 253.49$$
$$\bar{x} = 0.3192; \bar{y} = 31.688$$

Table 3.1 Maximum temperature and velocity component data

Dataset	P5_U (m/s)	P5_V (m/s)	Maximum temperature (°C)
1	0.0866	0.0006	29.2900
2	0.1498	0.8393	32.8321
3	0.4612	0.2654	35.7261
4	0.2085	0.4358	37.4637
5	0.2741	0.4124	35.4984
6	0.1776	0.4254	34.7860
7	0.2575	0.6898	29.8706
8	0.3961	0.3365	30.8523
9	0.1344	0.8079	30.5450
10	0.4620	0.6341	30.8277
11	0.3486	0.5329	29.0377
12	0.5922	0.5403	26.6316
13	0.4589	0.6936	28.4584
14	0.7945	0.2295	31.6171
15	0.2470	0.5656	34.1303
16	0.3151	0.0797	36.3697
17	0.4231	0.1995	38.2229
18	0.6213	0.1268	32.1490
19	0.1849	0.1248	30.4881
20	0.2417	0.6392	29.6145
21	0.3178	0.2921	30.8463
22	0.5320	0.4352	31.1348
23	0.0423	0.1007	29.5313
24	0.2335	0.0754	27.7787
25	0.0201	0.6405	28.5168

$$b_1 = \frac{n\sum xy - [(\sum x)(\sum y)]}{n\sum x^2 - (\sum x)^2} = \frac{25 * 253.49 - 7.9808 * 792.2191}{25 * 3.4202 - 7.9808 * 7.9808} = 0.674$$

$$b_0 = \bar{y} - b_1\bar{x} = 31.688 - 0.674 * 0.3192 = 31.47$$

$$S_{xx} = \frac{\left[n\sum x^2 - (\sum x)^2\right]}{n} = \frac{25 * 3.4202 - 7.9808 * 7.9808}{25} = 0.8724$$

$$S_{yy} = \frac{\left[n\sum y^2 - (\sum y)^2\right]}{n} = \frac{25 * 25341.90 - 792.2191 * 792.2191}{25} = 237.45$$

$$S_{xy} = \frac{[n\sum xy - (\sum x)(\sum y)]}{n} = \frac{25 * 253.49 - 7.9808 * 792.2191}{25} = 0.5883$$

Table 3.2 Computation of various intermediate outputs for case (a)

Dataset	P5_U (m/s) (x)	Maximum temperature (°C) (y)	x^2	y^2	xy
1	0.0866	29.2900	0.0075	857.9041	2.5365
2	0.1498	32.8321	0.0224	1077.9468	4.9182
3	0.4612	35.7261	0.2127	1276.3542	16.4769
4	0.2085	37.4637	0.0435	1403.5288	7.8112
5	0.2741	35.4984	0.0751	1260.1364	9.7301
6	0.1776	34.7860	0.0315	1210.0658	6.1780
7	0.2575	29.8706	0.0663	892.2527	7.6917
8	0.3961	30.8523	0.1569	951.8644	12.2206
9	0.1344	30.5450	0.0181	932.9970	4.1052
10	0.4620	30.8277	0.2134	950.3471	14.2424
11	0.3486	29.0377	0.1215	843.1880	10.1225
12	0.5922	26.6316	0.3507	709.2421	15.7712
13	0.4589	28.4584	0.2106	809.8805	13.0596
14	0.7945	31.6171	0.6312	999.6410	25.1198
15	0.2470	34.1303	0.0610	1164.8774	8.4302
16	0.3151	36.3697	0.0993	1322.7551	11.4601
17	0.4231	38.2229	0.1790	1460.9901	16.1721
18	0.6213	32.1490	0.3860	1033.5582	19.9742
19	0.1849	30.4881	0.0342	929.5242	5.6372
20	0.2417	29.6145	0.0584	877.0186	7.1578
21	0.3178	30.8463	0.1010	951.4942	9.8030
22	0.5320	31.1348	0.2830	969.3758	16.5637
23	0.0423	29.5313	0.0018	872.0977	1.2492
24	0.2335	27.7787	0.0545	771.6562	6.4863
25	0.0201	28.5168	0.0004	813.2079	0.5732

Correlation $R = \frac{S_{xy}}{\sqrt{S_{xx}S_{yy}}} = \frac{0.5883}{\sqrt{0.8724 * 237.45}} = 0.0408$ (very low value); preferred value is 1.

Regression equation is

$$y = 31.47 + 0.674\,x \tag{3.13}$$

(b) Polynomial relationship between predictand (temperature) and predictor (Zonal Velocity Component at 500 Hpa Height in m/s) in the form of $y = b_0 + b_1 x + b_2 x^2$ (Table 3.3).
n = number of datasets = 25

Table 3.3 Computation of various intermediate outputs for case (b)

Dataset	P5_U (m/s) (x)	Maximum temperature (°C) (y)	x^2	x^3	x^4	xy	x^2y
1	0.0866	29.2900	0.0075	0.0006	0.0001	2.5365	0.2197
2	0.1498	32.8321	0.0224	0.0034	0.0005	4.9182	0.7368
3	0.4612	35.7261	0.2127	0.0981	0.0452	16.4769	7.5991
4	0.2085	37.4637	0.0435	0.0091	0.0019	7.8112	1.6286
5	0.2741	35.4984	0.0751	0.0206	0.0056	9.7301	2.6670
6	0.1776	34.7860	0.0315	0.0056	0.0010	6.1780	1.0972
7	0.2575	29.8706	0.0663	0.0171	0.0044	7.6917	1.9806
8	0.3961	30.8523	0.1569	0.0621	0.0246	12.2206	4.8406
9	0.1344	30.5450	0.0181	0.0024	0.0003	4.1052	0.5517
10	0.4620	30.8277	0.2134	0.0986	0.0456	14.2424	6.5800
11	0.3486	29.0377	0.1215	0.0424	0.0148	10.1225	3.5287
12	0.5922	26.6316	0.3507	0.2077	0.1230	15.7712	9.3397
13	0.4589	28.4584	0.2106	0.0966	0.0443	13.0596	5.9930
14	0.7945	31.6171	0.6312	0.5015	0.3985	25.1198	19.9577
15	0.2470	34.1303	0.0610	0.0151	0.0037	8.4302	2.0823
16	0.3151	36.3697	0.0993	0.0313	0.0099	11.4601	3.6111
17	0.4231	38.2229	0.1790	0.0757	0.0320	16.1721	6.8424
18	0.6213	32.1490	0.3860	0.2398	0.1490	19.9742	12.4100
19	0.1849	30.4881	0.0342	0.0063	0.0012	5.6372	1.0423
20	0.2417	29.6145	0.0584	0.0141	0.0034	7.1578	1.7300
21	0.3178	30.8463	0.1010	0.0321	0.0102	9.8030	3.1154
22	0.5320	31.1348	0.2830	0.1506	0.0801	16.5637	8.8119
23	0.0423	29.5313	0.0018	0.0001	0.0000	1.2492	0.0528
24	0.2335	27.7787	0.0545	0.0127	0.0030	6.4863	1.5146
25	0.0201	28.5168	0.0004	0.0000	0.0000	0.5732	0.0115
Sum	7.9808	792.2191	3.4202	1.7436	1.0022	253.491	107.9448

$$\sum x = 7.9808; \sum y = 792.2191; \sum x^2 = 3.4202; \sum x^3 = 1.7436; \sum x^4 = 1.0022$$
$$\sum xy = 253.49; \sum x^2 y = 107.9448$$

$$nb_0 + b_1 \sum_{i=1}^{n} x_i + b_2 \sum_{i=1}^{n} x_i^2 = \sum_{i=1}^{n} y_i$$

$$b_0 \sum_{i=1}^{n} x_i + b_1 \sum_{i=1}^{n} x_i^2 + b_2 \sum_{i=1}^{n} x_i^3 = \sum_{i=1}^{n} x_i y_i$$

$$b_0 \sum_{i=1}^{n} x_i^2 + b_1 \sum_{i=1}^{n} x_i^3 + b_2 \sum_{i=1}^{n} x_i^4 = \sum_{i=1}^{n} x_i^2 y_i$$

$$25b_0 + 7.9808b_1 + 3.4202b_2 = 792.2191 \tag{3.14}$$

$$7.9808b_0 + 3.4202b_1 + 1.7436b_2 = 253.49 \tag{3.15}$$

$$3.4202b_0 + 1.7436b_1 + 1.0022b_2 = 107.9448 \tag{3.16}$$

Solving simultaneous equations yields:

$$\begin{aligned} &b_0 = 29.59; b_1 = 14.49;\ b_2 = -18.49 \\ &y = b_0 + b_1x + b_2x^2 = 29.59 + 14.49x - 18.49x^2 \end{aligned} \tag{3.17}$$

(c) Multiple linear relationship between predictand (temperature) and predictors (Zonal Velocity Component at 500 Hpa Height in m/s and Meridoinal Velocity

Table 3.4 Computation of various intermediate outputs for case (c)

Dataset	P5_U (m/s) (x_{i1})	P5_V (m/s) (x_{2i})	Maximum temperature (°C) (y_i)	$x_{i1} * x_{2i}$	$x_{i1} * y_i$	x_{i1}^2	x_{2i}^2	$x_{2i} * y_i$
1	0.0866	0.0006	29.2900	0.0001	2.5365	0.0075	0.0000	0.0176
2	0.1498	0.8393	32.8321	0.1257	4.9182	0.0224	0.7044	27.5560
3	0.4612	0.2654	35.7261	0.1224	16.4769	0.2127	0.0704	9.4817
4	0.2085	0.4358	37.4637	0.0909	7.8112	0.0435	0.1899	16.3267
5	0.2741	0.4124	35.4984	0.1130	9.7301	0.0751	0.1701	14.6395
6	0.1776	0.4254	34.7860	0.0756	6.1780	0.0315	0.1810	14.7980
7	0.2575	0.6898	29.8706	0.1776	7.6917	0.0663	0.4758	20.6047
8	0.3961	0.3365	30.8523	0.1333	12.2206	0.1569	0.1132	10.3818
9	0.1344	0.8079	30.5450	0.1086	4.1052	0.0181	0.6527	24.6773
10	0.4620	0.6341	30.8277	0.2930	14.2424	0.2134	0.4021	19.5478
11	0.3486	0.5329	29.0377	0.1858	10.1225	0.1215	0.2840	15.4742
12	0.5922	0.5403	26.6316	0.3200	15.7712	0.3507	0.2919	14.3891
13	0.4589	0.6936	28.4584	0.3183	13.0596	0.2106	0.4811	19.7387
14	0.7945	0.2295	31.6171	0.1823	25.1198	0.6312	0.0527	7.2561
15	0.2470	0.5656	34.1303	0.1397	8.4302	0.0610	0.3199	19.3041
16	0.3151	0.0797	36.3697	0.0251	11.4601	0.0993	0.0064	2.8987
17	0.4231	0.1995	38.2229	0.0844	16.1721	0.1790	0.0398	7.6255
18	0.6213	0.1268	32.1490	0.0788	19.9742	0.3860	0.0161	4.0765
19	0.1849	0.1248	30.4881	0.0231	5.6372	0.0342	0.0156	3.8049
20	0.2417	0.6392	29.6145	0.1545	7.1578	0.0584	0.4086	18.9296
21	0.3178	0.2921	30.8463	0.0928	9.8030	0.1010	0.0853	9.0102
22	0.5320	0.4352	31.1348	0.2315	16.5637	0.2830	0.1894	13.5499
23	0.0423	0.1007	29.5313	0.0043	1.2492	0.0018	0.0101	2.9738
24	0.2335	0.0754	27.7787	0.0176	6.4863	0.0545	0.0057	2.0945
25	0.0201	0.6405	28.5168	0.0129	0.5732	0.0004	0.4102	18.2650
Sum	7.9607	9.4825	763.702	3.0982	252.9178	3.4198	5.1662	299.1569

Component at 500 Hpa Height in m/s) in the form of $y = b_0 + b_1x_1 + b_2x_2$ (Table 3.4).

$$nb_0 + b_1\sum_{i=1}^{n} x_{1i} + b_2\sum_{i=1}^{n} x_{2i} = \sum_{i=1}^{n} y_i$$

$$b_0\sum_{i=1}^{n} x_{1i} + b_1\sum_{i=1}^{n} x_{1i}^2 + b_2\sum_{i=1}^{n} x_{1i}x_{2i} = \sum_{i=1}^{n} x_{i1}y_i$$

$$b_0\sum_{i=1}^{n} x_{2i} + b_1\sum_{i=1}^{n} x_{2i}x_{1i} + b_2\sum_{i=1}^{n} x_{2i}^2 = \sum_{i=1}^{n} x_{2i}y_i$$

$$25b_0 + 7.9607b_1 + 9.4825b_2 = 763.702 \tag{3.18}$$

$$7.9607b_0 + 3.4198b_1 + 3.0982b_2 = 252.9178 \tag{3.19}$$

$$9.4825b_0 + 3.0982b_1 + 5.1662b_2 = 299.1569 \tag{3.20}$$

$$\begin{aligned} &b_0 = 25.11;\ b_1 = 10.51;\ b_2 = 5.52 \\ &y = b_0 + b_1x_{1i} + b_2x_{2i} = 25.11 + 10.51x_{1i} + 5.52x_{2i} \end{aligned} \tag{3.21}$$

In the present section, multiple linear regression is discussed. Similar study can be extended to multiple non-linear regression.

3.2.2 Artificial Neural Networks

Artificial Neural Networks (ANN) are emerging as familiar and prominent due to their efficacy to establish non-linear relationships between inputs and outputs. In the present chapter, Feed Forward with Back Propagation (FFBP) category neural networks are described in brief. Mathematically, neural networks are composed of numerous layers of neurons consisting of input layer; number of hidden layers (minimum 1 is desirable) whereas it may extend to any number depending on the chosen problem; output layer. Each layer of neurons gets its input from the preceding layer or from the network input layer. Output of each neuron provides the information to the next layer/output of the network. Figure 3.1 presents ANN architecture consisting of three layers. Note that input layer consists of a, b, and c neurons, hidden layer (one in this case) consists of d, e, f, and g whereas output layer consists of neuron h. Neurons can also be termed as nodes. The lines connecting the neurons represent the weights.

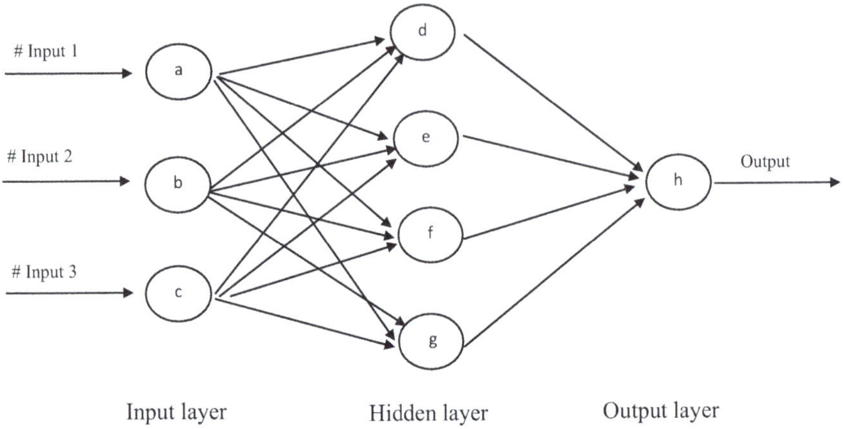

Fig. 3.1 Architecture of feed forward with back propagation

Terminology

Some of the important terminology relevant to FFBP is described to explain the methodology effectively (Rumelhart and McClelland 1986; ASCE Task Committee on Application of Artificial Neural Networks in Hydrology 2000; Nagesh Kumar 2004; Ross 2011).

- Preprocessing is the procedure for data processing before applying to the network.
- Weights: Each connection link has strength and is represented in the form of weights and may vary from $(-\infty, \infty)$ and update these weights continuously in the training process.
- Epoch is iteration and iteratively alters the network weights.
- Activation function estimates the output of a node/neuron from the total inputs they receive. Sigmoid function is one of the frequently used activation functions with characteristics of non-decreasing, bounded, and monotonic that provides graded non-linear surface. Mathematical expression of sigmoid function is as follows:

$$f(t) = \frac{1}{\left(1 + e^{\sum -r_i x_i}\right)} \tag{3.22}$$

Here, r_i and x_i are weights and corresponding input values. Selection of activation function has significance on the output. Other forms of activation functions are bipolar sigmoid and hyperbolic tangent.

- Training is a process through which network connection strengths/weights are improved continuously. In other words, training or learning process is to determine optimal weights that minimize the error. Error is the squared difference between the simulated values predicted by the network at output layer, and the observed/targeted values (Eq. 3.23).

$$E = \sum (O - P)^2 \tag{3.23}$$

  Here, P and O represent simulated output from ANN and observed/targeted value, respectively. Analogous analysis can also be made for cross training, testing, and validation using the optimized weights obtained from training.
- Learning rate increases the opportunity for the training process to reach global minima instead of trapping at local minima which may result in erroneous outputs. Learning rate varies from 0 to 1. The network learns and understands slowly with low learning rate whereas weights and objective function diverge with high learning rate.
- Momentum factor minimizes fluctuations in weights and improves training process.

FFBP technique consists of two phases (Rumelhart and McClelland 1986; ASCE Task Committee on Application of Artificial Neural Networks in Hydrology, 2000), forward and backward phases: Phase 1 is related to forward movement/pass where the impact of the input(s) is transmitted in forward direction to reach the last layer, i.e., output layer. In phase 2, error/errors (Eq. 3.23; depending on the single output or multiple outputs) are backpropagated toward the first/input layer with the weights being modified progressively. The training initiates with weights which are random in nature. Later they are modified throughout the process to achieve minimum error.

Weight adjustment is as follows: Change in weight at epoch n = −learning rate * rate of variation of error with reference to weight + momentum rate * Change in weight at epoch (n − 1). Mathematically, it can be expressed as:

$$\Delta r_{ij}(n) = -L_r * \frac{\partial E}{\partial r_{ij}} + M_r * \Delta r_{ij}(n-1) \tag{3.24}$$

where L_r and M_r are learning rate and momentum rate, respectively. $\Delta r_{ij}(n)$ and $\Delta r_{ij}(n-1)$ are increments of weights between node i and node j during nth and (n − 1)th epochs.

$$r_{ij}(new) = r_{ij}(old) + \Delta r_{ij} \tag{3.25}$$

This process is necessary to determine the final weights and the process will be terminated after reaching error tolerance/targeted number of epochs whichever is earlier.

Establishment of non-linear relationship between input and outputs and adaptivity are always helpful to achieve better results. However, major challenges such as hidden layers and corresponding neurons, initialization of momentum and learning rates, initialization of weights and their ranges, scaling the parameters in the form of normalization for better compatibility, and reasonably good training are to be handled with more caution. In addition, number of input and output parameters is to be chosen efficiently. Fewer inputs and more outputs or vice versa without physical significance and relevance to the chosen problem may not provide the desired results. In this case, output cannot be transformed for implementation.

Numerical Problem 3.2 Three inputs of rainfall of magnitude, 1, 2, and 3 units are producing runoff equivalent of 0.6996 units. Relationship between inputs and outputs are expected to be established through ANN with backpropagation methodology. Three-layered network is proposed. Input layer consists of three nodes representing rainfall (named as a, b, and c), hidden layer of four nodes (named as d, e, f, and g), and output layer for runoff of one node (named as h). Figure 3.1 is presenting the architecture of the ANN. Notation for weights is represented as r. Table 3.5 presents initial weights between the inputs to hidden layer as well as hidden to output layer. Assume learning rate as 0.4. Choose sigmoid as the activation function. Estimate the error at the output node and update the weights. Perform for one epoch.

Solution

Input values x_1, x_2, x_3 are 1, 2, 3; Observed value O is 0.6996

Epoch 1

1. Predicted outputs from the second layer

$$O_d = \frac{1}{1+e^{-(r_{ad}*x_1+r_{bd}*x_2+r_{cd}*x_3)}} = \frac{1}{1+e^{-(0.03*1+0.04*2+0.05*3)}} = \frac{1}{1+e^{-0.26}} = 0.5646$$

$$O_e = \frac{1}{1+e^{-(r_{ae}*x_1+r_{be}*x_2+r_{ce}*x_3)}} = \frac{1}{1+e^{-(0.13*1+0.07*2+0.03*3)}} = \frac{1}{1+e^{-0.36}} = 0.5890$$

$$O_f = \frac{1}{1+e^{-\left(r_{af}*x_1+r_{bf}*x_2+r_{cf}*x_3\right)}} = \frac{1}{1+e^{-(0.12*1+0.06*2+0.09*3)}} = \frac{1}{1+e^{-0.51}} = 0.6248$$

$$O_g = \frac{1}{1+e^{-\left(r_{ag}*x_1+r_{bg}*x_2+r_{cg}*x_3\right)}} = \frac{1}{1+e^{-(0.08*1+0.06*2+0.05*3)}} = \frac{1}{1+e^{-0.35}} = 0.5866$$

Table 3.5 Node assignment data and assumed weights

From node	To node	Notation for weight	Assumed weight
a	d	r_{ad}	0.03
b	d	r_{bd}	0.04
c	d	r_{cd}	0.05
a	e	r_{ae}	0.13
b	e	r_{be}	0.07
c	e	r_{ce}	0.03
a	f	r_{af}	0.12
b	f	r_{bf}	0.06
c	f	r_{cf}	0.09
a	g	r_{ag}	0.08
b	g	r_{bg}	0.06
c	g	r_{cg}	0.05
d	h	r_{dh}	0.22
e	h	r_{eh}	0.24
f	h	r_{fh}	0.11
g	h	r_{gh}	0.12

2. Predicted output from the third layer

$$O_h = \frac{1}{1+e^{-\left(r_{dh}*o_d + r_{eh}*o_e + r_{fh}*o_f + r_{gh}*o_g\right)}}$$
$$= \frac{1}{1+e^{-(0.22*0.5646 + 0.24*0.5890 + 0.11*0.6248 + 0.12*0.5866)}}$$
$$= \frac{1}{1+e^{-0.4047}} = 0.5998$$

3. Error E_h = Observed runoff − Predicted runoff = 0.6996 − 0.5998 = 0.0998
4. Distribution of errors is as follows:
   $E_{j-1} = O_n(1 - O_n)\sum r_{nj}E_j$ [If j is h, j − 1 is d, e, f, g]

   Assigning errors to elements in second layer

$$E_d = O_d * (1 - O_d) * (r_{dh} * E_h)$$
$$= 0.5646(1 - 0.5646) * (0.22 * 0.0998) = 0.00539$$
$$E_e = O_e * (1 - O_e) * (r_{eh} * E_h)$$
$$= 0.5890(1 - 0.5890) * (0.24 * 0.0998) = 0.005798$$
$$E_f = O_f * \left(1 - O_f\right) * (r_{fh} * E_h)$$
$$= 0.6248 * (1 - 0.6248) * (0.11 * 0.0998) = 0.00257$$
$$E_g = O_g * \left(1 - O_g\right) * (r_{gh} * E_h)$$
$$= 0.5866 * (1 - 0.5866) * (0.12 * 0.0998) = 0.0029$$

Update the weights based on the associated errors for all the elements in the network using the following equation:

$$r^i_{jk}new = r^i_{jk}old + L_r * E^{i+1}_k * x_{jk}$$

where r^i_{jk} is associated weight relating jth neuron of ith layer to the kth neuron of (i + 1)th layer; L_r is learning rate (0.4). E^{i+1}_k error associated with the kth neuron of (i + 1)th layer; x_{jk} is input from the jth neuron in the ith layer to the kth neuron in the (i + 1)th layer.

Table 3.6 presents updated weights connecting neurons in first and second layers whereas Table 3.7 presents updated weights connecting neurons in second and third layers.

Epoch 2

$$O_d = 0.5720,\ O_e = 0.5969,\ O_f = 0.6282,\ O_g = 0.6254$$
$$O_h = 0.6154$$

Table 3.6 Updated weights connecting neurons in first and second Layers

Notation for updated weight	Updated equation	Substitution	Updated weight
$r_{ad.new}$	$r_{ad.old} + L_r * E_d * x_1$	0.03 + 0.4 * 0.0053 * 1	0.0321
$r_{bd.new}$	$r_{bd.old} + L_r * E_d * x_2$	0.04 + 0.4 * 0.0053 * 2	0.0443
$r_{cd.new}$	$r_{cd.old} + L_r * E_d * x_3$	0.05 + 0.4 * 0.0053 * 3	0.0565
$r_{ae.new}$	$r_{ae.old} + L_r * E_e * x_1$	0.13 + 0.4 * 0.00579 * 1	0.1323
$r_{be.new}$	$r_{be.old} + L_r * E_e * x_2$	0.07 + 0.4 * 0.00579 * 2	0.0746
$r_{ce.new}$	$r_{ce.old} + L_r * E_e * x_3$	0.03 + 0.4 * 0.005798 * 3	0.0369
$r_{af.new}$	$r_{af.old} + L_r * E_f * x_1$	0.12 + 0.4 * 0.00257 * 1	0.1210
$r_{bf.new}$	$r_{bf.old} + L_r * E_f * x_2$	0.06 + 0.4 * 0.00257 * 2	0.0620
$r_{cf.new}$	$r_{cf.old} + L_r * E_f * x_3$	0.09 + 0.4 * 0.00257 * 3	0.0931
$r_{ag.new}$	$r_{ag.old} + L_r * E_g * x_1$	0.08 + 0.4 * 0.00290 * 1	0.0812
$r_{bg.new}$	$r_{bg.old} + L_r * E_g * x_2$	0.06 + 0.4 * 0.00290 * 2	0.0623
$r_{cg.new}$	$r_{cg.old} + L_r * E_g * x_3$	0.05 + 0.4 * 0.00290 * 3	0.0535

Table 3.7 Updated weights connecting neurons in second and third layers

Notation for updated weight	Updated equation	Substitution	Updated weight
$r_{dh.new}$	$r_{dh.old} + L_r * E_h * E_d$	0.22 + 0.4 * 0.0998 * 0.5646	0.2425
$r_{eh.new}$	$r_{eh.old} + L_r * E_h * E_e$	0.24 + 0.4 * 0.0998 * 0.5890	0.2635
$r_{fh.new}$	$r_{fh.old} + L_r * E_h * E_f$	0.11 + 0.4 * 0.0998 * 0.6248	0.1349
$r_{gh.new}$	$r_{gh.old} + L_r * E_h * E_g$	0.12 + 0.4 * 0.0998 * 0.5866	0.1434

Error E = Observed runoff − Predicted runoff = 0.6996 − 0.6154 = 0.0842

Process continues until number of epochs is completed or error termination criterion is reached whichever occurs earlier.

Numerical Problem 3.3 Three inputs of rainfall of magnitude, 0.8, 0.4, and 0.3 are producing runoff equivalent of 1.0 units. Relationship between inputs and outputs are established through ANN with backpropagation methodology. Three-layered network is proposed. Input layer consists of three nodes representing rainfall (named as 1, 2, and 3), hidden layer of three nodes (named as 4, 5, 6), and output layer for runoff of one node (named as 7). Notation for weights is represented as r. Table 3.8 presents initial weights between the inputs to hidden layer as well as hidden to output layer. Assume learning rate as 0.45. Choose sigmoid as the activation function. Estimate the error at the output node and update the weights after incorporating the error. Perform for one epoch.

Solution

Input values x_1, x_2, and x_3 are 0.8, 0.4, and 0.3; Observed value O is 1.0

Epoch 1

1. Predicted outputs from the second layer

$$O_4 = \frac{1}{1+e^{-(r_{14}*x_1+r_{24}*x_2+r_{34}*x_3)}} = \frac{1}{1+e^{-(045*0.8+0.55*0.4+0.87*0.3)}} = \frac{1}{1+e^{-0.841}} = 0.6987$$

$$O_5 = \frac{1}{1+e^{-(r_{15}*x_1+r_{25}*x_2+r_{35}*x_3)}} = \frac{1}{1+e^{-(0.57*0.8+0.62*0.4+0.92*0.3)}} = \frac{1}{1+e^{-0.98}} = 0.7271$$

$$O_6 = \frac{1}{1+e^{-(r_{16}*x_1+r_{26}*x_2+r_{36}*x_3)}} = \frac{1}{1+e^{-(0.73*0.8+0.88*0.4+0.94*0.3)}} = \frac{1}{1+e^{-1.218}} = 0.7717$$

Table 3.8 Node assignment data and assumed weights

From node	To node	Notation for weight	Assumed weight
1	4	r_{14}	0.45
2	4	r_{24}	0.55
3	4	r_{34}	0.87
1	5	r_{15}	0.57
2	5	r_{25}	0.62
3	5	r_{35}	0.92
1	6	r_{16}	0.73
2	6	r_{26}	0.88
3	6	r_{36}	0.94
4	7	r_{47}	0.8
5	7	r_{57}	0.9
6	7	r_{67}	1.0

2. Predicted output of the third layer

$$O_7 = \frac{1}{1+e^{-(r_{47}*o_4+r_{57}*o_5+r_{67}*o_6)}} = \frac{1}{1+e^{-(0.8*0.6987+0.90*0.7271+1.0*0.7717)}}$$
$$= \frac{1}{1+e^{-1.985}} = 0.8792$$

3. Error E_7 = Observed runoff − Predicted runoff = 1 − 0.8792 = 0.1208
4. Distribution of errors is as follows:
   $E_{j-1} = O_n(1 - O_n)\sum r_{nj}E_j$ [If j is h, j − 1 is 4, 5, 6]

   Assigning errors to elements in second layer

$$E_4 = O_4 * (1 - O_4) * (r_{47} * E_7)$$
$$= 0.6987(1 - 0.6987) * (0.80 * 0.1208) = 0.0203$$
$$E_5 = O_5 * (1 - O_5) * (r_{57} * E_7)$$
$$= 0.7271(1 - 0.7271) * (0.90 * 0.1208) = 0.0215$$
$$E_6 = O_6 * (1 - O_6) * (r_{67} * E_7)$$
$$= 0.7717 * (1 - 0.7717) * (1.0 * 0.1208) = 0.0212$$

Update the weights based on the associated errors for all the elements in the network using the following equation:

$$r^i_{jk}new = r^i_{jk}old + L_r * E^{i+1}_k * x_{jk}$$

where r^i_{jk} is associated weight relating jth neuron of ith layer to the kth neuron of (i + 1)th layer; L_r is learning rate (0.45). E^{i+1}_k error associated with the kth neuron of (i + 1)th layer; x_{jk} is input from the jth neuron in the ith layer to the kth neuron in the (i + 1)th layer. Updated weights are presented in Table 3.9.

After the updated weights are incorporated output value is 0.8886 and error is 1 − 0.8886 = 0.1114 which is less than previous, i.e., 0.1208. However, the process can continue until error is within tolerance limit.

3.2.3 Statistical Downscaling Model

Statistical Downscaling Model (SDSM) is a combination of regression and conditional weather generator techniques. SDSM software handles downscaling of daily weather series effectively. Calibrated modeling option formulates regression model by taking predictand and predictor data. Coefficients of the regression equations were determined using simplex algorithm. Screening option helps to keep the predictor variables that are relevant and SDSM has a provision of handling 12

Table 3.9 Updated weights

From node	To node	Notation for updated weight	Substitution	Updated weight
1	4	$r_{14.new}$	0.45 + 0.45 * 0.0203 * 0.8	0.4573
2	4	$r_{24.new}$	0.55 + 0.45 * 0.0215 * 0.4	0.5536
3	4	$r_{34.new}$	0.87 + 0.45 * 0.0212 * 0.3	0.8728
1	5	$r_{15.new}$	0.57 + 0.45 * 0.0203 * 0.8	0.5773
2	5	$r_{25.new}$	0.62 + 0.45 * 0.0215 * 0.4	0.6227
3	5	$r_{35.new}$	0.92 + 0.45 * 0.0212 * 0.3	0.9228
1	6	$r_{16.new}$	0.73 + 0.45 * 0.0203 * 0.8	0.7373
2	6	$r_{26.new}$	0.88 + 0.45 * 0.0215 * 0.4	0.8838
3	6	$r_{36.new}$	0.94 + 0.45 * 0.0212 * 0.3	0.9425
4	7	$r_{47.new}$	0.8 + 0.45 * 0.1208 * 0.6987	0.8379
5	7	$r_{57.new}$	0.9 + 0.45 * 0.1208 * 0.7271	0.9395
6	7	$r_{67.new}$	1.0 + 0.45 * 0.1208 * 0.7717	1.0419

predictors at a given time. Numerous authors suggested to enhance predictors' selection procedure. Brief features of SDSM are as follows: (Wilby et al. 2002; Wilby and Dawson 2007, 2013):

- Quality control identifies data errors, missing data and outliers.
- Transform provides selected data transformation.
- Screen variables operation is to select relevant downscaling predictor variables.
- Calibration of model establishes relationship between predictand and set of predictor variables.
- Definition of model structure such as monthly, seasonal, or annual sub-models.
- Conditional or unconditional processing of data.
- Weather and Scenario Generators compute ensembles of synthetic daily weather series with NCEP reanalysis/observed data and climate model atmospheric predictor variables rather than observed predictors, respectively.

3.2.4 Change Factor Technique

Change factor technique is another simple but effective downscaling technique which is explained as follows: (Hay et al. 2000; Anandhi et al. 2011; Sylwia and Emilie 2014):

1. Computation of average values of GCM simulated baseline GCMBL (20C3M) and future climates GCMFC

$$AVGCMBL = \frac{\sum GCMBL_i}{NBL} \tag{3.26}$$

$$AVGCMFC = \frac{\sum GCMFC_i}{NFC} \tag{3.27}$$

where AVGCMBL and AVGCMFC are average of GCM values of baseline and future; NBL and NFC are number of values in baseline and future scenarios. For example, for duration 1991–2010, NBL values are 20 on a monthly time step, i.e., there are 20 values of March. Similarly, for 2021–2050, NFL values are 30 on a monthly time step.

2. Two types of change factors are explored in the present study, namely, additive and multiplicative change factors that are expressed as follows:

$$CF_{additive} = AVGCMFC - AVGCMBL \tag{3.28}$$

$$CF_{multiplica} = \frac{AVGCMFC}{AVGCMBL} \tag{3.29}$$

3. Estimation of local scale future values $(LSCF_{additive}, LSCF_{multiplica})$ using $CF_{additive}$ and $CF_{multiplica}$.

$$LSCF_{additive,j} = LOBS_j + CF_{additive} \tag{3.30}$$

$$LSCF_{multiplica,j} = LOBS_j * CF_{multiplica} \tag{3.31}$$

where $LOBS_j$ is observed values of the climate variables at jth time step for any location; $LSCF_{additive}, LSCF_{multiplica}$ are values of future scenario obtained using additive and multiplicative change factors.

Numerical Problem 3.4 Local temperature (in °Celsius) data for 1971–1995 obtained from India Meteorological Department for January for a location in India is presented in Table 3.10 (column 2). Corresponding historic (baseline) temperature data (20C3M) in °Kelvin obtained from ACCESS 1.3 is also presented in column 4. Future temperature for 2021–2045 in °Kelvin obtained from ACCESS 1.3 is presented in column 6. Compute local temperature in °Celsius for the period 2021–2045 using Change factor analysis.

Solution

Local temperature (in °Celsius) in column 2 is converted to °Kelvin (column 3)
NBL and NFC = Number of values in baseline and future scenario = 25 and 25
Average of baseline data (Average of values in column 4) $= AVGCMBL = \frac{\sum GCMBL_i}{NBL} = \frac{7754}{25} = 310.16$

Average of future series data (Average of values in column 6) $= AVGCMFC$ $= \frac{\sum GCMFC_i}{NFC} = \frac{7910}{25} = 316.4$

Change factor value in additive scenario $= CF_{additive} = 316.4 - 310.16 = 6.24$

Change factor value in multiplicative scenario $= CF_{multiplicative}$ $= \frac{AVGCMFC}{AVGCMBL} = \frac{316.40}{310.16} = 1.02$

$$LSCF_{additive\,j} = LOBS_j + CF_{additive}$$

$$LSCF_{multiplica,j} = LOBS_j * CF_{multiplica}$$

For example, local scale temperature in 2021 using additive change factor is

$LSCF_{2021additive} = 302$ °K (local observed temperature at 1971) + 6.24 = 308.24 °K or 35.24 °C

$LSCF_{2021multiplica} = 302$ °K (local observed temperature at 1971) * 1.02 = 308.04 °K or 35.04 °C

Local scale variables for 2021–2045 for additive (values in columns 7 and 8) and multiplicative (values in columns 9 and 10) change factors are presented in Table 3.10.

3.2.5 Support Vector Machine

Most of the traditional neural network models seek to minimize the training error by implementing the empirical risk minimization principle, whereas the Support Vector Machine (SVMs) implements the structural risk minimization principle, which attempts to minimize an upper bound of the generalization error, by creating a right balance between the training error and the capacity of the machine (i.e., the ability of the machine to learn any training set without error). The solution of traditional neural network models may tend to fall into a local optimal solution, whereas global optimum solution is guaranteed in SVM (Haykin 2003; Anandhi et al. 2008, 2009; Anandhi 2010). Further, the traditional ANNs have considerable subjectivity in model architecture whereas, SVM's learning algorithm automatically decides the model architecture (number of hidden units). Moreover, traditional ANN models do not give much emphasis on generalization performance, while SVMs seek to address this issue in a rigorous theoretical setting. The flexibility of the SVM is provided by the use of kernel functions that implicitly maps the data to a higher, possibly infinite, and dimensional space. A linear solution, in the higher dimensional feature space, corresponds to a non-linear solution in the original lower dimensional input space. This makes SVM a plausible choice for solving a variety of problems in hydrology, which are non-linear in nature (Vapnik 1995, 1998; Schölkopf et al. 1998; Suykens 2001; Sastry 2003).

The Least Square Support Vector Machine (LS-SVM) provides a computational advantage over standard SVM (Suykens 2001; Tripathi et al. 2006; Anandhi et al. 2008). Consider a finite training sample of N patterns $\{(\mathbf{x}_i, y_i),\ i = 1, \ldots, N\}$, where $\mathbf{x}_i$ represents the 'i-th' pattern in n-dimensional space (i.e. $\mathbf{x}_i = [x_{1i}, \ldots, x_{ni}] \in \Re^n$)

Table 3.10 Temperature information and velocity component data

Year (1)	IMD Local temperature (°C) (2)	IMD Local temperature (°K) (3)	Historic 20C3M ACCESS 1.3 (°K) (4)	Year (5)	Future ACCESS 1.3 (°K) from 2021 to 2045 (6)	Local temperature (°K) from 2021 to 2045 additive change factor (7)	Local temperature (°C) from 2021 to 2045 additive change factor (8)	Local temperature (°K) from 2021 to 2045 multiplicative change factor (9)	Local temperature (°C) from 2021 to 2045 multiplicative change factor (10)
1971	29	302	300	2021	327	308.24	35.24	308.04	35.04
1972	32	305	303	2022	323	311.24	38.24	311.1	38.1
1973	35	308	309	2023	319	314.24	41.24	314.16	41.16
1974	37	310	312	2024	315	316.24	43.24	316.2	43.2
1975	35	308	297	2025	299	314.24	41.24	314.16	41.16
1976	34	307	298	2026	327	313.24	40.24	313.14	40.14
1977	29	302	283	2027	288	308.24	35.24	308.04	35.04
1978	30	303	292	2028	297	309.24	36.24	309.06	36.06
1979	31	304	284	2029	328	310.24	37.24	310.08	37.08
1980	30	303	279	2030	291	309.24	36.24	309.06	36.06
1981	29	302	291	2031	295	308.24	35.24	308.04	35.04
1982	26	299	292	2032	298	305.24	32.24	304.98	31.98
1983	28	301	301	2033	305	307.24	34.24	307.02	34.02
1984	31	304	302	2034	306	310.24	37.24	310.08	37.08
1985	34	307	317	2035	319	313.24	40.24	313.14	40.14
1986	36	309	321	2036	311	315.24	42.24	315.18	42.18
1987	38	311	326	2037	338	317.24	44.24	317.22	44.22
1988	32	305	328	2038	332	311.24	38.24	311.1	38.1
1989	30	303	329	2039	335	309.24	36.24	309.06	36.06

(continued)

Table 3.10 (continued)

Year (1)	IMD Local temperature (°C) (2)	IMD Local temperature (°K) (3)	Historic 20C3M ACCESS 1.3 (°K) (4)	Year (5)	Future ACCESS 1.3 (°K) from 2021 to 2045 (6)	Local temperature (°K) from 2021 to 2045 additive change factor (7)	Local temperature (°C) from 2021 to 2045 additive change factor (8)	Local temperature (°K) from 2021 to 2045 multiplicative change factor (9)	Local temperature (°C) from 2021 to 2045 multiplicative change factor (10)
1990	29	302	331	2040	338	308.24	35.24	308.04	35.04
1991	30	303	342	2041	321	309.24	36.24	309.06	36.06
1992	31	304	352	2042	322	310.24	37.24	310.08	37.08
1993	29	302	312	2043	324	308.24	35.24	308.04	35.04
1994	27	300	325	2044	325	306.24	33.24	306	33
1995	28	301	328	2045	327	307.24	34.24	307.02	34.02

constitutes the input to LS-SVM, and $y_i \in \Re$ is the corresponding value of the desired model output. Further, let the learning machine be defined by a set of possible mappings $\mathbf{x} \mapsto f(\mathbf{x}, \mathbf{w})$, where $f(\cdot)$ is a deterministic function, for a given input pattern X and adjustable parameters $\mathbf{W}(\mathbf{w} \in \Re^n)$, always give the same output. The training phase of the learning machine involves adjusting the parameters w. These parameters are estimated by minimizing the cost function $\psi_{\mathrm{L}}(\mathbf{w}, e)$.

$$\psi_{\mathrm{L}}(\mathbf{w}, e) = \frac{1}{2}\mathbf{w}^{\mathrm{T}}\mathbf{w} + \frac{1}{2}C\sum_{i=1}^{N} e_i^2 \tag{3.32}$$

Subjected to the equality constraint

$$y_i - \hat{y}_i = e_i \quad i = 1, \ldots, N \tag{3.33}$$

$$\hat{y}_i = w^T \phi(x) + b \tag{3.34}$$

where C is a positive real constant and $\hat{y}_i$ is the actual model output. The first term of the cost function represents weight decay or model complexity. It is used to regularize the weight sizes and to penalize the large weights. This helps in improving the generalization performance. The second term of the cost function represents penalty function.

The solution of the optimization problem is obtained by considering the Lagrangian as:

$$L(\mathbf{w}, b, \mathbf{e}, \boldsymbol{\alpha}) = \frac{1}{2}\mathbf{w}^{\mathrm{T}}\mathbf{w} + \frac{1}{2}C\sum_{i=1}^{N} e_i^2 - \sum_{i=1}^{N} \alpha_i\{\hat{y}_i + e_i - y_i\} \tag{3.35}$$

where α_i are Lagrange multipliers and b is the bias term defined in Eq. 3.34. The conditions for optimality are given by

$$\begin{cases} \frac{\partial L}{\partial \mathbf{w}} = \mathbf{w} - \sum_{i=1}^{N} \alpha_i \phi(\mathbf{x}_i) = 0 & \\ \frac{\partial L}{\partial b} = \sum_{i=1}^{N} \alpha_i = 0 & \\ \frac{\partial L}{\partial e_i} = \alpha_i - Ce_i = 0 & i = 1, \ldots, N \\ \frac{\partial L}{\partial \alpha_i} = \hat{y}_i + e_i - y_i = 0 & i = 1, .., N \end{cases} \tag{3.36}$$

The above conditions of optimality can be expressed as the solution to the following set of linear equations after elimination of $\mathbf{w}$ and e_i.

$$\begin{bmatrix} 0 & \vec{\mathbf{1}}^{\mathrm{T}} \\ \vec{\mathbf{1}} & \Omega + \mathrm{C}^{-1}\mathbf{I} \end{bmatrix} \begin{bmatrix} b \\ \boldsymbol{\alpha} \end{bmatrix} = \begin{bmatrix} 0 \\ \mathbf{y} \end{bmatrix} \tag{3.37}$$

$$\text{where } \mathbf{y} = \begin{bmatrix} y_1 \\ y_2 \\ \vdots \\ y_N \end{bmatrix}; \vec{\mathbf{1}} = \begin{bmatrix} 1 \\ 1 \\ \vdots \\ 1 \end{bmatrix}_{N\times 1}; \boldsymbol{\alpha} = \begin{bmatrix} \alpha_1 \\ \alpha_2 \\ \vdots \\ \alpha_N \end{bmatrix}; \mathbf{I} = \begin{bmatrix} 1 & 0 & \dots & 0 \\ 0 & 1 & \dots & 0 \\ \vdots & \vdots & \vdots & \vdots \\ 0 & 0 & \dots & 1 \end{bmatrix}_{N\times N}$$

In Eq. (3.37), $\boldsymbol{\Omega}$ is obtained from the application of Mercer's theorem.

$$\Omega_{i,j} = K(\mathbf{x}_i, \mathbf{x}_j) = \phi(\mathbf{x}_i)^T \phi(\mathbf{x}_j) \quad \forall i,j \tag{3.38}$$

where $\phi(\cdot)$ represents non-linear transformation function defined to convert a non-linear problem in the original lower dimensional input space to linear problem in a higher dimensional feature space.

The resulting LS-SVM model for function estimation is:

$$f(\mathbf{x}) = \sum \alpha_i^* K(\mathbf{x}_i, \mathbf{x}) + b^* \tag{3.39}$$

where α_i^* and b^* are the solutions to Eq. (3.37) and $K(\mathbf{x}_i, \mathbf{x})$ is the inner product kernel function defined in accordance with Mercer's theorem (Courant and Hilbert 1970; Mercer 1909) and b^* is the bias.

3.3 Multisite Downscaling

Downscaling techniques discussed in this chapter are capable for single location. However, multisite downscaling is conducted using Multivariate Multiple Linear Regression (MMLR) (Jeong et al. 2012), Multisite multivariate statistical downscaling (Khalili et al. 2013), Modified Markov Model-Kernel Density Estimation (MMM-KDE) modeling framework (Mehrotra and Sharma 2010; Mehrotra et al. 2013) or Support Vector Machine for Multisite downscaling (Srinivas et al. 2014). Researchers can also refer research papers by Nagesh Kumar et al. (2000), Wilby et al. (2004), Tripathi et al. (2006), Anandhi et al. (2008, 2012), Johnson and Sharma (2012) for more information about multisite downscaling.

3.4 Nested Bias Correction

Nested Bias correction (NBC) (Johnson and Sharma 2011, 2012) is a technique that compensates some of the shortcomings of GCM predicted rainfall values. It corrects for systematic biases of GCM outputs (e.g., mean, standard deviation, lag-one correlation, etc.) at multiple timescales and allows the use of GCM outputs directly in hydrologic studies. When combined with spatial disaggregation, bias correction techniques can provide model inputs suitable for hydrological studies (Hashino et al. 2007; Johnson and Sharma 2009, 2011, 2012; Mehrotra and Sharma 2010;

Mehrotra et al. 2013). In the present study, technique suggested by Johnson and Sharma (2011, 2012) is employed with the provision that bias in the lag-one autocorrelation statistic at the various time scales is neglected.

The NBC technique represents a nested procedure which addresses bias across prespecified multiple timescales which is as follows: Denoting a variable for month i in year k as $y_{i,k}$:

1. Standardization to create $y'_{i,k}$ by subtracting the model monthly mean $(\mu_{\text{mod},i})$ and dividing by the standard deviation $(\sigma_{\text{mod},i})$ for that month as shown in (3.40).

$$y'_{i,k} = \frac{y_{i,k} - \mu_{\text{mod},i}}{\sigma_{\text{mod},i}} \tag{3.40}$$

2. Interposition of the mean $(\mu_{obs,i})$ and standard deviation $(\sigma_{obs,i})$ of reanalysis data to create a transformed time series $y^*_{i,k}$ at the monthly level.

$$y^*_{i,k} = y'_{i,k}\sigma_{obs,i} + \mu_{obs,i} \tag{3.41}$$

3. Aggregation of transformed monthly series $(y^*_{i,k})$ into the annual scale z_k. The standardization and transformation steps are repeated at the annual time step.
4. Transform the annual time series to z^*_k which exhibits the mean and standard deviation in the recorded annual data.

Subsequent to the above steps, raw GCM simulation at the monthly time step is transformed by the NBC to

$$Y_{i,k} = y_{i,k}\left[\frac{y^*_{i,k}}{y_{i,k}}\right]\left[\frac{z^*_k}{z_k}\right] \tag{3.42}$$

where $Y_{i,k}$ represents NBC transformed variable. Using the transformation Eq. (3.42), the corrections at monthly and annual scales can be applied to the monthly time series and also to create a one-step correction (Srikanthan 2009). In Eq. (3.42), $\left[\frac{y^*_{i,k}}{y_{i,k}}\right]\left[\frac{z^*_k}{z_k}\right]$ is a weighing factor, i.e., the ratio of the monthly corrected value to the raw GCM value for month i and year k, multiplied by the ratio of the yearly corrected value to the aggregated GCM rainfall for year k. Above equations were used to transform the GCM simulations for the present climate.

The present chapter discusses various downscaling techniques including Nested Bias Correction (NBC) that are very useful to climate modeling situations. Next chapter discusses various statistical and optimization techniques.

Software (Information as on 30.12.2016)

Multiple linear/Non-linear Regression: Statistics and Machine Learning Tool box of MATLAB (http://in.mathworks.com/products/statistics/)

- SPSS (http://www-03.ibm.com/software/products/en/spss-statistics)
- MINITAB; Statistical Analysis System (SAS) http://www.sas.com/en_in/home.html
- XLSTAT (https://www.xlstat.com/en/)

Artificial Neural Networks: Neural Network Tool Box of MATLAB (http://in.mathworks.com/products/neural-network/)

- Statistical Package for Social Sciences (http://www-03.ibm.com/software/products/en/spss-statistics)
- MINITAB (https://www.minitab.com/en-us/)
- You can also view some of the related software through this link https://en.softonic.com/s/neural-network-software

Statistical Downscaling Model: You can register at this site http://co-public.lboro.ac.uk/cocwd/SDSM/index.html for downloading the software. You can mail R.L.Wilby@lboro.ac.uk for any queries.

Support Vector Machine (SVM): Statistics and Machine Learning Tool box of MATLAB.

- You can also access SVM-related software through this link http://www.support-vector-machines.org/SVM_soft.html; http://www.svms.org/software.html

Revision Questions and Exercise Problems

3.1 What is the purpose of downscaling? How many types of downscaling are possible?

3.2 What are the various available Statistical downscaling techniques?

3.3 What is the difference between Statistical and Dynamical downscaling?

3.4 What are different types of regression models? On what basis these can be differentiated?

3.5 What are the various steps in (a) regression (b) artificial neural networks?

3.6 What is the purpose of activation function and epoch in ANN?

3.7 What is the difference between training and learning ANN?

3.8 What is the purpose of momentum factor in ANN?

3.9 What is the procedure for weight adjustment in ANN?

3.10 Maximum temperature (in °Celsius) data generated from India Meteorological Department is dependent on number of parameters that are available from National Centers for Environmental Prediction (NCEP). Notable among them are P5_U (Zonal Velocity Component at 500 Hpa Height in m/s), P5_V (Meridoinal Velocity Component at 500 Hpa Height in m/s). Data is presented in Table 3.1. Establish (a) linear relationship between predictand (temperature) and predictor (Meridoinal Velocity Component at

500 Hpa Height in m/s) (b) polynomial relationship between predictand (temperature) and predictor (Meridoinal Velocity Component at 500 Hpa Height in m/s. Take only 15 data sets for evaluation.

3.11 Solve Numerical Problem 3.2 related to Artificial Neural Networks using data presented in Table 3.5. Consider only three nodes (d, e, and f) in hidden layer. $x_1 = 0.8$, $x_2 = 1.6$, and $x_3 = 2.4$, Observed value O is 0.7.

3.12 Solve Numerical Problem 3.4 related to change factor technique. Analyze the Numerical Problem from 1971 to 1985 for baseline and 2021–2035 for future scenarios. Use data in Table 3.10.

3.13 What is the difference between Support Vector Machine and Neural Networks?

3.14 What is SDSM? What is its purpose?

3.15 What is the change factor technique? What is the procedure for downscaling using change factor technique?

3.16 What is Multisite downscaling? How it is advantageous? What are the relevant techniques falling in this category?

3.17 What is difference between predictand and predictors?

3.18 What is purpose of Nested Bias Correction and how it is useful for downscaling?

Advanced Review Questions

3.19 Is there any possibility such that output of regression can be input of artificial neural networks or vice versa? If so, how these techniques can be complimented to each other? Provide examples of how these techniques help in climate modeling with practical significance?

3.20 Discuss the limitations of regression techniques?

3.21 Mention any other techniques that are falling in regression category?

3.22 Is there any possibility to relate multiple linear regression with support vector machine?

3.23 Mention names of six researchers who applied statistical techniques in the field of climate modeling? Mention the technique of application and case study of application.

3.24 Name relevant software that can be used for (a) regression (b) artificial neural networks. Provide salient points of the software.

3.25 What are the limitations of GCM for implementing its output to impact assessment?

3.26 How SDSM is different from other statistical downscaling techniques?

3.27 In your opinion, which downscaling technique is suitable for water resources planning? Justify with discussion.

3.28 How extreme events can be handled while downscaling? Discuss the suitability of statistical and dynamical downscaling in this regard.

References

Anandhi A (2010) Assessing impact of climate change on season length in Karnataka for IPCC scenarios. J Earth Syst Sci 119(4):447–460

Anandhi A, Srinivas VV, Nanjundiah RS, Kumar DN (2008) Downscaling precipitation to river basin in India for IPCC SRES scenarios using support vector machine. Int J Climatol 28 (3):401–420

Anandhi A, Srinivas VV, Kumar DN, Nanjundiah RS (2009) Role of predictors in downscaling surface temperature to river basin in India for IPCC SRES scenarios using support vector machine. Int J Climatol 29(4):583–603

Anandhi A, Frei A, Pierson DC, Schneiderman EM, Zion MS, Lounsbury D, Matonse AH (2011) Examination of change factor methodologies for climate change impact assessment. Water Resour Res 47:W03501

Anandhi A, Srinivas VV, Nagesh Kumar D, Nanjundiah RS (2012) Daily relative humidity projections in an indian river basin for IPCC SRES scenarios. Theoret Appl Climatol 108:85–104

ASCE Task Committee on Application of Artificial Neural Networks in Hydrology (2000) Artificial neural networks in hydrology 1: preliminary concepts. J Hydrol Eng 5(2):115–123

Courant R, Hilbert D (1970) Methods of mathematical physics, vol I, II. Wiley Interscience, New York, USA

Fowler HJ, Blenkinsop S, Tebaldi C (2007) Linking climate change modelling to impacts studies: recent advances in downscaling techniques for hydrological modelling. Int J Climatol 27 (12):1547–1578

Hashino T, Bradley AA, Schwartz SS (2007) Evaluation of bias-correction methods for ensemble streamflow forecasts. Hydrol Earth Syst Sci 11:939–950

Hay LE, Wilby RL, Leavesley GH (2000) A comparison of delta change and downscaled GCM scenarios for three mountainous basins in the United States. J Am Water Resour Assoc 36 (2):387–397

Haykin S (2003) Neural networks: a comprehensive foundation. Pearson Education, Singapore

Jeong DI, St-Hilaire A, Ouarda TBMJ, Gachon P (2012) A multivariate multi-site statistical downscaling model for daily maximum and minimum temperatures. Climate Res 54:129–148

Johnson F, Sharma A (2009) Measurement of GCM skill in predicting variables relevant for hydro climatological assessments. J Clim 22:4373–4382

Johnson F, Sharma A (2011) Accounting for interannual variability: a comparison of options for water resources climate change impact assessments. Water Resour Res 47:W04508

Johnson F, Sharma A (2012) A nesting model for bias correction of variability at multiple time scales in general circulation model precipitation simulations. Water Resour Res 48:W01504

Karl T, Wang W, Schlesinger M, Knight R, Portman D (1990) A method of relating general circulation model simulated climate to the observed local climate part I: seasonal statistics. J Clim 3:1053–1079

Khalili M, Nguyenb VTV, Gachonc P (2013) A statistical approach to multi-site multivariate downscaling of daily extreme temperature series. Int J Climatol 33(1):15–32

Mercer J (1909) Functions of positive and negative type and their connection with the theory of integral equations. Philos Trans R Soc A 209:415–446

Mehrotra R, Sharma A (2010) Development and application of a multisite rainfall stochastic downscaling framework for climate change impact assessment. Water Resour Res 46:W07526

Mehrotra R, Sharma A, Nagesh Kumar D, Reshmidevi TV (2013) Assessing future rainfall projections using multiple GCMs and a multi-site stochastic downscaling model. J Hydrol 488:84–100

Milton JS, Arnold JC (2007) Introduction to probability and statistics. Tata McGraw Hill

Mujumdar PP, Nagesh Kumar D (2012) Floods in a changing climate: hydrologic modeling. International hydrology series. Cambridge University Press, Cambridge

Nagesh Kumar D, Lall U, Peterson MR (2000) Multi-site disaggregation of monthly to daily streamflow. Water Resour Res 36(7):1823–1833
Nagesh Kumar D (2004) ANN applications in hydrology-merits and demerits. In: Srinivasa Raju K, Sarkar AK, Dash M (eds) Integrated water resources planning and management. Jain Brothers, New Delhi, pp. 31–42
Ross TJ (2011) Fuzzy logic with engineering applications. Wiley
Rumelhart D, McClelland J (1986) Parallel distributed processing. MIT Press, Cambridge, Mass
Sastry P (2003) An introduction to support vector machines. In: Computing and information sciences: recent trends. Narosa Publishing House, New Delhi
Schölkopf B, Burges C, Smola A (1998) Advances in kernel methods—support vector learning. MIT Press
Srikanthan R (2009) A nested multisite daily rainfall stochastic generation model. J Hydrol 371:142–153
Srinivas VV, Basu B, Nagesh Kumar D, Jain SK (2014) Multi-site downscaling of maximum and minimum daily temperature using support vector machine. Int J Climatol 34:1538–1560
Suykens JAK (2001) Nonlinear modelling and support vector machines. In: Proceedings of IEEE instrumentation and measurement technology conference, Budapest, Hungary, pp 287–294
Sylwia T, Emilie S (2014) A review of downscaling methods for climate change projections. In: African and latin american resilience to climate change (ARCC). USAID http://www.ciesin.org/documents/Downscaling_CLEARED_000.pdf. Accessed 31 Jan 2017)
Tripathi S, Srinivas VV, Nanjundiah RS (2006) Downscaling of precipitation for climate change scenarios: a support vector machine approach. J Hydrol 330:621–640
Vapnik V (1995) The nature of statistical learning theory. Springer, New York
Vapnik V (1998) Statistical learning theory. Wiley, New York
Wang Y, Leung LR, McGregor JL, Lee DK, Wang WC, Ding Y, Kimura F (2004) Regional climate modeling: progress challenges, and prospects. J Meteorol Soc Jpn 82:1599–1628
Wigley T, Jones P, Briffa K, Smith G (1990) Obtaining subgrid scale information from coarse-resolution general circulation model output. J Geophys Res-Atmosp 95:1943–1953
Wilby RL, Dawson CW, Barrow EM (2002) SDSM—a decision support tool for the assessment of regional climate change impacts. Environ Model Softw 17(2):147–159
Wilby RL, Charles SP, Zorita E, Timbal B, Whetton P, Mearns LO (2004) The guidelines for use of climate scenarios developed from statistical downscaling methods: supporting material of the intergovernmental panel on climate change (IPCC), prepared on behalf of task group on data and scenario support for impacts and climate analysis (TGICA). www.ipcc-data.org/guidelines/dgm_no2_v1_09_2004.pdf. Accessed 31 Jan 2017
Wilby RL, Dawson CW (2007) SDSM 4.2: a decision support tool for the assessment of regional climate change impacts—user manual. http://co-public.lboro.ac.uk/cocwd/SDSM/SDSMManual.pdf. Accessed 31 Jan 2017
Wilby RL, Dawson CW (2013) The statistical downscaling model (SDSM): insights from one decade of application. Int J Climatol 33(7):1707–1719

Suggested Further Reading

Benestad RE (2008) Empirical-statistical downscaling. World Scientific Publishing Co., Pte. Ltd, Singapore
Bishop CM (1995) Neural networks for pattern recognition. Oxford University Press, New York
Burden FR, Winkler DA (2015) Relevance vector machines: sparse classification methods for QSAR. J Chem Inf Model 55(8):1529–1534
Flood I, Kartam N (1994) Neural networks in civil engineering. I: principles and understandings. J Comput Civil Eng 8(2):131–148

Hastie T, Tibshirani (2011) The elements of statistical learning: data mining, inference, and prediction. Series in Statistics. Springer

Helsel DR, Hirsch RM (2002) Statistical methods in water resources techniques of water resources investigations, book 4, chapter A3, US Geological Survey, p. 522

Rojas R (1996) Neural networks: a systematic introduction. Springer, Berlin

Wilks DS (1999) Multi-site downscaling of daily precipitation with a stochastic weather generator. Climate Res 11:125–136

Statistical and Optimization Techniques in Climate Modeling

4

Abstract

This chapter presents data compression techniques, namely, cluster and fuzzy cluster analysis, Kohonen neural networks for clustering GCMs and principal component analysis for transforming a set of observations of possible correlation into a set of linearly uncorrelated variables applying an orthogonal transformation. F–statistic test which can be used as the basis for finding optimal clusters is also discussed. Trend detection techniques, namely, Kendall's rank correlation and turning point test along with mathematical background are also briefed with the objective to ascertain the quality of the hydrological or climatological records. In addition, optimization techniques, namely, linear and non-linear programming and genetic algorithms along with mathematical description are also discussed. The reader is expected to understand various statistical and optimization techniques along with their applicability by studying this chapter.

Keywords

Cluster · Data compression techniques · Fuzzy · Genetic algorithms · Kohonen neural networks · Linear · Non-linear · Optimization · Principal component analysis · Trend detection

Electronic supplementary material The online version of this chapter (doi:10.1007/978-981-10-6110-3_4) contains supplementary material, which is available to authorized users.

K. Srinivasa Raju and D. Nagesh Kumar, *Impact of Climate Change on Water Resources*, Springer Climate, https://doi.org/10.1007/978-981-10-6110-3_4

4.1 Introduction

Statistical and optimisation techniques are gaining importance in the field of climate modeling due to their capability of analyzing the data in an effective manner. Data compression techniques include Cluster and fuzzy cluster analysis, Kohonen Neural Networks (KNN), Principal Component Analysis (PCA) as well as trend detection and optimization techniques are discussed in this chapter.

4.2 Data Compression Techniques

Data compression techniques are explained with reference to GCMs and are as follows:

4.2.1 Cluster Analysis

K-means algorithm is used for clustering GCMs. Procedural steps to conduct K-means cluster analysis are presented in Fig. 4.1 (Jain and Dubes 1988; Raju and Nagesh Kumar 2014; Bezdek 1981):

Numerical Problem 4.1 Nine GCMs are evaluated on three indicators and presented in Table 4.1. Use K-means algorithm to cluster the GCMs. Consider number of clusters as 3.

Solution:

Goal: To group A1 to A9 into G1, G2, G3 based on the three indicators C_1, C_2, and C_3.

Iteration 0

- Number of clusters are 3. Initially, each GCM is assigned randomly to each cluster (column 5) in Table 4.1.
- Knowing the GCMs in each group (3 in G1, 2 in G2, 4 in G2), compute the mean of each group. Mean values are presented in Table 4.2.
- Column 2 of Table 4.3 corresponds to the Euclidean distance (error) of each GCM to the mean of first group G1, and the column 3 is the distance of each GCM to group G2. The process continues up to the last chosen group. For example, Euclidean distance between GCM A1 (Table 4.1) and group G1 (Table 4.2) can be computed as (Table 4.3).

$$\sqrt{(20-33.33)^2+(30-53.33)^2+(50-40)^2}=28.67$$

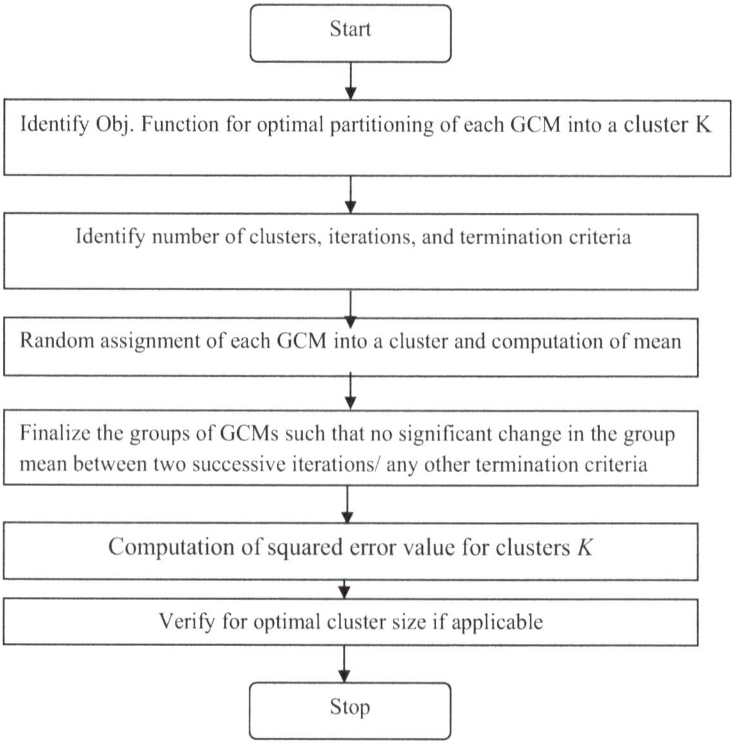

Fig. 4.1 Flowchart of procedural steps for K-means cluster analysis

Table 4.1 Data matrix of 9 GCMs and random assignment of clusters

GCMs (1)	C_1 (2)	C_2 (3)	C_3 (4)	Assigned random clusters/groups iteration 0 (5)
A1	20	30	50	G1
A2	60	70	40	G1
A3	20	60	30	G1
A4	20	70	32	G2
A5	30	42	22	G2
A6	72	43	31	G3
A7	22	11	12	G3
A8	16	18	21	G3
A9	21	17	19	G3

Table 4.2 Mean values of G1, G2, G3 based on random clusters for iteration 0

	C_1	C_2	C_3	Remark
G1	33.33 (20 + 60 + 20)/3	53.33 (30 + 70 + 60)/3	40.00 (50 + 40 + 30)/3	Average of A1, A2, A3 (G1)
G2	25.00 (20 + 30)/2	56.00 (70 + 42)/2	27.00 (32 + 22)/2	Average of A4, A5 (G2)
G3	32.75 (72 + 22 + 16 + 21)/4	22.25 (43 + 11 + 18 + 17)/4	20.75 (31 + 12 + 21 + 19)/4	Average of A6, A7, A8, A9 (G3)

Table 4.3 Euclidean distance of each GCM to the means for iteration 0

GCM (1)	G1 (2)	G2 (3)	G3 (4)	Minimum value among columns 2, 3, 4 (5)	Assigned group based on the minimum value iteration 1 (6)
A1	28.67 (1)	35.07 (0)	32.84 (0)	28.67	G1
A2	31.45 (1)	39.87 (0)	58.25 (0)	31.45	G1
A3	17.95 (0)	7.07 (1)	40.90 (0)	7.07	G2
A4	22.79 (0)	15.68 (1)	50.69 (0)	15.68	G2
A5	21.53 (0)	15.68 (1)	19.98 (0)	15.68	G2
A6	41.02 (1)	48.93 (0)	45.57 (0)	41.02	G1
A7	52.01 (0)	47.53 (0)	17.85 (1)	17.85	G3
A8	43.70 (0)	39.51 (0)	17.28 (1)	17.28	G3
A9	43.74 (0)	40.01 (0)	12.99 (1)	12.99	G3

Iteration 1

Assign A1 to A9 to groups G1 to G3 based on the minimum Euclidean distance (column 6, Table 4.3). Thus, A1 is assigned to G1 due to least distance among 28.67 (G1), 35.07 (G2), 32.84 (G3). Similar inferences can be made for other GCMs (column 6) of Table 4.3.

It is also to be noted that membership of A1, A2, A6 in G1 is 1, whereas their membership in G2 and G3 are 0. These memberships for each GCM are presented in Table 4.3 in the form of parenthesis. Similar inferences can be made for other GCMs (Table 4.3).

Iteration 2

Mean values and Euclidean distances are presented in Tables 4.4 and 4.5, respectively. It is observed that A2, A6 falls into G1; A1, A3, A4, A5 falls into G2; A7, A8, A9 falls into G3 (column 6 of Table 4.5).

Iteration 3

Mean values are presented in Table 4.6. It is observed that A2, A6 falls into G1; A1, A3, A4, A5 falls into G2; A7, A8, A9 falls into G3 (Table 4.7). There is no change in the grouping of GCMs as compared to iteration 2. The final grouping of the GCMs is presented in Table 4.8.

Determination of an optimal number of clusters for a group of datasets is very important for effective decision-making and notable techniques are Davies–Bouldin,

Table 4.4 Mean values of G1, G2, G3 based on assigned clusters for iteration 1

	C_1	C_2	C_3	Remark
G1	50.67	47.67	40.33	Average of A1, A2, A6 (G1)
G2	23.33	57.33	28.00	Average of A3, A4, A5 (G2)
G3	19.67	15.33	17.33	Average of A7, A8, A9 (G3)

Table 4.5 Euclidean distance of each GCM to the means for iteration 1

GCM (1)	G1 (2)	G2 (3)	G3 (4)	Minimum value among columns 2, 3, 4 (5)	Assigned group based on the minimum value iteration 2 (6)
A1	36.69 (0)	35.25 (1)	35.81 (0)	35.25	G2
A2	24.21 (1)	40.61 (0)	71.62 (0)	24.21	G1
A3	34.63 (0)	4.71 (1)	46.43 (0)	4.71	G2
A4	38.84 (0)	13.70 (1)	56.60 (0)	13.70	G2
A5	28.20 (0)	17.76 (1)	28.98 (0)	17.76	G2
A6	23.75 (1)	50.82 (0)	60.75 (0)	23.75	G1
A7	54.49 (0)	49.04 (0)	7.26 (1)	7.26	G3
A8	49.55 (0)	40.62 (0)	5.83 (1)	5.83	G3
A9	47.70 (0)	41.39 (0)	2.71 (1)	2.71	G3

Table 4.6 Mean values of G1, G2, G3 based on assigned clusters for iteration 2

	C_1	C_2	C_3	Remark
G1	66.00	56.50	35.50	Average of A2, A6 (G1)
G2	22.50	50.50	33.50	Average of A1, A3, A4, A5 (G2)
G3	19.67	15.33	17.33	Average of A7, A8, A9 (G3)

Table 4.7 Euclidean distance of each GCM to the means for iteration 2

GCM (1)	G1 (2)	G2 (3)	G3 (4)	Minimum value among columns 2, 3, 4 (5)	Assigned group based on the minimum value iteration 3 (6)
A1	55.03 (0)	26.43 (1)	35.81 (0)	26.43	G2
A2	15.44 (1)	42.76 (0)	71.62 (0)	15.44	G1
A3	46.46 (0)	10.43 (1)	46.43 (0)	10.43	G2
A4	48.07 (0)	19.72 (1)	56.60 (0)	19.72	G2
A5	41.09 (0)	16.15 (1)	28.98 (0)	16.15	G2
A6	15.44 (1)	50.13 (0)	60.75 (0)	15.44	G1
A7	67.52 (0)	44.97 (0)	7.26 (1)	7.26	G3
A8	64.75 (0)	35.42 (0)	5.83 (1)	5.83	G3
A9	62.11 (0)	36.53 (0)	2.71 (1)	2.71	G3

Dunn's indices (Davies and Bouldin 1979; Dunn 1974; Raju and Nagesh Kumar 2014) and F-Statistic test (Burn 1989). Brief details of F-Statistic test is presented in sub-Sect. 4.1.5 and information about Davies–Bouldin and Dunn's indices is available in Davies and Bouldin (1979), Dunn (1974), Raju and Nagesh Kumar (2014).

Table 4.8 Data matrix of 9 GCMs and assignment of clusters

GCM	C_1	C_2	C_3	Assigned random clusters/groups	Final partition
A1	20	30	50	G1	G2
A2	60	70	40	G1	G1
A3	20	60	30	G1	G2
A4	20	70	32	G2	G2
A5	30	42	22	G2	G2
A6	72	43	31	G3	G1
A7	22	11	12	G3	G3
A8	16	18	21	G3	G3
A9	21	17	19	G3	G3

4.2.2 Fuzzy Cluster Analysis

The procedure of fuzzy cluster analysis remains same as that of cluster analysis except that each dataset belongs to a cluster to degree (varies between zero and one). Procedural steps to conduct fuzzy cluster analysis are presented in Fig. 4.2 (Ross 2011; Bezdek 1981).

Numerical Problem 4.2 Nine GCMs are evaluated on three indicators and presented in Table 4.9. Data is following triangular membership function. Use fuzzy C-means algorithm to cluster the GCMs. The number of clusters can be taken as 2. Constant α value can be taken as 2. Compute membership function values after first iteration.

Solution:

Goal: To group A1 to A9 into G1 and G2 based on the three indicators C1, C2, and C3.

Iteration 0

- A number of clusters is 2. Initially, each GCM is assigned randomly to each cluster (column 5 in Tables 4.9 and 4.10).
- Knowing the GCMs in each group, compute the mean of each group, mean values are presented in Table 4.11.

  Distance between two fuzzy triangular numbers is

$$\sqrt{\frac{\left[\left(p_{aj}-p_j\right)^2+\left(q_{aj}-q_j\right)^2+\left(s_{ia}-s_j\right)^2\right]}{3}}$$

where (p, q, s) are elements of Triangular fuzzy number.

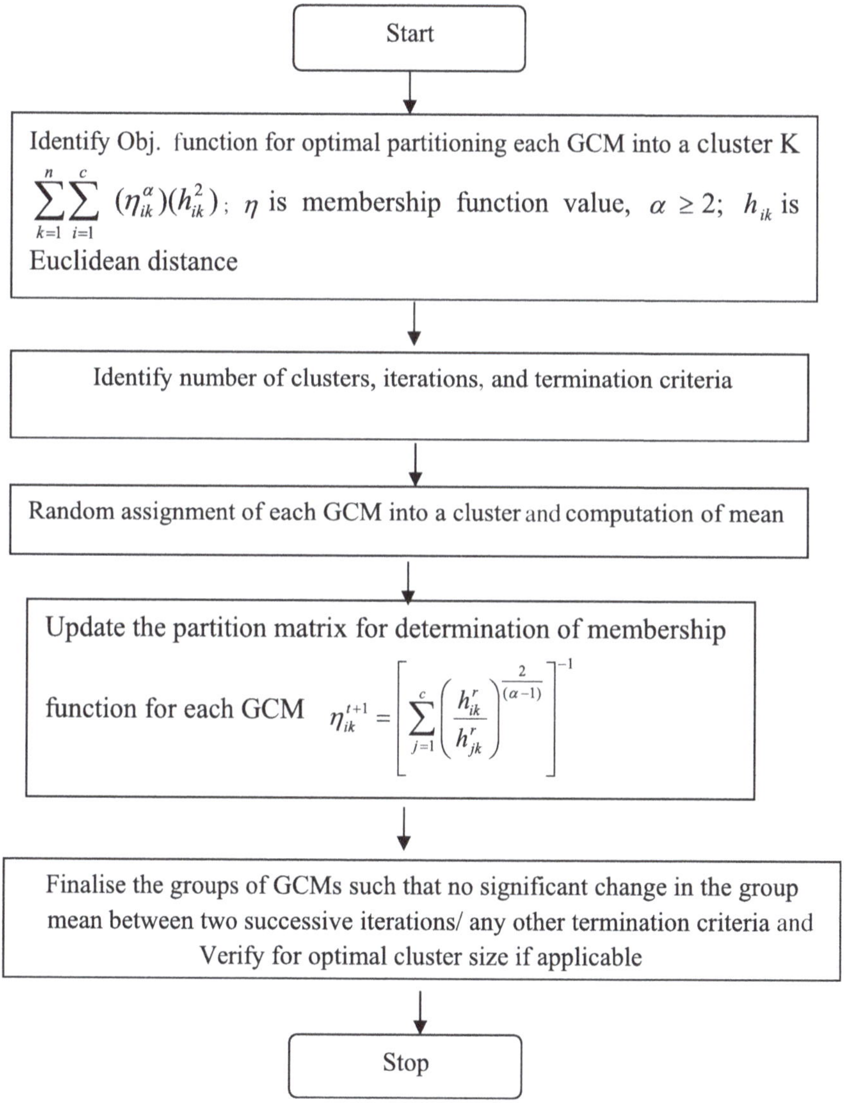

Fig. 4.2 Flow chart of procedural steps for fuzzy C-means

- Table 4.12 incorporates the Euclidean distance (error) of each GCM to the mean of first group G1 (columns 2, 3, 4), columns 6, 7, 8 correspond to the distance of each GCM to group G2 mean. Process continues up to the last chosen group. For example, Euclidean distance between GCM A1 (Table 4.9) and group G1 (Table 4.11) can be computed as (Table 4.12):

Table 4.9 Data matrix of 9 GCMs and random assignment of clusters

GCMs (1)	C_1 (2)	C_2 (3)	C_3 (4)	Assigned random clusters/groups iteration 0 (5)
A1	(2, 3, 4)	(4, 5, 6)	(7, 8, 10)	G1
A2	(5, 6, 7)	(8, 9, 12)	(4, 6, 8)	G1
A3	(3, 6, 9)	(8, 10, 12)	(6, 8, 10)	G1
A4	(6, 10, 12)	(1, 2, 4)	(4, 5, 8)	G2
A5	(5, 6, 7)	(4, 6, 8)	(1, 4, 6)	G2
A6	(3, 6, 9)	(2, 4, 6)	(1, 3, 5)	G2
A7	(2, 3, 5)	(4, 8, 12)	(3, 7, 8)	G1
A8	(2, 5, 8)	(1, 4, 8)	(2, 4, 8)	G2
A9	(1, 4, 7)	(5, 8, 9)	(6, 8, 12)	G1

Table 4.10 Random partition matrix/membership η of each GCM in groups G1 and G2

	A1	A2	A3	A4	A5	A6	A7	A8	A9
G1	1	1	1	0	0	0	1	0	1
G2	0	0	0	1	1	1	0	1	0

Table 4.11 Mean coordinate for G1 and G2 based on random clusters for iteration 0

	C1	C2	C3	Remark
G1	[2.6, 4.4, 6.4] [2 + 5 + 3 + 2 + 1)/5, (3 + 6 + 6 + 3 + 4)/5, (4 + 7 + 9 + 5 + 7)/5]	[5.8, 8, 10.2] [4 + 8 + 8 + 4 + 5)/5, (5 + 9 + 10 + 8 + 8)/5, (6 + 12 + 12 + 12 + 9)/5	[5.2, 7.4, 9.6] [7 + 4 + 6 + 3 + 6)/5, (8 + 6 + 8 + 7 + 8)/5, (10 + 8 + 10 + 8 + 12)/5	Average of A1, A2, A3, A7, A9 (G1)
G2	[4, 6.8, 9] [6 + 5 + 3 + 2)/4, (10 + 6 + 6 + 5)/4, (12 + 7 + 9 + 8)/4]	[2, 4, 6.5] [1 + 4 + 2 + 1)/4, (2 + 6 + 4 + 4)/4, (4 + 8 + 6 + 8)/4]	[2, 4, 6.75] [4 + 1 + 1 + 2)/4, (5 + 4 + 3 + 4)/4, (8 + 6 + 5 + 8)/4]	Average of A4, A5, A6, A8 (G2)

Table 4.12 Euclidean distance of each GCM from the group mean G1 and G2

GCM (1)	Distance from group mean of G1				Distance from group mean of G2			
	C1 (2)	C2 (3)	C3 (4)	Total error from group 1 (5)	C1 (6)	C2 (7)	C3 (8)	Total error from group 2 (9)
A1	1.64	3.16	1.12	5.92 (h_{11})	3.79	1.32	4.15	9.26 (h_{21})
A2	1.7	1.74	1.409	4.85 (h_{12})	1.36	5.52	1.79	8.67 (h_{22})
A3	1.78	2.01	0.622	4.41 (h_{13})	0.72	5.84	3.77	10.33 (h_{23})
A4	1.19	1.47	1.587	4.25 (h_{14})	3.37	4.09	1.96	9.42 (h_{24})
A5	1.01	0.83	1.501	3.34 (h_{15})	2.62	3.23	4.46	10.31 (h_{25})
A6	4.98	5.7	1.804	12.48 (h_{16})	2.8	1.94	1.48	6.22 (h_{26})
A7	1.7	2.01	3.749	7.46 (h_{17})	1.36	1.85	0.72	3.93 (h_{27})
A8	1.78	4	4.403	10.18 (h_{18})	0.72	0.29	1.3	2.31 (h_{28})
A9	1.05	3.82	2.85	7.72 (h_{19})	1.64	1.04	0.72	3.40 (h_{29})

$$\sqrt{\frac{[(p_{aj}-p_j)^2+(q_{aj}-q_j)^2+(s_{ia}-s_j)^2]}{3}}$$

$$D^{+}_{A1G1}=\sqrt{\frac{[(p_{aj}-p_j^*)^2+(q_{aj}-q_j^*)^2+(s_{aj}-s_j^*)^2]}{3}}=\sqrt{\frac{[(2-2.6)^2+(3-4.4)^2+(4-6.4)^2]}{3}}\text{ for C1}$$
$$+\sqrt{\frac{[(4-5.8)^2+(5-8)^2+(6-10.2)^2]}{3}}\text{ for C2}+\sqrt{\frac{[(7-5.2)^2+(8-7.4)^2+(10-9.6)^2]}{3}}\text{ for C3}$$
$$=1.64+3.16+1.12=5.92$$

For example, membership function for GCM A1 is

$$\eta_{A1.1}=\left[\left(\frac{h_{11}}{h_{11}}\right)^2+\left(\frac{h_{11}}{h_{21}}\right)^2\right]^{-1}=\left[\left(\frac{5.92}{5.92}\right)^2+\left(\frac{5.92}{9.26}\right)^2\right]^{-1}=[1^2+0.6393^2]^{-1}=[1.4087]^{-1}$$
$$=0.7099$$
$$\eta_{A1.2}=1-0.7099=0.2901$$

Using similar computations, membership functions of other GCMs can be computed (Table 4.13).

4.2.3 Kohonen Neural Networks

Kohonen Neural Networks (KNN) consists of input and output layers (Fig. 4.3) (Kohonen 1989; Yegnanarayana 1998; Raju and Nagesh Kumar 2014) and Procedural steps to conduct KNN are presented in Fig. 4.4.

Numerical Problem 4.3 The following normalized data presents information about three GCMs, namely, A1, A2, A3 for three evaluation indicators C1, C2, C3. Climate expert felt that A1, A2, A3 can be classified into two categories (Fig. 4.3). Weights of category G1 and G2 are (0.1, 0.2, 0.3) and (0.4, 0.5, 0.6). Assume learning rate L_r as 0.45. Using KNN, cluster the GCM into two categories G1 and G2. Use Data from Table 4.14.

Solution:

For classifying GCM A1

Input vector (GCM A1)			Output neuron G1			Output neuron G2		
C_1	C_2	C_3	r_{11}	r_{21}	r_{31}	r_{12}	r_{22}	r_{32}
0.2	0.3	0.5	0.1	0.2	0.3	0.4	0.5	0.6

Learning rate $L_r=0.45$

Table 4.13 Membership function information of GCMs

GCM	Membership function in G1 (η)	Membership function in G2 = 1 − η	Membership in group g1 crisp clustering	Membership in group G2 crisp clustering
A1	$\left[\left(\frac{h_{11}}{h_{11}}\right)^2+\left(\frac{h_{11}}{h_{21}}\right)^2\right]^{-1}=[1.4087]^{-1}=0.7099$	0.2901	1	0
A2	$\left[\left(\frac{h_{12}}{h_{12}}\right)^2+\left(\frac{h_{12}}{h_{22}}\right)^2\right]^{-1}=[1.3129]^{-1}=0.7617$	0.2383	1	0
A3	$\left[\left(\frac{h_{13}}{h_{13}}\right)^2+\left(\frac{h_{13}}{h_{23}}\right)^2\right]^{-1}=[1.1822]^{-1}=0.8458$	0.1542	1	0
A4	$\left[\left(\frac{h_{14}}{h_{14}}\right)^2+\left(\frac{h_{14}}{h_{24}}\right)^2\right]^{-1}=[1.2035]^{-1}=0.8309$	0.1691	1	0
A5	$\left[\left(\frac{h_{15}}{h_{15}}\right)^2+\left(\frac{h_{15}}{h_{25}}\right)^2\right]^{-1}=[1.1049]^{-1}=0.905$	0.0950	1	0
A6	$\left[\left(\frac{h_{16}}{h_{16}}\right)^2+\left(\frac{h_{16}}{h_{26}}\right)^2\right]^{-1}=[5.0257]^{-1}=0.199$	0.8010	0	1
A7	$\left[\left(\frac{h_{17}}{h_{17}}\right)^2+\left(\frac{h_{17}}{h_{27}}\right)^2\right]^{-1}=[4.6032]^{-1}=0.2172$	0.7828	0	1
A8	$\left[\left(\frac{h_{18}}{h_{18}}\right)^2+\left(\frac{h_{18}}{h_{28}}\right)^2\right]^{-1}=[20.421]^{-1}=0.0490$	0.9510	0	1
A9	$\left[\left(\frac{h_{19}}{h_{19}}\right)^2+\left(\frac{h_{19}}{h_{29}}\right)^2\right]^{-1}=[6.1555]^{-1}=0.1625$	0.8375	0	1

Computation of squared Euclidean distance:

$$h_i=\sum_{j=1}^{M}\left[I_j(t)-r_{ij}(t)\right]^2,\ \text{for } i=1,2,\ldots,N$$

$h_1=[(0.2-0.1)^2+(0.3-0.2)^2+(0.5-0.3)^2]=0.06$ [Squared distance from A1 to G1].

$h_2=[(0.2-0.4)^2+(0.3-0.5)^2+(0.5-0.6)^2]=0.09$ [Squared distance from A1 to G2].

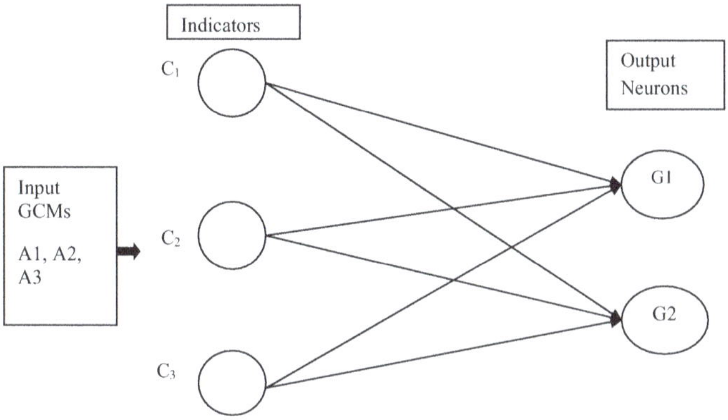

Fig. 4.3 Architecture of Kohonen neural networks

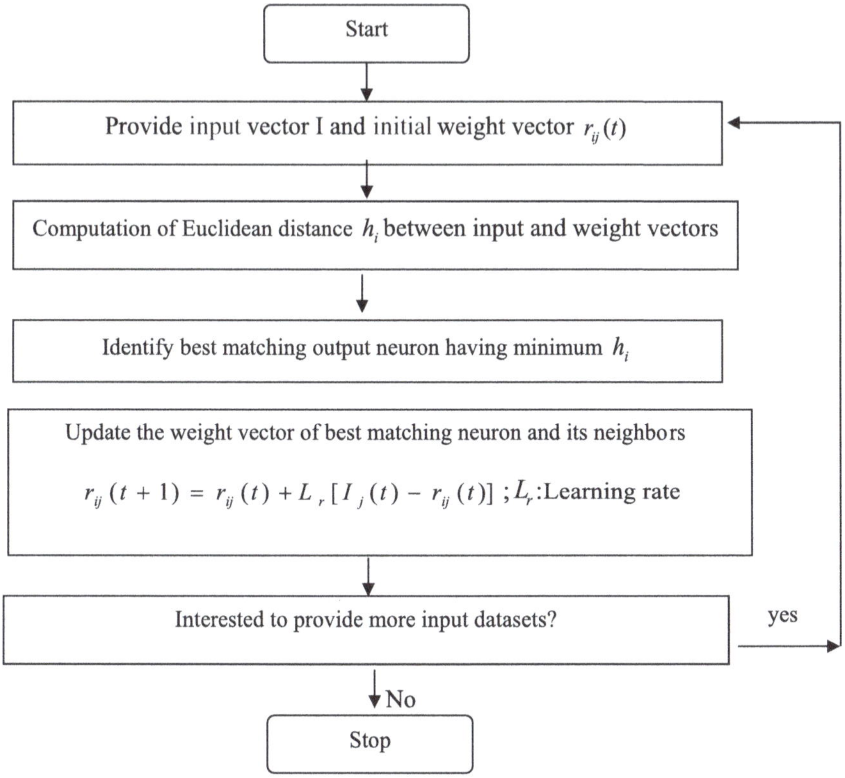

Fig. 4.4 Flowchart of procedural steps for Kohonen neural networks

Table 4.14 Normalized Data

GCM	C_1	C_2	C_3
A1	0.20	0.30	0.50
A2	0.60	0.77	0.40
A3	0.20	0.60	0.30

Best matching neuron is G1 with a minimum h_I of 0.06.
Weights updation for output neuron G1 (where distance is minimum)

$$r_{11updated} = r_{11} + L_r[C_1 - r_{11}] = 0.1 + 0.45[0.2 - 0.1] = 0.145$$
$$r_{21updated} = r_{21} + L_r[C_2 - r_{21}] = 0.2 + 0.45[0.3 - 0.2] = 0.245$$
$$r_{31updated} = r_{31} + L_r[C_3 - r_{31}] = 0.3 + 0.45[0.5 - 0.3] = 0.39$$

For classifying GCM A2

Input vector (GCM A2)			Updated output neuron G1			Output neuron G2		
C_1	C_2	C_3	r_{11}	r_{21}	r_{31}	r_{12}	r_{22}	r_{32}
0.6	0.77	0.40	0.145	0.245	0.39	0.4	0.5	0.6

$h_I = [(0.60 - 0.145)^2 + (0.77 - 0.245)^2 + (0.40 - 0.39)^2] = 0.48275$ [Squared distance from A2 to G1]

$h_2 = [(0.60 - 0.4)^2 + (0.77 - 0.5)^2 + (0.40 - 0.60)^2] = 0.1529$ [Squared distance from A2 to G2]

The best matching neuron is G2 with a minimum h_2 of 0.1529.
Weights updation for output neuron G2 (where distance is minimum)

$$r_{12updated} = r_{12} + L_r[C_1 - r_{12}] = 0.40 + 0.45[0.60 - 0.40] = 0.49$$
$$r_{22updated} = r_{22} + L_r[C_2 - r_{22}] = 0.50 + 0.45[0.77 - 0.50] = 0.6215$$
$$r_{32updated} = r_{32} + L_r[C_3 - r_{32}] = 0.60 + 0.45[0.40 - 0.60] = 0.51$$

For classifying GCM A3

Input vector (GCM A3)			Output neuron G1			Updated output neuron G2		
C_1	C_2	C_3	r_{11}	r_{21}	r_{31}	r_{12}	r_{22}	r_{32}
0.20	0.60	0.30	0.145	0.245	0.39	0.49	0.6215	0.51

$h_I = [(0.20 - 0.145)^2 + (0.60 - 0.245)^2 + (0.30 - 0.39)^2] = 0.1372$ [Squared distance from A3 to G1].

$h_2 = [(0.20 - 0.49)^2 + (0.60 - 0.6215)^2 + (0.30 - 0.51)^2] = 0.1286$ [Squared distance from A3 to G2].

Best matching neuron is G2 with a minimum h_2 of 0.1286.

Weights updation for output neuron G2 (where distance is minimum)

$$r_{12updated} = r_{12} + L_r[C_1 - r_{12}] = 0.49 + 0.45[0.20 - 0.49] = 0.3595$$
$$r_{22updated} = r_{22} + L_r[C_2 - r_{22}] = 0.6215 + 0.45[0.60 - 0.6215] = 0.6118$$
$$r_{32updated} = r_{32} + L_r[C_3 - r_{32}] = 0.51 + 0.45[0.30 - 0.51] = 0.4155$$

Updated weights for output neuron G2 = [0.3595, 0.6118, 0.4155].

GCMs A1, A2, and A3 are falling into Group G1, G2, G2, respectively.

4.2.4 Principal Component Analysis

Principal Component Analysis (PCA) transforms a set of observations of possible correlation into a set of linearly uncorrelated variables applying an orthogonal transformation (Mujumdar and Nagesh Kumar 2012; Raschka 2015). Procedural steps to conduct PCA are presented in Fig. 4.5.

PCA is also termed as Empirical Orthogonal Function (EOF) (The Climate Data Guide: Empirical Orthogonal Function (EOF) Analysis and Rotated EOF Analysis 2017).

Numerical Problem 4.4 Precipitation data simulated from the 4 GCMs for various runs are presented in Table 4.15. Use principal component analysis for data reduction. Compute (a) correlation matrix (b) eigen values and eigen vectors for each component (c) Final component matrix. Consider all four components.

Solution:

Step 1: See Tables 4.16, 4.17.

Step 2: Here correlation matrix is equal to the covariance matrix due to the standardization of data (Table 4.18).

Step 3: See Fig. 4.6, Table 4.19.

Step 4: The user can decide the number of components that can be considered. It depends on the computed eigenvalues or percentage of variance. However, the percentage of variance is the ratio of eigenvalues and number of components (Table 4.20).

Step 5: See Table 4.21.

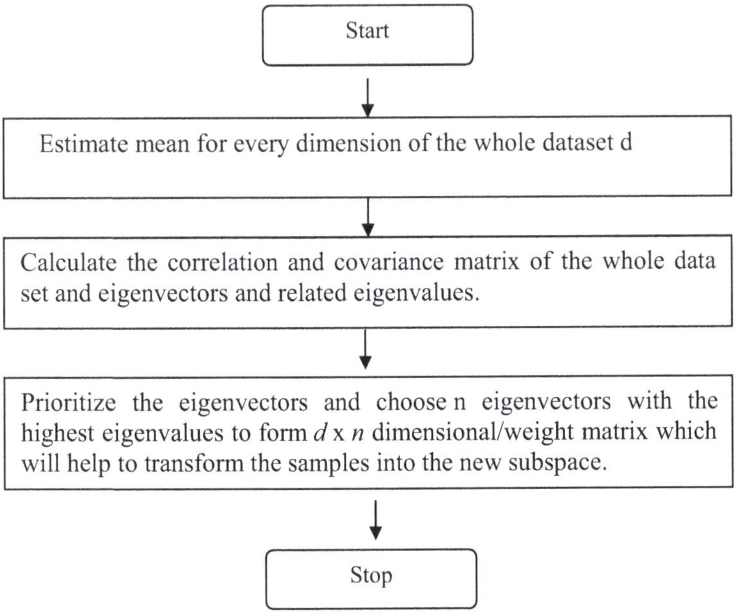

Fig. 4.5 Flowchart of procedural steps for principal component analysis

Table 4.15 Given precipitation data

Runs	GCM1	GCM2	GCM3	GCM4
1	3	7	13	5
2	4	8	2	8
3	5	9	5	17
4	6	10	8	3
5	7	11	16	9
6	8	12	12	18
7	9	10	14	2
8	12	15	11	1

Table 4.16 Estimation of mean, variance, and standard deviation

Parameter	GCM1	GCM2	GCM3	GCM4
Mean	6.75	10.25	10.125	7.875
Variance	8.5	6.214	22.696	42.98
Standard deviation	2.915	2.493	4.764	6.555

Table 4.17 Standardised matrix (x-mean)/standard deviation [X]

	GCM1	GCM2	GCM3	GCM4
1	−1.286	−1.304	0.6	−0.44
2	−0.943	−0.9	−1.7	0.019
3	−0.6	−0.5	−1.07	1.392
4	−0.257	−0.1	−0.44	−0.744
5	0.0858	0.3	1.233	0.17
6	0.428	0.7	0.394	1.544
7	0.772	−0.1	0.81	−0.896
8	1.801	1.9	0.18	−1.05

Table 4.18 Correlation matrix CM $[(X^TX)/(n-1)]$

	GCM1	GCM2	GCM3	GCM4
GCM1	1.000	0.934	0.404	−0.286
GCM2	0.934	1.000	0.322	−0.120
GCM3	0.404	0.322	1.000	−0.224
GCM4	−0.286	−0.120	−0.224	1.000

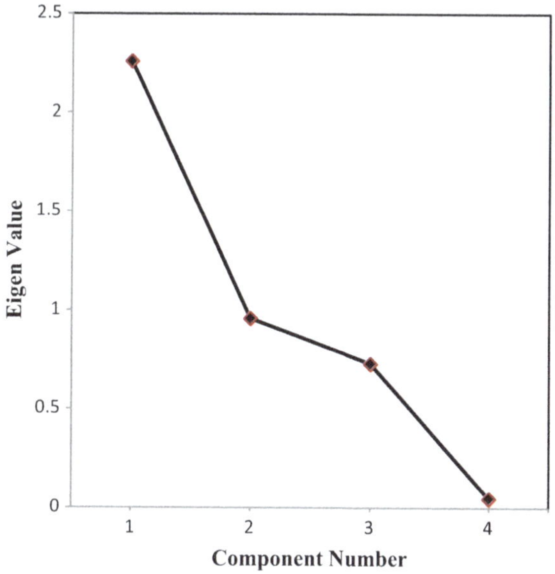

Fig. 4.6 Pictorial representation of eigenvalues

Table 4.19 Eigenvalues of each component and corresponding percentage of variance

Component	Initial eigenvalues		
	Total	% of variance [Eigen values/number of components]	Cumulative %
1	2.261	56.529	56.529
2	0.959	23.979	80.508
3	0.730	18.261	98.769
4	0.049	1.231	100.000

Table 4.20 Eigenvalues and corresponding eigenvector (in the decreasing order of eigenvalues)

	Component			
	1	2	3	4
Eigen value	2.261	0.959	0.73	0.049
GCM1	0.955	0.193	−0.160	0.161
GCM2	0.901	0.375	−0.159	−0.151
GCM3	−0.410	0.833	0.370	0.027
GCM4	0.608	−0.296	0.737	−0.011

Table 4.21 Transformed data using projection [XW]

Runs	GCM1	GCM2	GCM3	GCM4
1	−2.917	−0.107	0.311	0.011
2	−1.003	−1.941	−0.321	−0.062
3	0.262	−1.607	0.806	−0.065
4	−0.607	−0.233	−0.654	−0.03
5	−0.05	1.106	0.52	0
6	1.817	0.216	1.104	−0.043
7	−0.23	1.051	−0.468	0.171
8	2.72	1.521	−1.298	0.019

4.2.5 F-statistic Test

F-statistic is a statistical test which measures the variance reduction from clusters K to K + 1 (Burn 1989) and expressed as

$$F_K = \left(\frac{ER_K}{ER_{K+1}} - 1\right)(T - K + 1) \quad (4.1)$$

where F_K = F-statistic value for cluster K, T = Number of GCMs and ER_{K+1} = Squared error value for (K + 1). Optimal K corresponds to F_K value greater than 10 (Burn 1989).

Numerical Problem 4.5 Squared error values for clusters 2 and 3 are 10.00 and 9.15. Number of GCMs are 10. Compute F-statistic value.

Solution:

F-statistic value $F_K = \left(\frac{ER_K}{ER_{K+1}} - 1\right)(T - K + 1)$

F-statistic value for cluster 2 $= F_2 = \left(\frac{10}{9.15} - 1\right)(10 - 2 + 1) = 0.836$

4.3 Trend Detection Techniques

It is necessary to check the availability and quality of the hydrological or climatological records as erroneous and short length data records can lead to spurious inferences in detecting trends (Sonali and Nagesh Kumar 2013). In the present study, Kendall's rank correlation and turning point test are discussed with numerical examples. More information about trend detection techniques are available in Mann (1945), Kendall (1975), and Patra (2010).

4.3.1 Kendall's Rank Correlation Test

1. Analyze the data y_i regarding number of times it is greater than the remaining values sequentially in the data set (i = 1, 2, … I). For example, y_1 is greater than successive values Q1 times, y_2 is greater than successive values Q2 times, etc. Process of superiority of y_i computation is continued until end. Total superiority value Q is the summation of Q1 + Q2 + Q3 + ⋯ + Qn.
2. Kendall's rank correlation τ and variance of τ are described as $\left[\frac{4Q}{T(T-1)} - 1\right]$ and $\left[\frac{2(2T+5)}{9T(T-1)}\right]$, respectively, where number of the data is T samples.
3. Compute standard normal variate $Z = \frac{\tau}{\sqrt{Var(\tau)}}$ and test the same at desired level of significance. If Z is in the range (for example, value of the range at 5% significance is between ±1.96) it can be concluded that there is no trend in the data series.

Numerical Problem 4.6 Conduct the Kendall's rank correlation test for the following rainfall data: 760, 840, 120, 684, 345, 231, 134, 110, 451, 138, 521, 615. Verify existence of trend at 5% significance level.

Solution:

Total number of data = T = 12.

Superiority of 760 over other successive data = It is superior than 120, 684, 345, 231, 134, 110, 451, 138, 521, 615 (total number of times of superiority Q1 = 10)

Similarly, number of times of superiority of 840, 120, 684, 345, 231, 134, 110, 451, 138, 521, and 615 over successive data is 10, 1, 8, 4, 3, 1, 0, 1, 0, 0, 0

$$\text{Total Q} = 10 + 10 + 1 + 8 + 4 + 3 + 1 + 0 + 1 + 0 + 0 + 0 = 38$$

$$\text{Kendall's rank correlation } \tau = \left[\frac{4Q}{T(T-1)} - 1\right] = \left[\frac{4*38}{12*11} - 1\right] = 0.1515$$

$$\text{Variance of } \tau = \left[\frac{2(2T+5)}{9T(T-1)}\right] = \left[\frac{2*(2*12+5)}{9*12(12-1)}\right] = 0.04882$$

Value of Z $= \frac{\tau}{\sqrt{Var(\tau)}} = \frac{0.1515}{\sqrt{0.04882}} = 0.6865$; since value of Z is within the range of ± 1.96 (at 5% significance level), indicating no presence of trend at the chosen significance level.

4.3.2 Turning Point Test

Objective of the test is to ascertain the number of turning points possible in a given data (Patra 2010). Turning point value is 1 if the chosen data y_i satisfies any one of the following conditions (1) $y_{i-1} \prec y_i \succ y_{i+1}$ or (2) $y_{i-1} \succ y_i \prec y_{i+1}$ and otherwise turning point value is zero. Procedural steps are as follows:

1. Analyze the data y_i for turning point conditions and check for number of turning points t.
2. Compute expected number of turning points $E(t)$ and variance $Var(t)$ being $\frac{2(T-2)}{3}$ and $\frac{(16T-29)}{90}$, respectively, where the number of data is T.
3. Compute standard normal variate Z $= \frac{(t-E(t))}{\sqrt{Var(t)}}$ and test the same at desired level of significance. If Z is in the range (for example, value of the range at 5% significance is between ± 1.96) indicated, thus there is no trend in the data series.

Numerical Problem 4.7

Verify existence of trend at 5% significance level for the following data: 760, 840, 120, 684, 345, 231, 134, 110, 451, 138, 521, 615. Use turning point test for analysis.

Solution:

Data was arranged as follows for verifying the conditions 1 and 2.

Number of data points T = 12.

Sequencing of data and turning point value (which is in parenthesis): 760, 840, 120 (1); 840, 120, 684 (1); 120, 684, 345 (1); 684, 345, 231 (0); 345, 231, 134 (0); 231, 134, 110 (0); 134, 110, 451 (1); 110, 451, 138 (1); 451, 138, 521 (1); 138, 521, 615 (0)

Total number of turning points t = 6.

Expected number of turning points $E(t)$ and variance $Var(t)$ are $\frac{2(T-2)}{3}$ and $\frac{(16T-29)}{90}$. By substituting T values $\frac{2(12-2)}{3}$ and $\frac{(16*12-29)}{90}$, resulting expectation, and variance are 6.66, 1.811.

Standard normal variate Z $= \frac{(t-E(t))}{\sqrt{Var(t)}} = \frac{(6-6.66)}{\sqrt{1.811}} = -0.4904$

Since value of Z is within the range of ±1.96 (at 5% significance level), indicating no presence of trend at the chosen significance level.

4.4 Optimization Techniques

Optimization is a process of finding the optimum value of a function for a given condition (Loucks et al. 1981; Jain and Singh 2003; Vedula and Mujumdar 2005). Detailed information about optimization techniques is available in Ravindran et al. (2001), Rao (2003), Taha (2005), and Nagesh Kumar (2017).

4.4.1 Linear Programming Problem

If the objective function and all the constraints are "linear" functions of the design variables, the optimization problem is termed as Linear Programming Problem (LPP). Objective of the model is to determine X $(x_1, x_2, \ldots x_n)$ and model formulation is as follows:

Objective function:

$$Z = \sum_{i=1}^{n} c_i x_i \tag{4.2}$$

Constraints:

$$\sum_{i=1}^{n} a_{ij} x_i \le b_j, \quad j = 1, 2, \ldots m \tag{4.3}$$

$$x_i \ge 0 \quad i = 1, 2, \ldots n \tag{4.4}$$

where c_i, a_{ij}, and b_j are constants.

Numerical Problem 4.8 Australian Water Resources expert interested to know the cropping pattern of four crops A1, A2, A3, A4 for the year 2030 in Murray–Darling Basin, Australia. GCM employed is ACCESS 1.0 and the chosen pathway is RCP 2.6. It is estimated from the GCM-RCP analysis that 15 ha-m from surface water resources, 1.5 ha-m from groundwater resources and 1.5 ha-m from reuse of treat wastewater is expected to be available in the command area. Keeping inflation and expected fluctuations in the market, expected net benefits from crops A1 to A4 are AUS$ 1200/ha, 1400/ha, 2200/ha, and 1600/ha. Expected land availability obtained from Cellular Automata and other analyses are 140 ha. Fertilizer requirement for crops A1 to A4 are 0.1 kg/ha, 0.12 kg/ha, 0.16 kg/ha, 0.20 kg/ha, respectively, and

Table 4.22 Linear programming model formulation

Information	LINGO syntax
Objective function	Max Z = 1200 * A1 + 1400 * A2 + 2200 * A3 + 1600 * A4;
Area constraint	A1 + A2 + A3 + A4 <= 140;
Fertilizer constraint	0.10 * A1 + 0.12 * A2 + 0.16 * A3 + 0.2 * A4 <= 16;
Water constraint	0.14 * A1 + 0.22 * A2 + 0.16 * A3 + 0.22 * A4 <= 18;
Labour constraint	45 * A1 + 67 * A2 + 35 * A3 + 72 * A4 < = 5000
Crop area 1 bounds	A1 >= 25; A1 <= 45;
Crop area 2 bounds	A2 >= 24; A2 <= 30;
Crop area 3 bounds	A3 >= 16; A3 <= 24;
Crop area 4 bounds	A4 >= 20; A4 <= 22;
	END

availability of fertiliser is expected to be 16 kg. Penman–Monteith technique estimated water requirement of crops as 0.14 m, 0.22 m, 0.16 m, 0.22 m per hectare for crops A1 to A4. Man-days required per hectare for the crops A1 to A4 are 45, 67, 35, 72 and total man power availability is 5000 Man-days. Lower and upper limits for crops in hectare for A1, A2, A3, A4 are [25, 45], [24, 30], [16, 24], [20, 22]. Determine optimum cropping pattern for achieving maximum net benefits using linear programming framework.

Solution:

Total water expected to be available = 15 + 1.5 + 1.5 = 18 ha-m.

A1, A2, A3, A4 represent crop acreage in hectares that can be optimally determined (Table 4.22).

Outputs obtained from Linear Programming Solution solver LINGO (http://www.lindo.com/index.php/products/lingo-and-optimization-modeling). Crop acreages of A1, A2, A3, A4 are 25 ha, 24 ha, 23.628 ha, 20 ha and objective function value is Aus $ 147582.9.

4.4.2 Non-linear Programming Problem

Non-linear Programming (NLP) can be applied when some or all of the objective functions and/or constraints are non-linear in nature. A quadratic programming, sub-case of non-linear programming problem with a quadratic objective function and linear constraints is demonstrated with Numerical Problem 4.9.

Numerical Problem 4.9 Solve quadratic programming below using LINGO.

$$MaxZ = 1200 * x_1^2 + 1400 * x_2^2$$

Subject to

$$x_1 + x_2 \leq 100$$
$$0.1 * x_1 + 0.12 * x_2 \leq 10$$
$$45 * x_1 + 67 * x_2 \leq 4000$$
$$x_1 \geq 25$$
$$x_1 \leq 40$$
$$x_2 \geq 20$$
$$x_2 \leq 100$$

Solution:

$x_1 = 25$ units; $x_2 = 21.43$ units.
Objective function Z = 1392857 units.

4.4.3 Evolutionary Algorithms

Most of the real-world optimization problems involve complexities like discrete, continuous or mixed variables, multiple conflicting objectives, non-linearity, discontinuity, and non-convex region are difficult to solve using existing linear or non-linear techniques. Such problems can be solved using other techniques which make use of Evolutionary Algorithms (EA) such as Genetic Algorithms (GA) (Deb 2002), Differential Evolution (Price et al. 2005; Das et al. 2016), Particle Swarm Optimization (Khare and Rangnekar 2013), Firefly Optimization Algorithms (Garousi-Nejad et al. 2016) and Harmony Search (Bashiri-Atrabi et al. 2015). Extensive discussion on non-traditional optimization algorithms are available in Yang (2010). In the present chapter, a brief description of GA is presented.

GAs are based on the principle of natural selection and natural genetics and combine survival of the fittest with a randomized exchange of information to form a search algorithm (Goldberg 1989; Deb and Agarwal 1995; Deb 1999, 2002). Selection, crossover and mutation are the parameters that facilitate the generation of a new population of points from the old parent population. A flowchart indicating the steps of a simple GA is shown in Fig. 4.7.

Numerical Problem 4.10 Solve the following problem using GA (Raju and Nagesh Kumar 2014)

$$\text{Max } z = -x_1 + 2x_2 + 3x_3$$

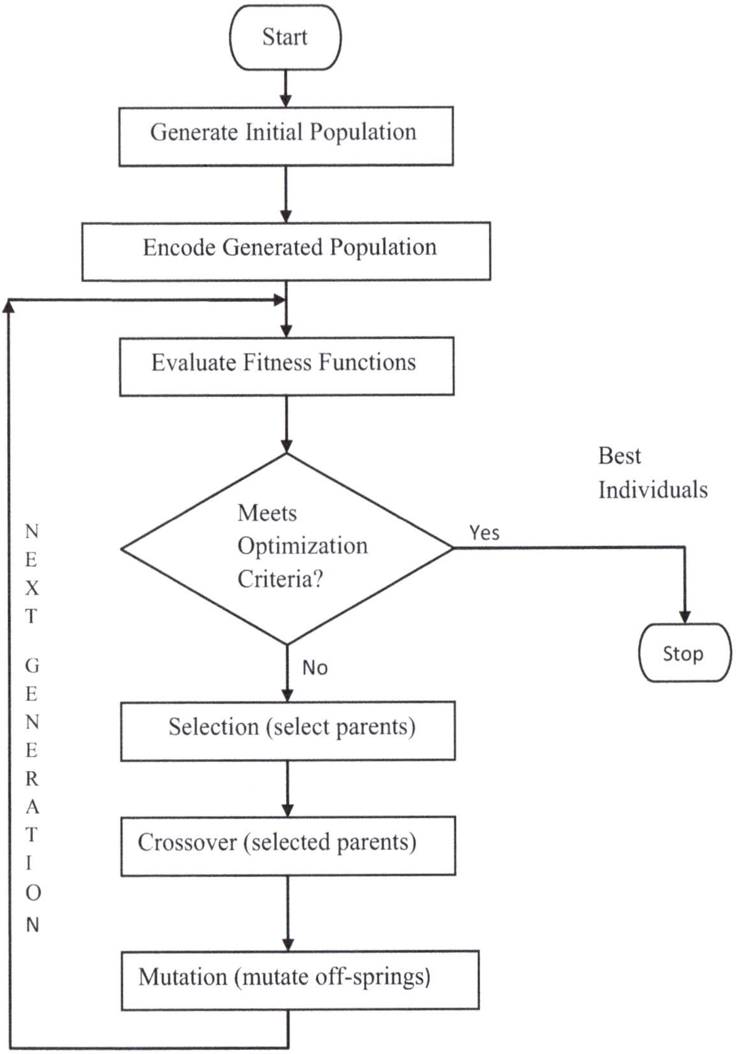

Fig. 4.7 Schematic Process of GA

Subject to

$$2x_1 + 3x_2 - 1.5x_3 \geq 26$$
$$-2.5x_1 + 4.5x_2 + 3x_3 \leq 65$$
$$2x_1 + 3.5x_2 + 1.5x_3 \leq 57$$
$$x_1 \geq 0;\ x_2 \geq 0;\ x_3 \geq 0$$

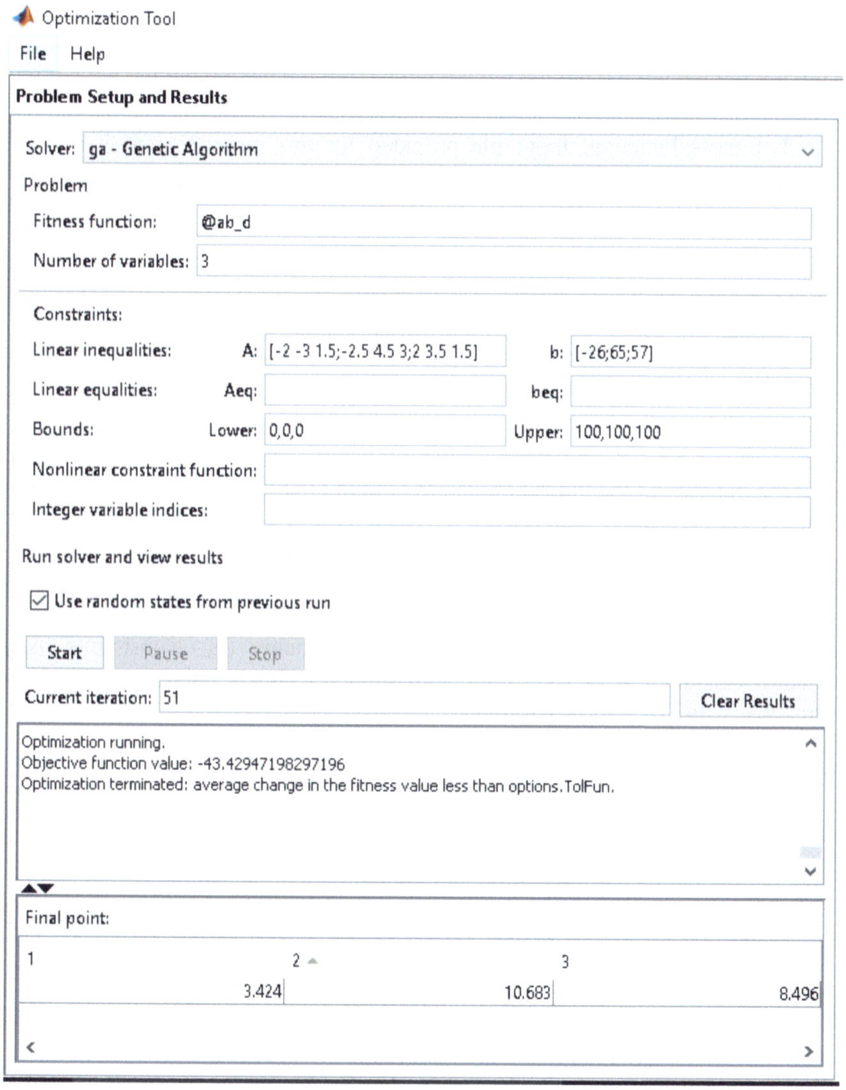

Fig. 4.8 Basic screen of the GA Tool Box

Solution:

GA tool box of MATLAB is employed to solve this problem (Genetic Algorithm 2017). Figure 4.8 presents appearance of the tool box where important information is to be provided as input. Detailed input procedures with reference to the given numerical problem are as follows:

Step 1: Choose the solver. A number of solvers are available in the Combo Box including linear programming. We chose solver GA for solving the given problem.

Step 2: Fitness Function: Input file provided for this purpose is @ab_d.m (as follows)

```
function f = ab_d(x)

x1 = double(x(1));
x2 = double(x(2));
x3 = double(x(3));
f = x1 - 2 * x2 - 3 * x3;
```

GA tool box handles only minimization problems. To make it compatible, given objective function of maximization nature is multiplied with negative sign to represent it as minimization. You can see the equation for f in the minimization format.

Step 3: Provide number of variables; In this case 3 (x1, x2, x3).

Step 4: GA tool box handles equality and inequality ($\leq$) constraints. Any inequality constraint of $\geq$ nature can be multiplied with a negative sign to make it compatible with GA tool box format.

See the first constraint $2x_1 + 3x_2 - 1.5x_3 \geq 26$; Transformation is as follows:

$$-2x_1 - 3x_2 + 1.5x_3 \leq -26 \quad \text{Constraint 1}$$
$$-2.5x_1 + 4.5x_2 + 3x_3 \leq 65 \quad \text{Constraint 2}$$
$$2x_1 + 3.5x_2 + 1.5x_3 \leq 57 \quad \text{Constraint 3}$$

Linear Inequalities Box:
Fill the information against block **A**: [−2−3 1.5; -2.5 4.5 3; 2 3.5 1.5] representing coefficients of x_1, x_2, x_3 for constraints 1, 2, and 3, respectively.

Fill the information against block **B** [−26; 65; 57] representing right-hand side values of constraint 1, 2, and 3, respectively.

Linear Equalities Box:
Not applicable for the present problem as there are no equality constraints.

$$\text{Bounds}: \text{ Lower } 0,\ 0,\ 0 \quad \text{Upper } 100,\ 100,\ 100$$

After filling the above information, click on **start** button. After certain number of iterations, it provides the value of the fitness function as well as the decision variables. As observed, value of the fitness function is −43.4294 (negative sign is representation of minimum). As per practical understanding of the problem, value of objective function is 43.4294 and corresponding x1, x2, x3 values are 3.424, 10.683, and 8.496, respectively.

Same problem was solved using linear programming and solution obtained was [$x_1 = 3.4418$; $x_2 = 10.6511; x_3 = 8.5581; z = 43.5349$] (Raju and Nagesh Kumar 2014).

Present chapter discusses about statistical and optimization techniques. Next chapter presents hydrological models.

Software (Information as on 30.12.2016)

Cluster Analysis:

- Statistics and Machine Learning Tool box of MATLAB
- SPSS (http://www-03.ibm.com/software/products/en/spss-statistics)
- MINITAB (https://www.minitab.com)
- Statistical Analysis System (SAS) (http://www.sas.com/en_in/home.html)
- CVAP: Cluster Validity Analysis Platform (cluster analysis and validation tool) (http://in.mathworks.com/matlabcentral/fileexchange/14620-cvap–cluster-validity-analysis-platform–cluster-analysis-and-validation-tool-)

Fuzzy Cluster Analysis: Fuzzy Logic tool box of MATLAB (http://in.mathworks.com/products/fuzzy-logic/)

Kohonen Neural Networks (KNN): Neural Network Toolbox of MATLAB (http://in.mathworks.com/products/neural-network/)

Principal Component Analysis:

- Statistics and Machine Learning Tool box of MATLAB
- SPSS
- MINITAB
- SAS
- XLSTAT
- You can also access Statistics related software through the link http://statpages.org/javasta2.html

Optimization Techniques (Handles linear, non-linear, and quadratic programming):

- General Algebraic Modeling System (GAMS) (https://www.gams.com/)
- LINGO: LINGO 16.0—Optimization Modeling Software for Linear, Non-linear and Integer Programming (http://www.lindo.com/index.php/products/lingo-and-optimization-modeling)
- Global Optimization Tool box (http://in.mathworks.com/products/global-optimization/).
- Optimization Tool Box: (http://in.mathworks.com/products/optimization/)
- Genetic Algorithm Tool box: (https://in.mathworks.com/help/gads/genetic-algorithm.html)

Revision Questions and Exercise Problems

4.1 What are different types of data compression techniques? On what basis they can be differentiated?

4.2 What are the various steps in (a) principal component analysis (b) cluster analysis, and (c) Kohonen Neural Networks?

4.3 Solve Numerical Problem 4.1 related to cluster analysis using only C1 and C2. Weights of C1 and C2 are 0.4 and 0.6, respectively. Identify the group of each GCM. Use data in Table 4.1.

4.4 Nine GCMs are evaluated on three indicators as presented in Table 4.23. Data is following Trapezoidal membership function. Use fuzzy C-means algorithm to cluster the GCMs. A number of clusters can be taken as 2. Constant m' value can be taken as 2. Compute membership function values after first iteration?

Hint: Distance between two fuzzy trapezoidal numbers is

$$\sqrt{\frac{\left[\left(p_{aj}-p_j\right)^2+\left(q_{aj}-q_j\right)^2+\left(s_{ia}-s_j\right)^2+\left(t_{ia}-t_j\right)^2\right]}{4}}$$

Where (p, q, s, t) are elements of Trapezoidal fuzzy number.

4.5 Table 4.24 presents normalized data information about four GCMs for three evaluation criteria C1, C2, C3. Climate expert felt that 4 GCMs can be classified into two categories. Weights of category 1 and 2 are (0.2, 0.4, 0.8), (0.2, 0.6, 0.8). Assume learning rate as 0.60. Using KNN, cluster the GCMs into the two categories?

Table 4.23 Data matrix of 9 GCMs and random assignment of clusters

GCMs	C_1	C_2	Assigned random clusters/groups iteration 0
(1)	(2)	(3)	(4)
A1	(0.2, 0.3, 0.4, 0.5)	(0.4, 0.55, 0.66, 0.88)	G2
A2	(0.5, 0.6, 0.7, 0.8)	(0.3, 0.5, 0.6, 0.8)	G2
A3	(0.3, 0.6, 0.9, 1.0)	(0.2, 0.40, 0.6, 0.8)	G1
A4	(0.6, 0.7, 0.8, 0.9)	(0.1, 0.2, 0.4, 0.8)	G2
A5	(0.5, 0.7, 0.8, 0.9)	(0.4, 0.6, 0.8, 1)	G1
A6	(0.2, 0.4, 0.6, 0.8)	(0.22, 0.44, 0.66, 0.88)	G1
A7	(0.2, 0.3, 0.5, 0.8)	(0.44, 0.8, 0.9,1)	G2
A8	(0.2, 0.4, 0.8, 1.0)	(0.11, 0.44, 0.8, 0.9)	G1
A9	(0.1, 0.4, 0.7, 0.9)	(0.3, 0.5, 0.9, 1)	G2

Table 4.24 Normalised data

GCM	C_1	C_2	C_3
A1	0.40	0.80	0.40
A2	0.30	0.82	0.55
A3	0.24	0.66	0.34
A4	0.32	0.76	0.42

4.6 Solve Numerical Problem 4.4 related to principal component analysis using data presented in Table 4.15. Consider only three GCMs for evaluation.

4.7 What are various trend detection techniques available? How expectation and variance are computed for each individual technique?

4.8 Verify existence of trend at 1% significance level for the following rainfall data (mm): 860, 930, 320, 584, 445, 341, 154, 220, 481, 158, 541, 645. Use turning point test for analysis.

4.9 Conduct the Kendall's rank correlation test for the following rainfall data (mm): 720, 810, 130, 674, 355, 241, 144, 180, 461, 178, 531, 625 to verify existence of trend at 1% significance level.

4.10 Formulate linear/non-linear programming/ genetic algorithm in your domain of interest (in similar aspects of Numerical Problem 4.8)? Mention mathematically objective function, constraints, and bounds for the chosen problem. Discuss the outcomes from the optimization problem and challenges for implementing the outcome in the field. You can assume suitable relevant numerical data wherever applicable (For example, greenhouse gases (GHGs) are increasing day by day resulting in an increase of global warming and health hazards due to the growing population and pollution from industries and vehicles. However, it is not possible to minimise the GHGs emissions as sustainable industrial and societal development is necessary. Solve for the hypothetical data, assuming an industrial environment is emitting Carbon dioxide (CO_2) and Methane (CH_4)).

Advanced Review Questions

4.11 Is there any possibility that output of principal component analysis can be input to cluster analysis or vice versa? If so, how these techniques can be complemented to each other? Provide examples how these techniques help in climate modeling with practical significance.

4.12 Discuss the limitations of data compression techniques?

4.13 Mention any other techniques that are falling in data compression category?

4.14 Name relevant software that can be used for (a) principal component analysis (b) cluster analysis? Provide salient features of the software.

4.15 In your opinion, which trend detection technique is suitable for rainfall in Indian conditions?

4.16 Mention various detection techniques that can be employed in climate change studies.

4.17 Mention two case studies in India where trend detection techniques were employed. Name the trend detection technique employed?

References

Bashiri-Atrabi H, Qaderi K, Rheinheimer DE, Sharifi E (2015) Application of harmony search algorithm to reservoir operation optimization. Water Resour Manage 29:5729–5748

Bezdek J (1981) Pattern recognition with fuzzy objective function algorithms. Plenum, NewYork

Burn DH (1989) Cluster analysis as applied to regional flood frequency. J Water Resour Plann Manag 115:567–582

Das S, Mullick SS, Suganthan PN (2016) Recent advances in differential evolution—an updated survey. Swarm Evol Comput 27:1–30

Davies DL, Bouldin DW (1979) A cluster separation measure. IEEE Trans Pattern Anal Mach Intell 1(4):224–227

Deb K, Agarwal RB (1995) Simulated binary crossover for continuous search space. Complex Syst 9:115–148

Deb K (1999) An introduction to genetic algorithms. Sadhana 24(4):293–315

Deb K (2002) Multi-objective optimization using evolutionary algorithms, 1st edn. Wiley

Dunn JC (1974) Well separated clusters and optimal fuzzy partitions. J Cybern 4:95–104

Garousi-Nejad I, Bozorg-Haddad O, Loáiciga HA, Marino MA (2016) Application of the firefly algorithm to optimal operation of reservoirs with the purpose of irrigation supply and hydropower production. J Irrig Drain Eng 142(10):04016041-1–04016041-12. doi:10.1061/(ASCE)IR.1943-4774.0001064

Genetic Algorithm (2017) https://in.mathworks.com/help/gads/genetic-algorithm.html. Accessed on 31 Jan 2017

Goldberg DE (1989) Genetic algorithms in search. Optimization and Machine Learning, Addison-Wesley, New York

Jain AK, Dubes RC (1988) Algorithms for clustering data. Prentice-Hall, Englewood Cliffs, New Jersey

Jain SK, Singh VP (2003) Water resources systems planning and management. Elsevier B.V, The Netherlands

Kendall MG (1975) Rank correlation methods. Charless Griffin, London

Khare A, Rangnekar S (2013) A review of particle swarm optimization and its applications I in solar photovoltaic system. Appl Soft Comput 13:2997–3006

Kohonen T (1989) Self organization and associative memory. Springer, Berlin

Loucks DP, Stedinger JR, Haith DA (1981) Water resources systems planning and analysis. Prentice–Hall, NJ

Mann HB (1945) Nonparametric tests against trend. Econometrica 13:245–259

Mujumdar PP, Nagesh Kumar D (2012) Floods in a changing climate: hydrologic modeling. Cambridge University Press, International Hydrology Series

Nagesh Kumar D (2017) Optimization methods. http://nptel.ac.in/syllabus/105108127/. Accessed on 31 Jan 2017

Patra KC (2010) Hydrology and water resources engineering. Narosa Publishing House

Price K, Storn R, Lampinen J (2005) Differential evolution—a practical approach to global optimization. Springer, Germany

Raju KS, Nagesh Kumar D (2014) Multicriterion analysis in engineering and management. Prentice Hall of India, New Delhi

Rao SS (2003) Engineering optimization: theory and practice. New Age International (P) Limited, New Delhi

Raschka (2015) Implementing a principal component analysis (PCA) in Python step by step http://sebastianraschka.com/Articles/2014_pca_step_by_step.html. Accessed on 31 Jan 2017

Ravindran A, Phillips DT, Solberg JJ (2001) Operations research—principles and practice. Wiley, New York

Ross TJ (2011) Fuzzy logic with engineering applications. Wiley

Sonali P, Nagesh Kumar D (2013) Review of trend detection methods and their application to detect temperature changes in India. J Hydrol 476:212–227

Taha HA (2005) Operations research—an introduction. Prentice-Hall of India Pvt. Ltd., New Delhi

The Climate Data Guide: Empirical Orthogonal Function (EOF) Analysis and Rotated EOF Analysis (2017). https://climatedataguide.ucar.edu/climate-data-tools-and-analysis/empirical-orthogonal-function-eof-analysis-and-rotated-eof-analysis#sthash.RBGslgxN.dpuf. Accessed on 31 Jan 2017

Vedula S, Mujumdar PP (2005) Water resources systems: modelling techniques and analysis. Tata McGraw Hill, New Delhi

Yang XS (2010) Nature-inspired metaheuristic algorithms. Lunvier Press, United Kingdom

Yegnanarayana B (1998) Artificial neural networks. Prentice Hall of India, New Delhi

Suggested Further Reading

Hamed KH, Rao AR (1998) A modified Mann-Kendall trend test for auto correlated data. J Hydrol 204:219–246

Haan CT (1977) Statistical methods in hydrology. Iowa State University Press, Ames, Iowa, p 378

Hastie T, Tibshirani R (2011) The elements of statistical learning: data mining, inference, and prediction. Springer Series in Statistics

Helsel DR, Hirsch RM (2002) Statistical methods in water resources techniques of water resources investigations, book 4, chapter A3, US geological survey, pp 522

Khaliq MN, Ouarda TBMJ, Gachon P, Sushama L, St.-Hilaire A (2009) Identification of hydrologic trends in the presence of serial and cross correlations. A review of selected methods and their application to annual flow regimes of Canadian rivers. J Hydrol 368:117–130

Lettenmaier DP (1976) Detection of trend in water quality data from record with dependent observations. Water Resour Res 12:1037–1046

Milton JS, Arnold JC (2007) Introduction to probability and statistics. Tata McGraw-Hill

North GR, Bell TL, Cahalan RF, Moeng FJ (1982) Sampling errors in the estimation of empirical orthogonal functions. Mon Weather Rev 110:699–706

Shrestha S, Anal AK, Salam PA, van der Valk M (eds) (2015) Managing water resources under climate uncertainty: examples from Asia, Europe, Latin America, and Australia. Springer Water, pp 438

Hydrological Modeling

5

Abstract

This chapter describes basic definitions, classification of models into various categories, with procedures for solving water resources engineering problems using hydrological models. Storm Water Management Model (SWMM), Hydrologic Engineering Center-Hydrologic Modeling System (HEC-HMS), Soil and Water Assessment Tool (SWAT), and Variable Infiltration Capacity (VIC) are discussed with mathematical background. Brief information about MIKE-based models are also part of the chapter. The reader is expected to understand various hydrological models along with their applicability by studying this chapter.

Keywords

HEC-HMS · Hydrological models · MIKE · SWAT · SWMM · VIC

5.1 Introduction

Hydrological system is an action of biological/physical/chemical processes on input variable(s) so as to generate outputs (Dooge 1992). Some of the variables are precipitation, evaporation, soil moisture, and streamflow. Xu (2002) made an extensive review on hydrological models whereas Francini and Pacciani (1991) made comparative analysis of several conceptual rainfall–runoff models. Some of the terminologies according to Xu (2002) are as follows:

Electronic supplementary material The online version of this chapter (doi:10.1007/978-981-10-6110-3_5) contains supplementary material, which is available to authorized users.

K. Srinivasa Raju and D. Nagesh Kumar, *Impact of Climate Change on Water Resources*, Springer Climate, https://doi.org/10.1007/978-981-10-6110-3_5

- Parameter is a measurable quantity that characterises a system which may be a constant or varies with time or some other variable and to be quantified/deduced from actual field measurements.
- Hydrological model describes hydrological cycle and simplifies representation of complex real systems, with some assumptions (Dooge 1992) for assessing hydrologic phenomena in a catchment in the laws of energy, momentum, and continuity equations. Some of the specific functions include streamflow forecasting that supports reservoir operation, flood, and drought studies. Accuracy of streamflow simulations are constrained by hydrologic model approximations, availability of data of the catchment, and the quality of measurements. In addition, some of the challenges include, choosing the most suitable model structure for chosen catchment, and quantification of uncertainty in hydrologic model simulations. This requires a thorough knowledge of the role of individual components in each hydrologic model for better assessment of their applicability and also to know the differences between simulations of various hydrologic models.

Models are classified into various categories, which are as follows (Eldho 2017):

- Theoretical/physical models have logical structure as that of real hydrological system. Examples include models related to infiltration built on two-phase flow porous media, watershed runoff based on St. Venant equations, evaporation established on turbulence and diffusion and groundwater on fundamental transport equations. Empirical models can be computed by using concurrent input and output measurements. Conceptual models are intermediate state between theoretical and empirical models.
- Linear and non-linear models are built on the regression relationship between inputs and output.
- Model is termed as time invariant, if relationship between input and output does not vary with time.
- Lumped models treat the total basin as homogeneous whereas basin is divided into elementary unit blocks/areas in case of distributed models. Intermediate process is termed as semi-distributed.
- Models are said to be deterministic if all the variables are free from randomness and vice versa for stochastic models.
- Models which concentrate on single floods and runoff series are termed as event-based and continuous models, respectively.

Combinations of models are possible such as lumped conceptual models, distributed theoretical, distributed conceptual, etc. Some of the models of this category are WASMOD (Water and Snow Balance Modeling System) a conceptual, stochastic, lumped water balance model; HBV (Hydrologiska Byråns Vattenbalansavdelning) model; a conceptual, deterministic, lumped (semi-distributed) daily rainfall–runoff model; TOPMODEL (TOPology MODEL: a physically based, semi-distributed model, and SHE (Systeme Hydrologique Europeen) model: a physically based, deterministic, distributed model. Extensive information on hydrological models are available (Singh and Woolhiser 2002; Xu 2002). Procedure for solving a water resource engineering problem using hydrological models are as follows:

1. Definition of the problem: Identification of the domain problem such as groundwater, surface water, runoff and salt water intrusion.
2. Data availability: Precipitation and its duration, temperature, runoff guide the choice of hydrological model.
3. Choice of model: Model choice may vary depending on the domain, problem statement data availability, and procuring of software in terms of freeware or commercial.
4. Calibrating the model: Establishing the values of parameters such that reasonable agreement exists between simulated output and observed value.
5. Validation of the model with calibrated parameters (obtained from step 4) to assess the hydrologic model efficacy.
6. Model application for simulation of the output after satisfying steps 4 and 5.

In the present study, Storm Water Management Model (SWMM), continuous dynamic event model; Hydrologic Engineering Center-Hydrologic Modeling System (HEC-HMS), a physical semi-distributed event-based runoff model; Soil and Water Assessment Tool (SWAT), a conceptual continuous simulation model; Variable Infiltration Capacity (VIC), a physically distributed model and MIKE models are discussed.

5.2 Storm Water Management Model

SWMM is used for single event or long-term/continuous simulation of both quality and quantity of runoff from urban areas. Runoff component works on a group of sub-catchments, which receive precipitation, and generates pollutant loads and runoff. The routing portion transports generated runoff through channels, pumps, regulators, pipes, and storage/treatment devices whichever are applicable as shown in Fig. 5.1. SWMM tracks the flow depth and rate, and water quality in each channel and pipe during a multiple time step simulation. SWMM is also extended to model low impact development (LID) controls (Fig. 5.2) and facilitates hydrologic, hydraulic, and water quality simulations and can estimate pollutant loads associated with storm water runoff as shown in Fig. 5.3.

SWMM facilitates multiple modeling aspects including designing and sizing of detention facilities, drainage system components for flood control and detailed applications of SWMM are available in Barco et al. (2008). The Storm Water Management Model Climate Adjustment Tool (SWMM-CAT) is a new addition to

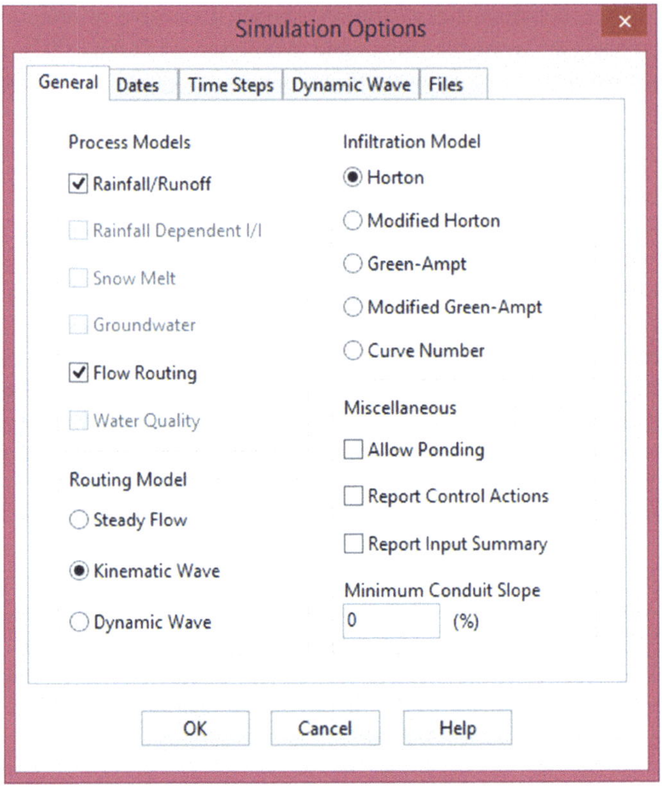

Fig. 5.1 Various methods in SWMM

SWMM in AR3 perspective (Fig. 5.4). More information on SWMM is available in Daniel et al. (2011) and SWMM (2017).

Numerical Problem 5.1 Demonstrate modeling in SWMM for a hypothetical storm system as shown in Fig. 5.5.

Solution:

EPA SWMM 5.1 (epaswmm5.1.exe) can be downloaded from EPA website. The. exe file should be RUN from the start menu. The path of the file can be browsed to desired location. Shortcut of SWMM can be pinned to the desktop for an easy access.

Steps to model in SWMM

1. Default set of options should be specified before starting a model and can be selected from project option. ID labels, sub-catchments, and node default options can be mentioned in the dialog box (Section 1).

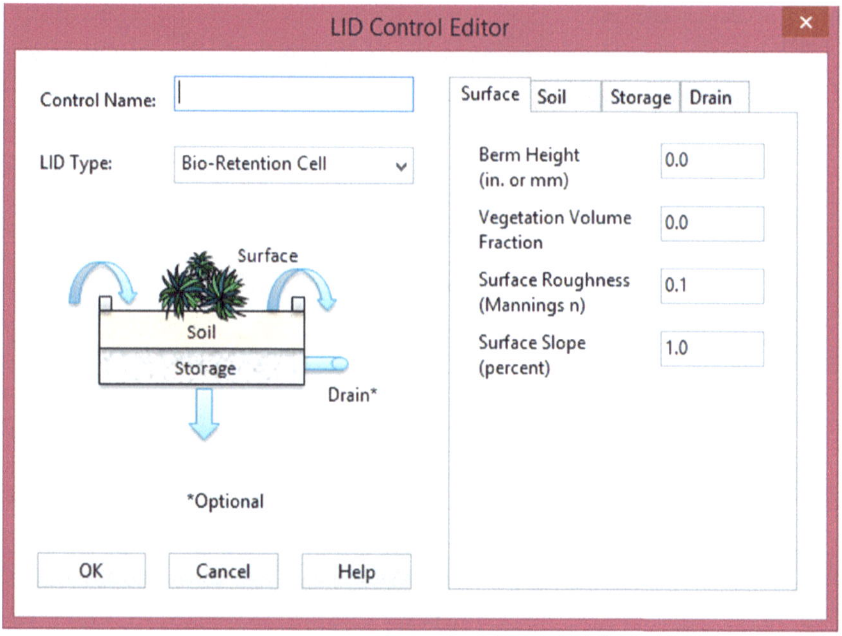

Fig. 5.2 LID control editor in SWMM

2. Network diagram of storm network of the study area should be drawn in the SWMM map window (Section 2).
3. Set of methods for each analysis have to be selected (Section 3).
4. Simulation and run option have to be selected for complete analysis of the model (Section 4).
5. Results can be viewed through summary report, profile, and scatter plots in the model (Section 5).

Section 1 (Fig. 5.6)

- Default set of options should be specified before starting a model.
- Default option can be selected from project option.
- Select Project ≫ Defaults ≫ Default dialog box.
- ID labels, sub-catchments, and node default options can be mentioned in the dialog box.
- ID labels include prefixes for various objects, viz., sub-catchments, nodes, junctions, outfall, etc.
- Sub-catchment properties include area, width, slope, etc.
- Nodes/links properties include geometry, roughness, routing methods, etc.

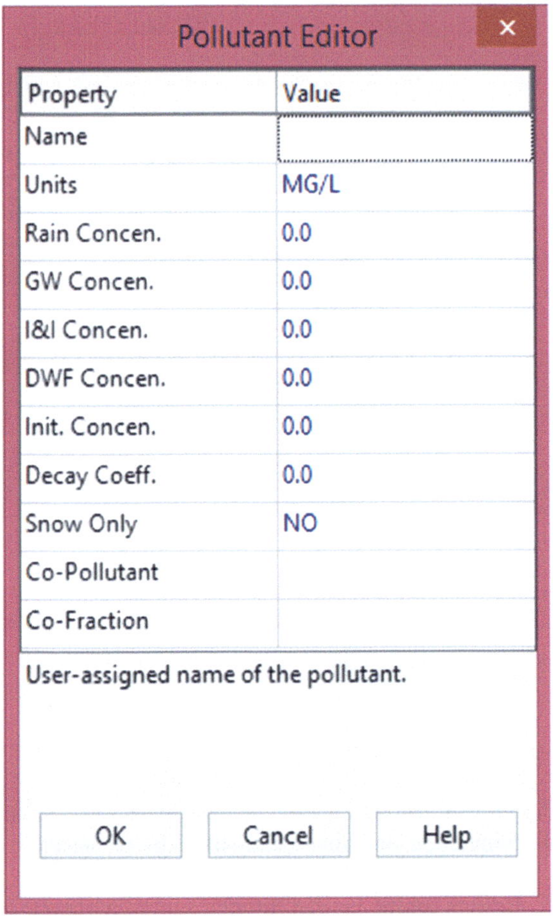

Fig. 5.3 Pollutant editor in SWMM

For numerical problem,

In the ID Labels dialog box, the following details are given as input.

Rain gauges—R; Sub-catchments—C; Junctions—J; Outfalls—O; Conduits—CO

In the sub-catchment dialog box, the following details can be given as input:

Area = 3; Width = 500; % Slope = 0.05; % Impervious = 20; N-Impervious = 0.02.

N-Pervious = 0.05; Dstore-Imperv = 0.03; Dstore-Perv = 0.04; % Zero-Imperv = 25

Infiltration Method—Horton

On the Nodes/Links default page

Node Invert

Node Max. Depth = 3

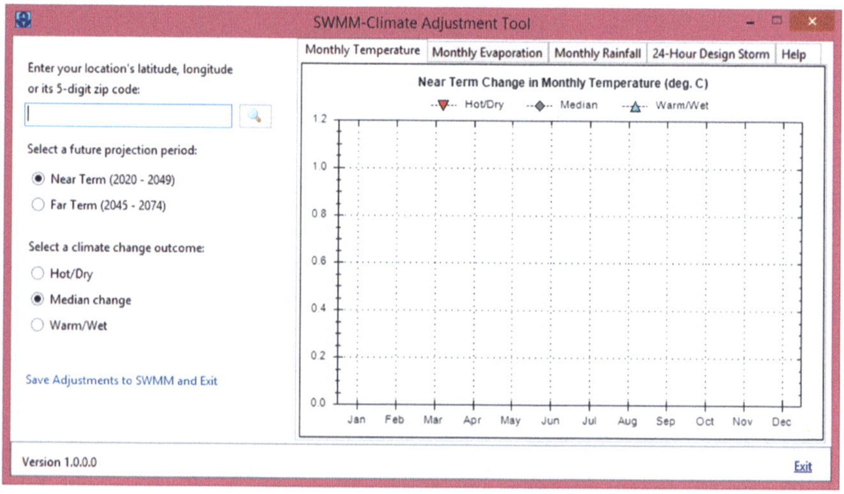

Fig. 5.4 Components in SWMM-CAT

Fig. 5.5 Example problem

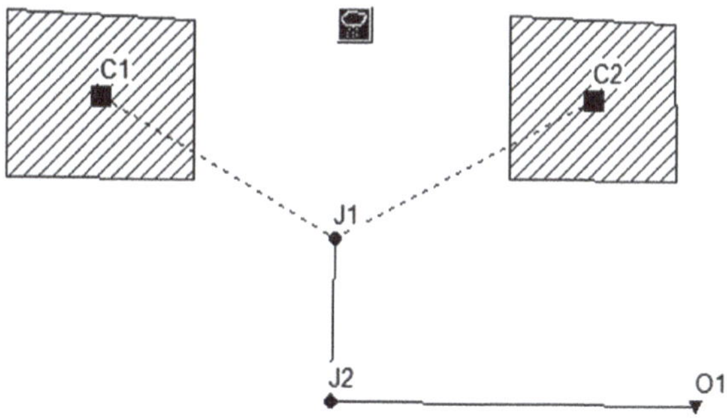

Node Ponded Area = 0
Conduit Length = 200
Conduit Geometry

- Barrels = 0
- Shape = Rectangular
- Max. Depth = 1.0

Conduit Roughness = 0.03
Flow Units = CMS
Link Offsets = DEPTH
Routing method Kinematic Wave

All the above details are input and OK is selected to apply them.

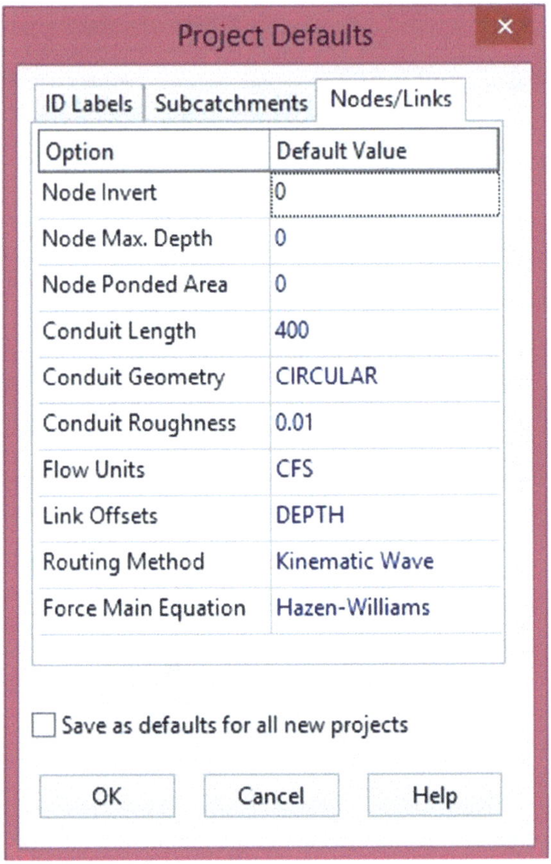

Fig. 5.6 Default options

Section 2 (Fig. 5.7)

- Network diagram of storm network of the study area should be drawn in the SWMM map window.
- Right click on the drawing window, select options, and set the annotations and font size to display names of the objects.
- Select View ≫ Dimensions to display the map dimensions dialog.
- To draw Sub-catchment in the drawing window.
- Select in the menu bar ≫ option.
- Draw the sub-catchment and right click to finish the sub-catchment drawing.
- Double click on the sub-catchment to display the dialog box showing sub-catchment properties.
- Input the sub-catchments, viz., area, width, slope, outlet, etc., as per the study area.

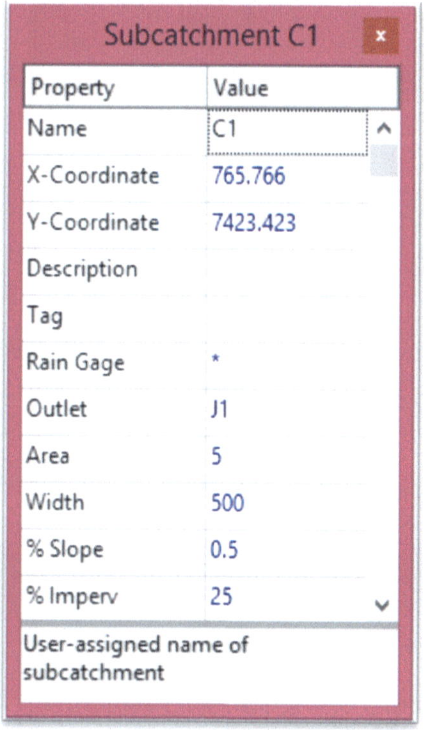

Fig. 5.7 Sub-catchment details

- If there are two or more sub-catchments, the data should be input in the sub-catchment dialog box for each sub-catchment, as these properties differ for each individual catchment.

Next junctions are assigned (Fig. 5.8)

- Select ≫ ○ option to draw junctions wherever required.
- Select the desired junction by double clicking on it and input properties of the junction in the dialog box displayed.
- Repeat the same, if multiple junctions exist.
- Draw the outfall, using Select ≫ ∇.
- Double click on the outfall to input its properties.

The storm system should be connected properly, viz connections between various junctions, junctions, outfalls, etc. This is done by connecting them with storm conduits.

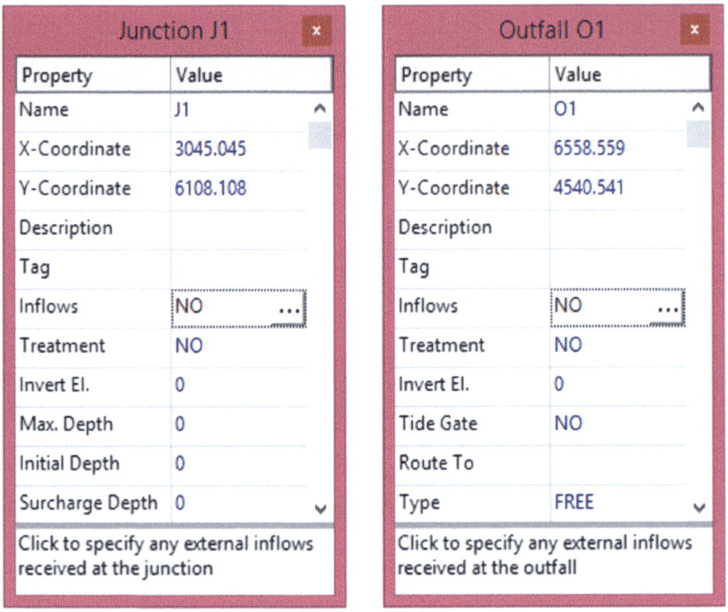

Fig. 5.8 Details of junction and outfall

- Conduit establishes connection between two junctions, for example, to connect junction J1 and J2 conduit CO1 is used
- Select ≫ option to draw conduits
- When the conduit option is selected, the cursor changes as a pencil, the conduit can be drawn at desired locations
- Repeat the procedure to draw conduits between various junctions

Now the study area drawing will be completed by assigning a rain gauge

- Select ≫ option to locate rain gauge in the study area
- Double click on the rain gauge to input time series data and rainfall units
- To input time series data, Project ≫ options ≫ curves ≫ time series
- The time series created above, should be further input into the rain gauge dialog box

For the numerical problem, the following data can be given as input

Node	Invert level
J1	98
J2	92
O1	85

For rain gauge
Rain format—Intensity
Rain Interval—0:30
Data Source—Time series
Series name—time series 1
Time series 1 data:

Time (HH:MM)	Value (mm)
0:00	0
0:30	10
1:00	6
1:30	2
2	0

Now after inputting all the above details, the project can be named by Data browser ≫ project title. To save the file select File ≫ Save as option and browse to the required path.

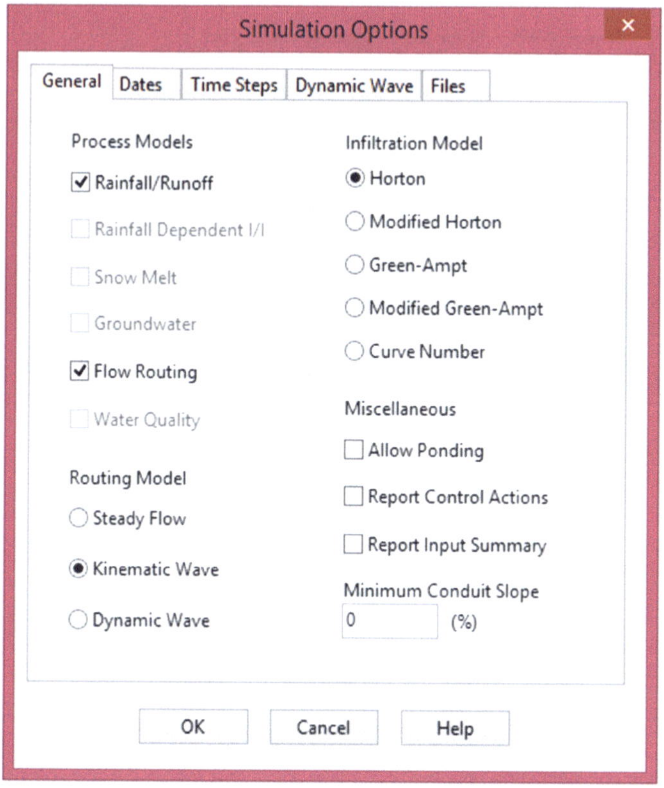

Fig. 5.9 Various available methods

Section 3

- There are several methods available in SWMM for infiltration, routing, etc.
- The user can chose the method based on his/her choice.
- From the side menu bar Project ≫ Options ≫ General.
- After selecting the models, click OK to apply.

Section 4 (Figs. 5.9, 5.10)

- Now as the complete network is created with all input values, the simulation can be RUN.
- The simulation period requires details of wet weather, dry weather durations, and analysis period can be given as input by the user.
- Select Project ≫ ≫ Run Simulation.
- If the models have errors and warnings, they would be displayed. On correcting the errors, the model will RUN successfully.

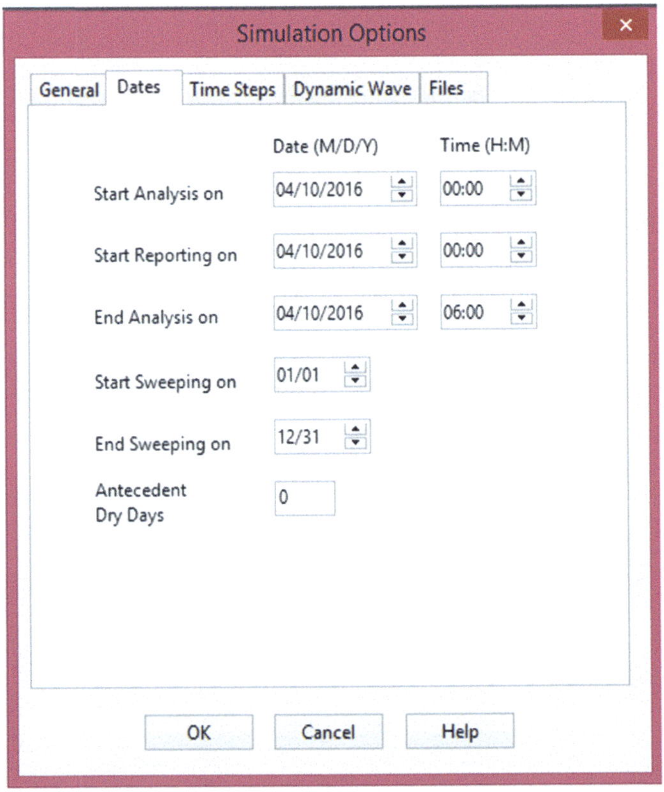

Fig. 5.10 Various simulation options

Section 5

To view results on map

- Select ≫ Reports ≫ Summary.
- To create Profile plot between various junctions Select ≫  from the menu bar.
- To compare two parameters for any two catchments, Scatter plot option can be selected from the menu bar.
- For colour coding a parameter, Select ≫View ≫ Legends.

5.3 Hydrologic Engineering Center-Hydrologic Modeling System

Hydrologic Engineering Center-Hydrologic Modeling System (HEC-HMS) simulates the precipitation runoff processes and can be applied to small urban or natural watershed, large river basin water supply, and flood hydrology. Components of HEC-HMS are as follows:

Watershed

Sub-basin, reach, junction, reservoir, diversion, source, and sink are the elements (Fig. 5.11). Initial constant, Soil Conservation Service (SCS), and gridded SCS curve number, exponential and Green Ampt options are available for computing infiltration losses. Unit Clark, Snyder, and SCS approaches are available for generating surface runoff from excess precipitation (Fig. 5.12). Several routing methods are also available (Fig. 5.13).

Meteorology

Meteorological model includes component of snowmelt, precipitation, and evapotranspiration (Fig. 5.14).

Hydrologic Simulation

Control specifications consist of start and end date of simulation with specified time intervals (Fig. 5.15). A simulation run is created by integrating relevant modules and control specifications (Fig. 5.16). Other components include parameter estimation, analyzing simulations, GIS connection. Researchers can refer Scharffenberg and Harris (2008), Merwade (2012), Silva et al. (2014) and HEC-HMS (2017) for more details about software, user's manual, and other relevant information.

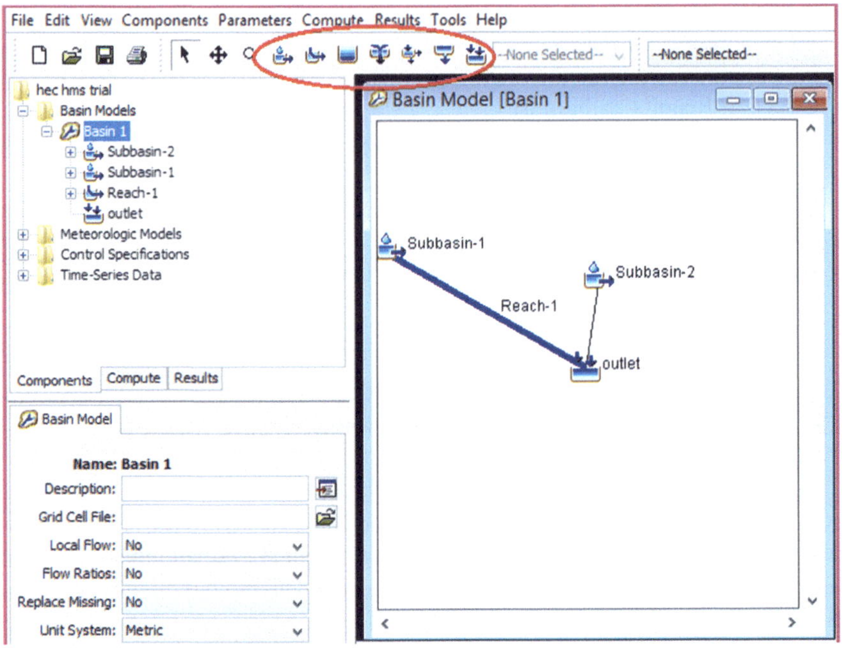

Fig. 5.11 Elements in HEC-HMS window

Numerical Problem 5.2 HEC-HMS Example Model

Data provided is hypothetical and is presented to demonstrate the software application.

Figure 5.17 is a storm water network consisting of three junctions connected by two conduits. Sub-basins 1, 2, and 3 have junction 1, junction 2, and junction 3 as outfalls respectively. Determine peak discharge in m^3/sec from each sub-basin, by using the following data:

Reach 1 and Reach 2: Dimensions of conduits: Rectangle section with width 5 m and length 50 m, Slope—0.03, roughness coefficient—0.02

$$\begin{aligned} \text{Sub}-\text{basin area}-\text{Sub}-\text{basin } 1 - 5\,\text{km}^2 \quad & \text{Lag time}: 60\,\text{min} \\ \text{Sub}-\text{basin } 2 - 3\,\text{km}^2 \quad & \text{Lag time}: 40\,\text{min} \\ \text{Sub}-\text{basin } 3 - 2\,\text{km}^2 \quad & \text{Lag time}: 20\,\text{min} \end{aligned}$$

The total area is 85% impervious

Use SCS Curve method as loss method, SCS Unit Hydrograph for transform method, Kinematic routing for flow routing.

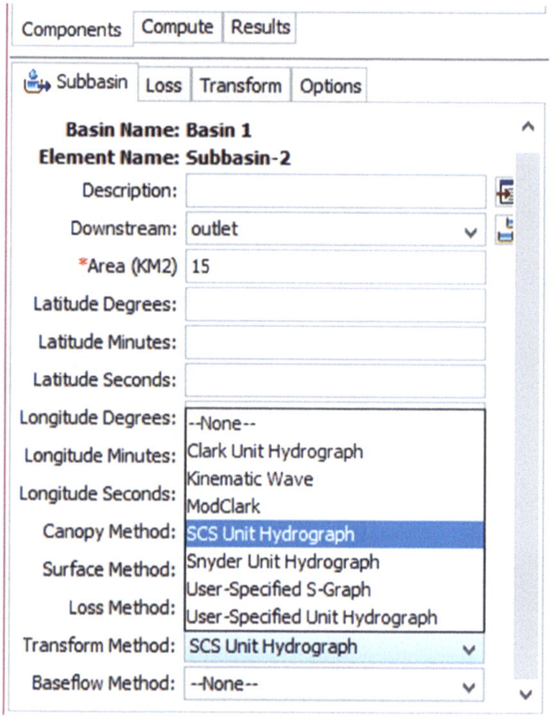

Fig. 5.12 Various surface runoff methods

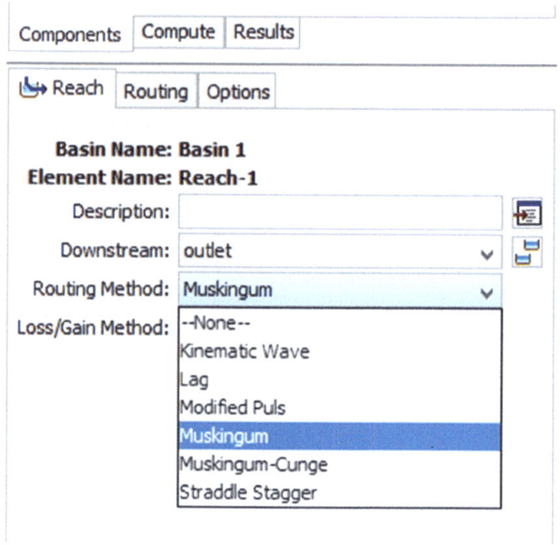

Fig. 5.13 Various routing methods

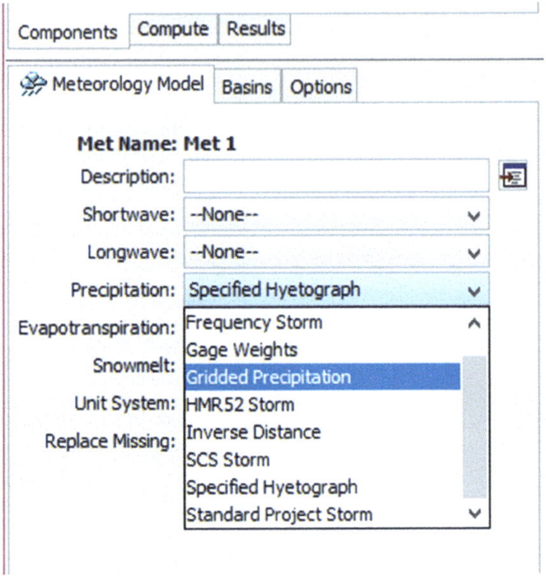

Fig. 5.14 Precipitation input forms

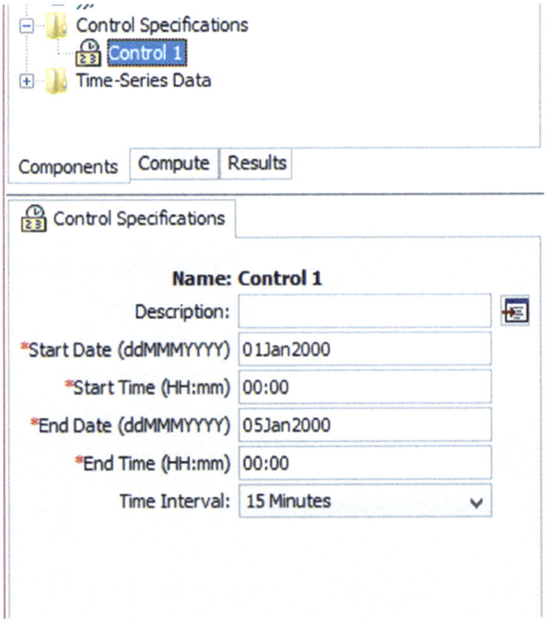

Fig. 5.15 Control specification components

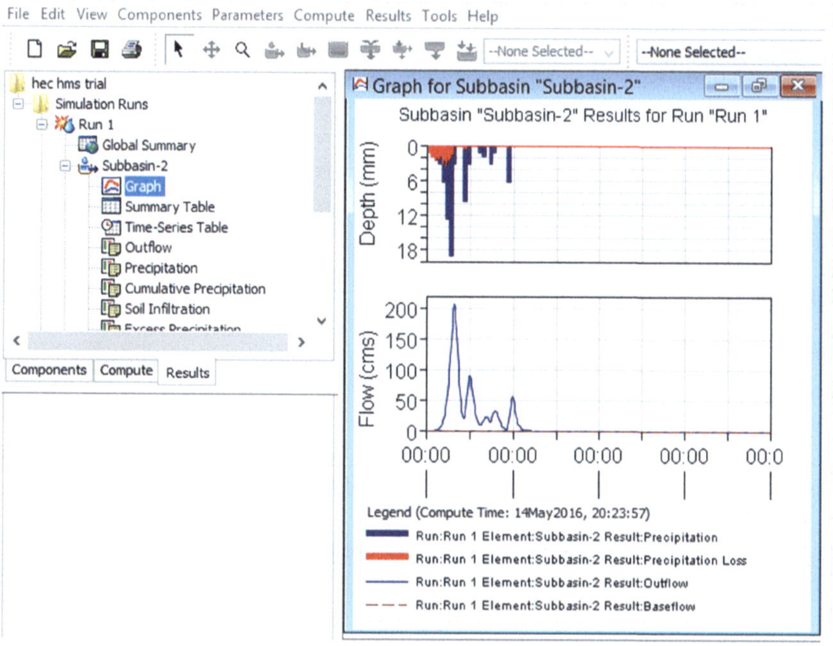

Fig. 5.16 Modeling results

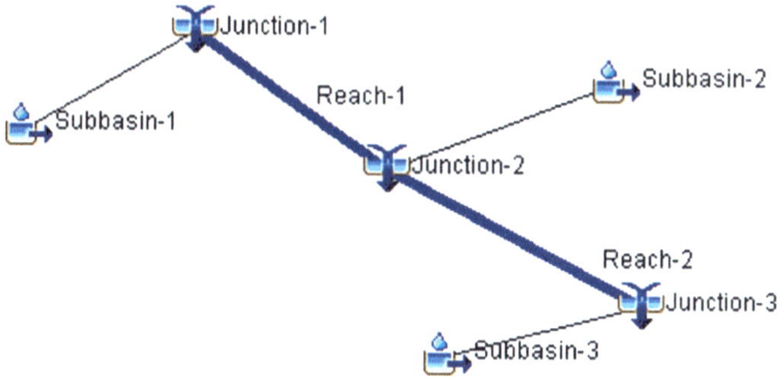

Fig. 5.17 Example problem

Rainfall can be mentioned as Specific Hyetograph with time series:

Day (time HH:MM)	Rainfall (mm)
January 01, 2000, 00:00	0.00
January 01, 2000, 06:00	5.00
January 01, 2000, 12:00	10.00
January 01, 2000, 18:00	3.00
January 02, 2000, 00:00	2.00

Solution:

Steps to modeling

Step 1:

- Before starting a model, a new project has to be created by the user and it should be given a name and saved at the required directory as shown in Fig. 5.18.
- File ≫ New
- HEC—HMS has several objects to represent the watershed system..

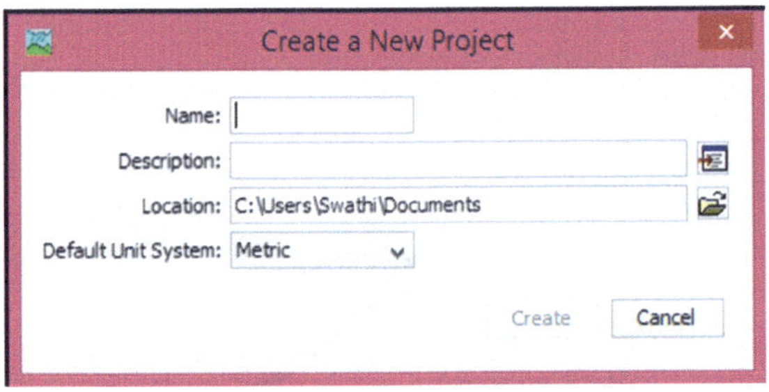

Fig. 5.18 Creating a new project in HEC-HMS

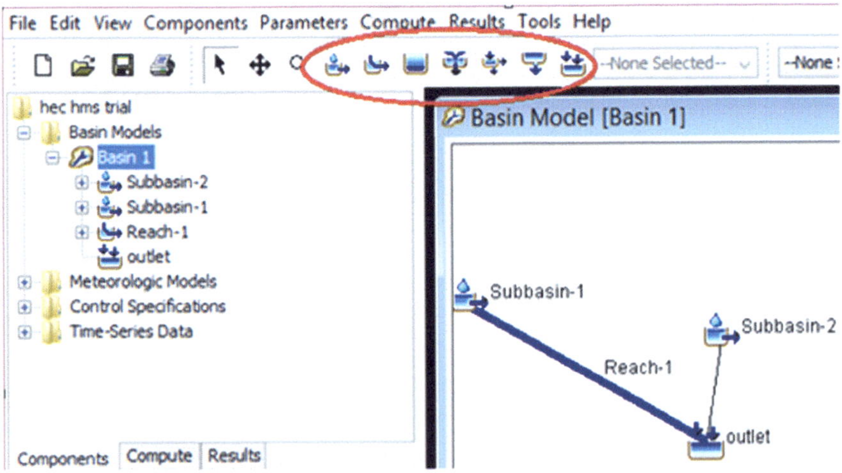

Fig. 5.19 Creating objects in HEC-HMS

- These objects are available on the main window of HEC-HMS which include sub-basin, reach, junction, reservoir, diversion, source, and sink as shown in Fig. 5.19.

Step 2:

- The watershed can be divided into required number of sub-basins.
- Each sub-basin should be created separately and should be assigned a name as in Figs. 5.20a, b, c.
- Components ≫ Basin Model Manager.
- All the required data for each sub-basin should input separately by selecting the desired sub-basin individually.

Draw the storm network in sequence using the objects on the menu (Fig. 5.20c). For the present example, three sub-basins are created, i.e., sub-basin 1, sub-basin 2 and sub-basin 3.

Step 3:
Input the details of junctions, reaches, and sub-basins

Step 4:

In the present numerical problem, SCS Curve number, SCS Unit Hydrograph, and Kinematic Routing are chosen as Loss, Transform (Fig. 5.21), and Routing (Fig. 5.22) methods, respectively, and necessary data is given as input.

Step 5:

- Once the user has successfully created the objects and assigned connections between objects, the user should then specify the precipitation details for which a meteorological model manager should be created.
- Components ≫ Meteorological Model Manager.
- There are many methods to define precipitation, here we have assigned specific Hyetograph method for both the sub-basins as shown in Figs. 5.23a, b.

Step 6:

- The precipitation that has occurred over a period of time and date of the rainfall event are to be input using time series manager, and the data should be assigned to the rain gauge, defined by the user.
- Components ≫ Time series manager.

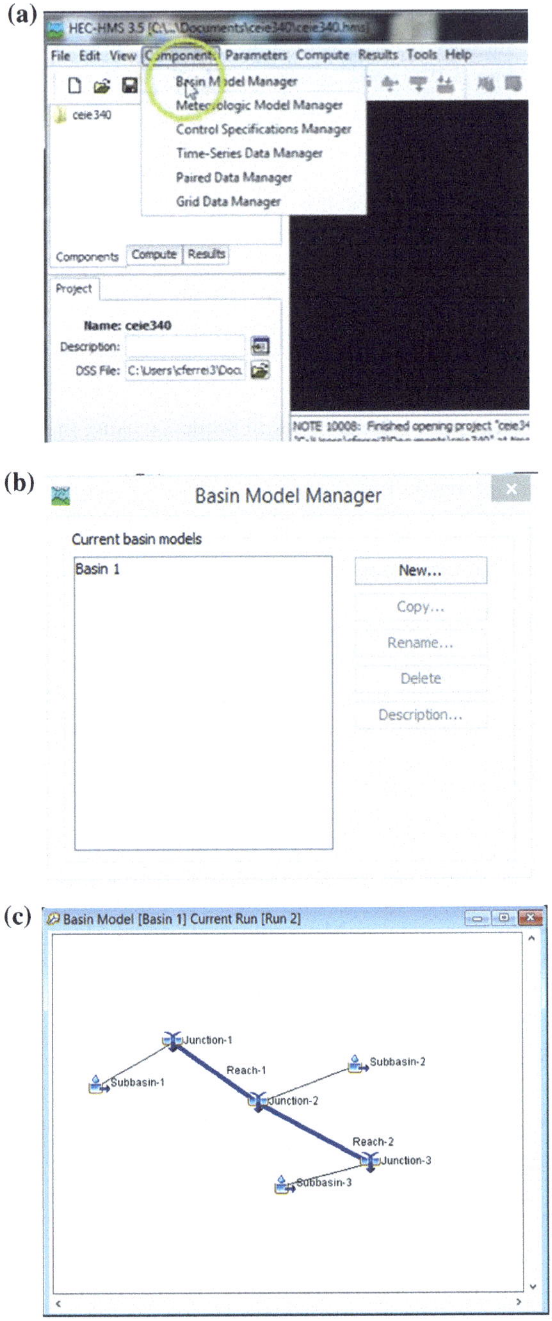

Fig. 5.20 **a** Basin model manager **b** Creating basin in HEC-HMS **c** Created basins in HEC-HMS

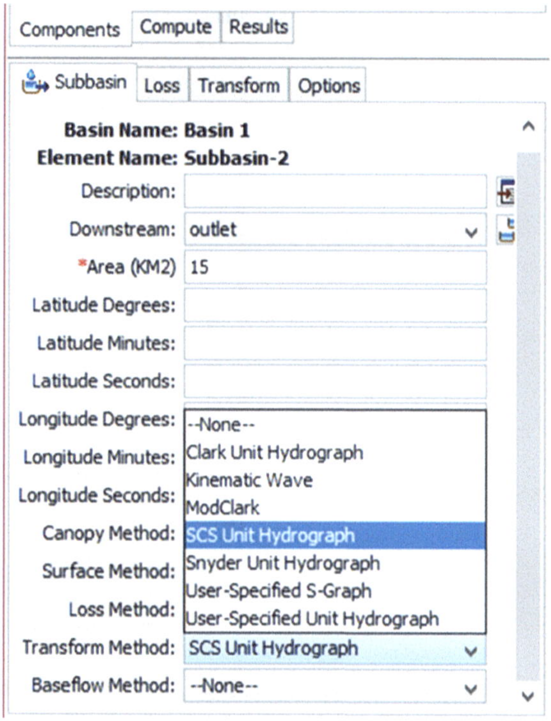

Fig. 5.21 Transform method selection

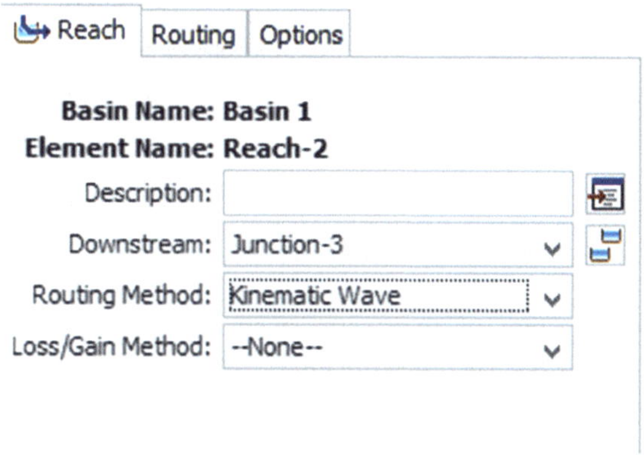

Fig. 5.22 Routing method selection

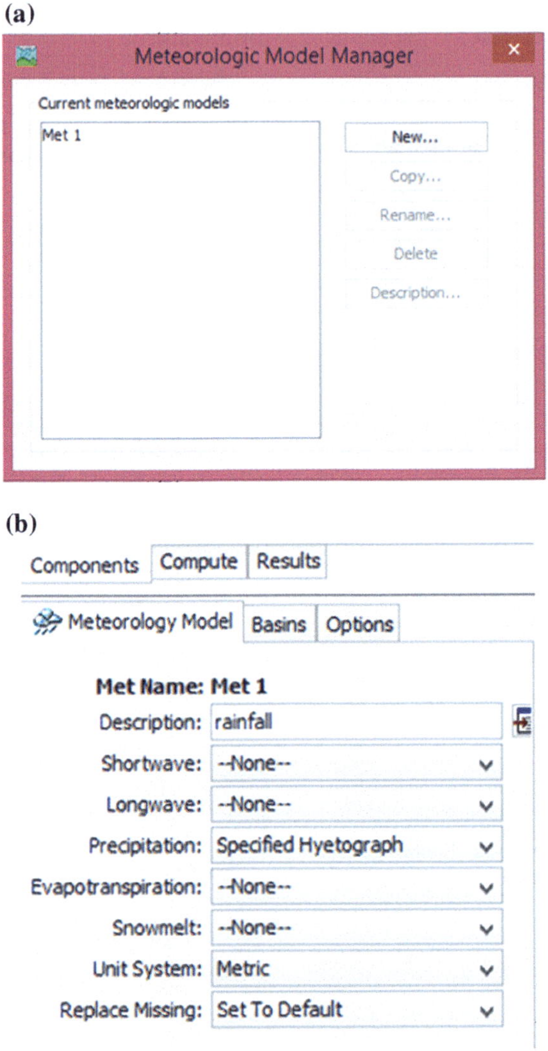

Fig. 5.23 **a** Meteorological model manager in HEC-HMS **b** Precipitation input in HEC-HMS

- Selection of time series manager > Precipitation gauge > gauge 1 > rainfall values. The precipitation gauge should be assigned to all sub-basins (Figs. 5.24a, b, c).

Step 7:

- After the precipitation data as input, the user should define control specifications.
- Components ≫ Control specifications.

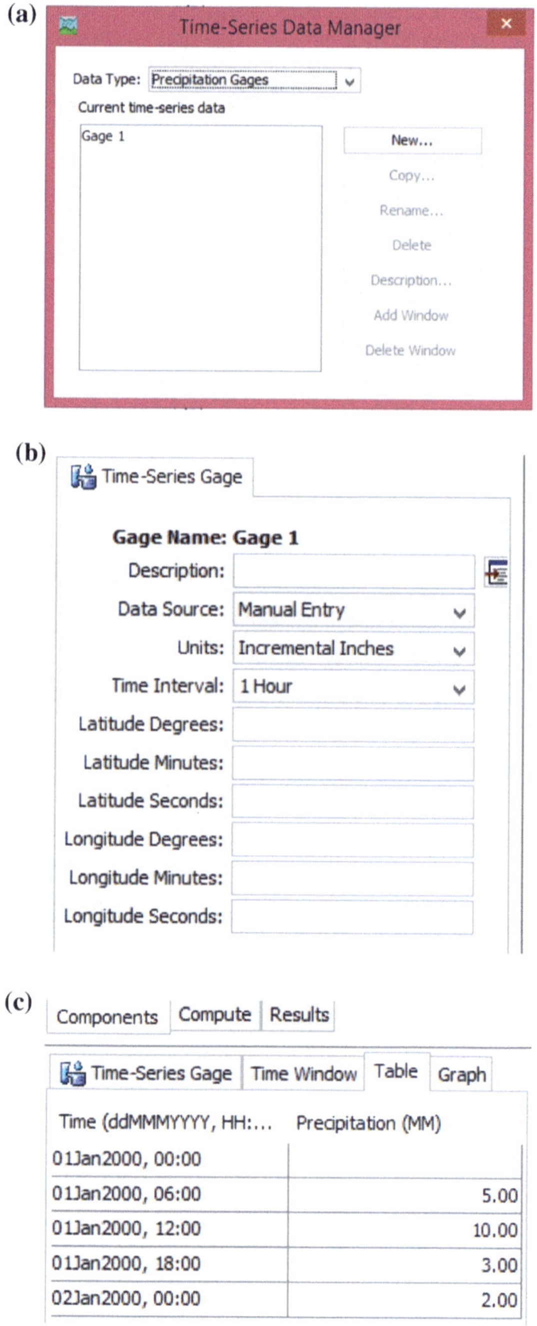

Fig. 5.24 **a** Creating time series manager in HEC-HMS **b** Time series gauge in HEC-HMS **c** Precipitation data entry in HEC-HMS

(a)

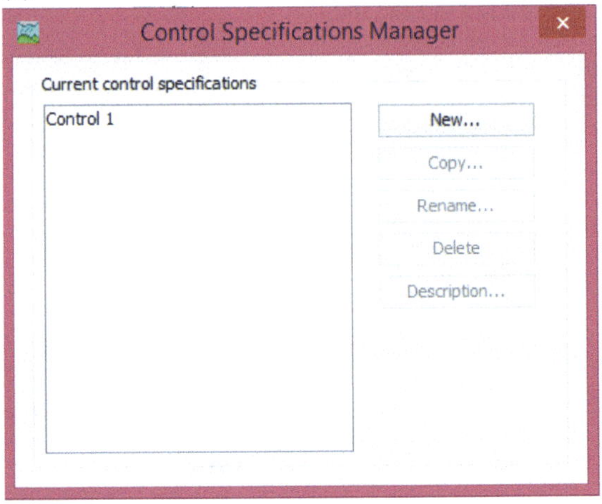

(b)

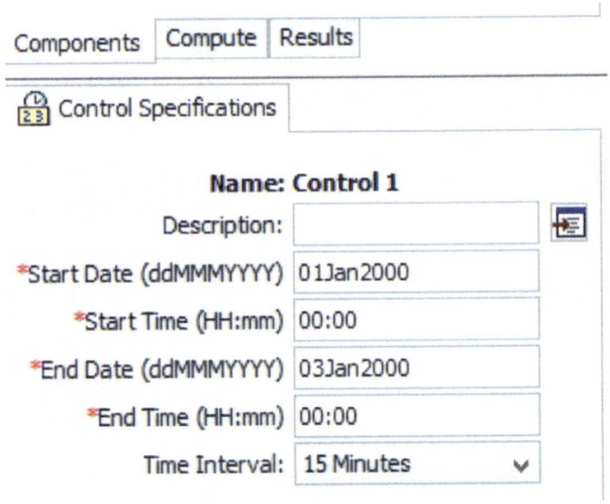

Fig. 5.25 **a** Creating control specifications manager in HEC-HMS **b** Start and end data of modeling in HEC-HMS

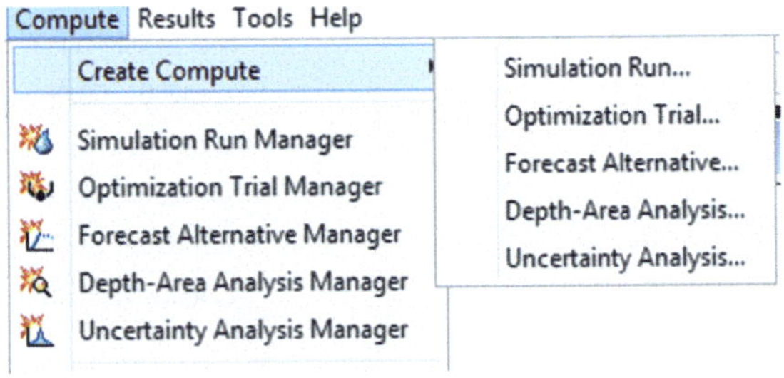

Fig. 5.26 Running the model

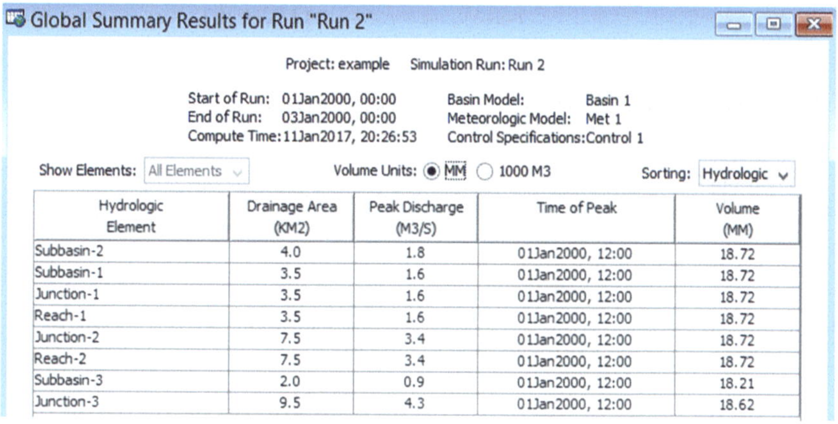

Hydrologic Element	Drainage Area (KM2)	Peak Discharge (M3/S)	Time of Peak	Volume (MM)
Subbasin-2	4.0	1.8	01Jan2000, 12:00	18.72
Subbasin-1	3.5	1.6	01Jan2000, 12:00	18.72
Junction-1	3.5	1.6	01Jan2000, 12:00	18.72
Reach-1	3.5	1.6	01Jan2000, 12:00	18.72
Junction-2	7.5	3.4	01Jan2000, 12:00	18.72
Reach-2	7.5	3.4	01Jan2000, 12:00	18.72
Subbasin-3	2.0	0.9	01Jan2000, 12:00	18.21
Junction-3	9.5	4.3	01Jan2000, 12:00	18.62

Fig. 5.27 Peak discharge of sub-basins

- Start time and end time of the model should be given as input to control specifications to run the model for simulation.
- Here, we have to run the model from the start date of precipitation till 2 days for every 15 min to know the runoff from sub-basins as in Figs. 5.25a, b.

Step 8:

- After all the above details are given as input, the model should be RUN, the warning messages and errors should be corrected, if any.
- Compute ≫ Create ≫ Simulation RUN ≫ Select required RUN.

Step 9:

- Running the model, Compute > Simulation run > Name and select the run (Fig. 5.26).

Step 10:

- Viewing Results, Results > Global summary table for complete results of all objects, Results > Element summary table for specific elements (Fig. 5.27).

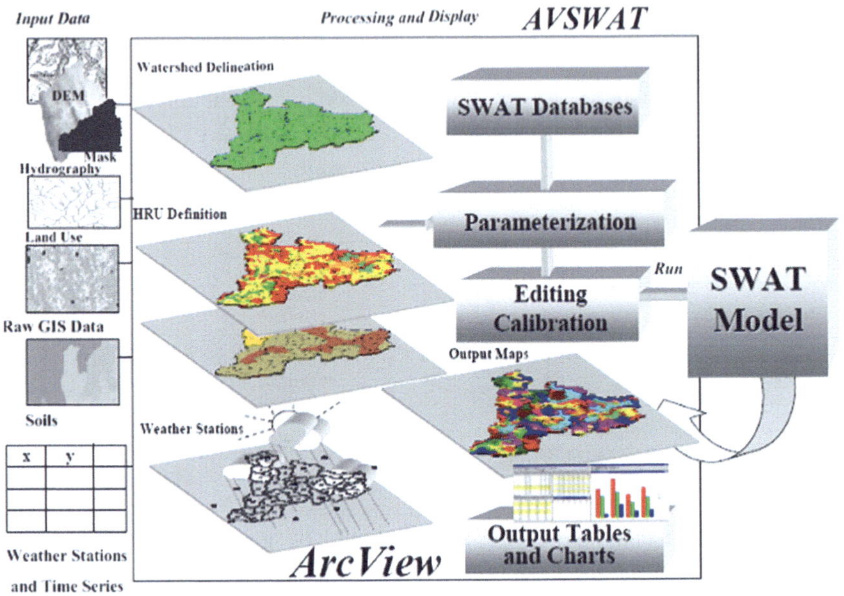

Fig. 5.28 Schematic of AVSWAT *Source* Di Luzio et al. 2002; Available at http://www.pcwp.tamu.edu/reports/2002/tr193.pdf

5.4 Soil and Water Assessment Tool

Soil and Water Assessment Tool (SWAT) is a river basin-scale model (Neitsch et al. 2002a, b). It can predict the impact of land management practices on water, agriculture, and sediment in complex watersheds with different soil categories, land use, and management aspects over long time periods (Anandhi 2007; Akhavan et al. 2010). Watershed is partitioned into Hydrological Response Units (HRUs). Simulation in SWAT is separated into two major parts with first and second parts handling land phase of the hydrologic cycle and routing phase respectively. AVSWAT-2000 is an ArcView extension along with Graphical User Interface (GUI) of the SWAT model (Di Luzio et al. 2002). Schematic view of AVSWAT is presented in Fig. 5.28. Maps of digital elevation, soil, land use/land cover, hydrography, and time series of weather variables along with their locations are the required inputs (Anandhi 2007). SWAT-CUP is another calibration/uncertainty or sensitivity program interface for SWAT, whereas QSWAT is a QGIS interface for SWAT. User manual/information about SWAT, SWAT-CUP, and QSWAT is available in SWAT (2017).

5.5 Other Modeling Techniques

Variable Infiltration Capacity (VIC) is another physical distributed model with components such as land cover, soil, snow model, meteorology/meteorological input data, frozen soil formulation, dynamic lake/wetland model, flow routing. Detailed information on VIC is available (Lohmann et al. 1996 and 1998). Researchers can refer VIC (2017) about software and manual.

MIKE has number of models, namely, MIKE URBAN (2017) covers water distribution and storm water drainage systems. MIKE FLOOD (2017) includes a specialised 1D and 2D flood simulation engines, enabling to model any flood problem. MIKE 21 (2017) handles data assessment for coastal and offshore structures. MIKE HYDRO Basin (2017) facilitates multisector solution alternatives to water allocation and water shortage problems, climate change impact assessments on water resources availability and quality. More detailed applications of MIKE are available in their respective homepages.

Other related software available is XPSWMM (2017). Detailed information about hydrological models are available in Singh and Woolhiser (2002), Devi et al. (2015) and Hydrological Models (2017).

The present chapter discusses various hydrological models relevant to climate change. Next chapter discusses about case studies.

Revision Questions and Exercise Problems

5.1 What is purpose of hydrological modeling?

5.2 What are the various features of SWMM?

5.3 What are the various features of HEC-HMS

5.4 What are the various features of SWAT?

5.5 Describe various MIKE-related hydrological models? Differentiate them with reference to their purpose?

5.6 Using SWMM, obtain the flooded areas of the drain and the maximum depth of flooding for maximum daily rainfall of 80 mm/day. Using Horton infiltration and dynamic routing for the storm network shown in Fig. 5.29.

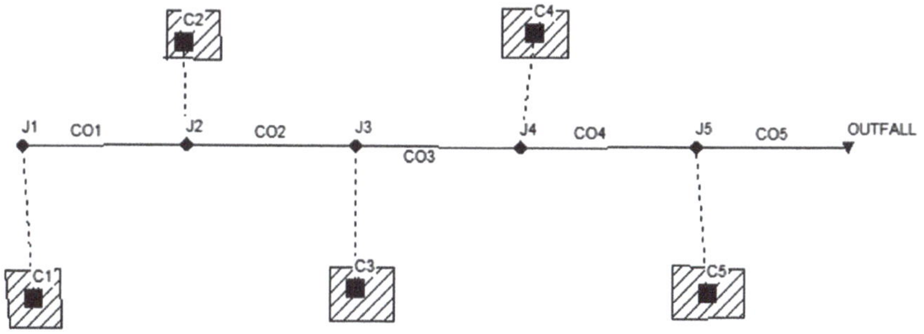

Fig. 5.29 Storm network

Table 5.1 Details about storm network: node information

S.no	Node no.	Junction name	Chainage (M)	Elevation (m)	Depth (m)
1	J1	Children's park	0	520.13	2.2
2	J2	Super market	300	516.67	3.5
3	J3	Temple road	600	508.53	1.8
4	J4	MKR street	900	506.29	2.4
5	J5	Shopping plaza	1200	500.33	4.1
6	Outfall	Lake	1500	498.22	3.0

Table 5.2 Details about storm network: conduits information

S.no	Conduit	Drain reach		Chainage (m)		Width (m)	Depth (m)	Manning's n
		From	To	From	To			
1	CO1	Children's park	Super market	0	300	3	2	0.003
2	CO2	Super market	Temple road	300	600	5	2	0.002
3	CO3	Temple road	MKR street	600	900	8	1.5	0.002
4	CO4	MKR street	Shopping plaza	900	1200	6	2.5	0.003
5	CO5	Shopping plaza	Lake (Outfall)	1200	1500	3.5	1.6	0.002

Table 5.3 Details about storm network: sub-catchments information

S. No	Sub-catchment	Place		Chainage (m)		Area (ha)	Slope of terrain	Weighted (%) impervious	Width (m)
		From	To	From	To				
1	C1	Children's park	Super market	0	300	86.23	0.013	50	320
2	C2	Super market	Temple road	300	600	56.25	0.025	43	430
3	C3	Temple Road	MKR street	600	900	72.65	0.016	28	350
4	C4	MKR street	Shopping plaza	900	1200	62.67	0.056	65	220
5	C5	Shopping plaza	Lake (Outfall)	1200	1500	92.45	0.016	73	184

Supporting information is also provided in Tables 5.1, 5.2, and 5.3. Also (a) obtain the flooded nodes of the storm network and the maximum depth of flooding for an hourly rainfall of 20 mm, use Green Ampt infiltration and dynamic routing, (b) Design the drainage network for no flooding scenario (Using detention ponds, placing pumps at necessary locations), and (c) Design the drainage network for no flooding scenario (Using low impact developments).

5.7 Explore the HEC-HMS model using Muskingum–Cunge routing method for the storm network data of numerical problem 5.2. Compare the difference in results using kinematic routing and Muskingum–Cunge routing?
5.8 Explore the HEC-HMS model for a rainfall of 45 mm occurring for a day by keeping the time series interval as 1 day and identify the critical reach locations, if any. Use data of numerical problem 5.2 wherever necessary, Consider impervious area of all sub-basins as 95%.
5.9 What are various components of VIC?

Advanced Review Questions

5.10 Compare SWMM, HEC-HMS, SWAT, and VIC on any three features?
5.11 In your opinion, which model is suitable for hydrological modeling? Why? If not so, can you describe ideal hydrologic model and its proposed components?
5.12 Can you provide idea of a new software/methodology that combines features of SWMM, HEC-HMS, SWAT, and VIC?
5.13 Can you suggest 10 softwares related to hydrological modeling that were not covered in this chapter?

References

Akhavan S, Abedi-Koupai J, Mousavi SF, Afyuni M, Eslamian SS, Abbaspour KC (2010) Application of SWAT model to investigate nitrate leaching in Hamadan-Bahar watershed Iran. Agr Ecosyst Environ 139:675–688
Anandhi A (2007) Impact assessment of climate change on hydrometeorology of Indian River Basin for IPCC SRES scenarios. PhD thesis, Indian Institute of Science, Bangalore
Barco J, Wong K, Stenstrom M (2008) Automatic calibration of the U.S. EPA SWMM model for a large urban catchment. J Hydraul Eng (ASCE) 134(4):466–474
Daniel EB, Camp JV, LeBoeuf EJ, Penrod JR, Dobbins JP, Abkowitz MD (2011) Watershed modeling and its applications: a state-of-the-art review. Open Hydrol J 5:26–50
Devi GK, Ganasri BP, Dwarakish GS (2015) A review on hydrological models. Aquatic Procedia 4:1001–1007
Di Luzio M, Srinivasan R, Arnold JG, Neitsch SL (2002) Soil and Water Assessment Tool: ArcView GIS Interface Manual: Version 2000. GSWRL Report 02-03, BRC Report 02-07, Texas Water Resources Institute TR-193, College Station, TX, pp. 346
Dooge JCI (1992) Hydrologic models and climate change. J Geophys Res 97(D3):2677–2686
Eldho TI (2017) Lecture Material on Watershed Management (Lecture number 16), Department of Civil Engineering, Indian Institute of Technology, Bombay (Accessed on 1.1.2017)
Francini M, Pacciani M (1991) Comparative analysis of several conceptual rainfall-runoff models. J Hydrol 122:161–219
HEC-HMS (2017). http://www.hec.usace.army.mil/software/hec-hms/. Accessed 31 Jan 2017
Hydrological Models (2017). https://csdms.colorado.edu/wiki/Hydrological_Models. Accessed 31 Jan 2017
Lohmann D, Nolte-Holube R, Raschke E (1996) A large-scale horizontal routing model to be coupled to land surface parametrization schemes. Tellus 48(A):708–721

Lohmann D, Raschke E, Nijssen B, Lettenmaier DP (1998) Regional scale hydrology: I. Formulation of the VIC-2L model coupled to a routing model. Hydrol Sci J 43(1):131–141
Merwade V (2012) Hydrologic Modeling using HEC-HMS. http://web.ics.purdue.edu/~vmerwade/tutorial.html
MIKE 21 (2017). https://www.mikepoweredbydhi.com/products/mike-21. Accessed 31 Jan 2017
MIKE FLOOD (2017). https://www.mikepoweredbydhi.com/products/mike-flood. Accessed 31 Jan 2017
MIKE HYDRO Basin (2017). https://www.mikepoweredbydhi.com/products/mike-hydro-basin. Accessed 31 Jan 2017
MIKE URBAN (2017). https://www.mikepoweredbydhi.com/products/mike-urban. Accessed 31 Jan 2017
Neitsch SL, Arnold JG, Kiniry JR, Williams JR, King KW (2002a) Soil and Water Assessment Tool: Theoretical Documentation: Version 2000, Texas Water Resources Institute (TWRI) Report TR-191. College Station, Texas
Neitsch SL, Arnold JG, Kiniry JR, Srinivasan R, Williams JR (2002b) Soil and Water Assessment Tool: User's Manual: Version 2000, Texas Water Resources Institute (TWRI) Report TR-192. College Station, Texas
Scharffenberg W, Harris J (2008) Hydrologic Engineering Center Hydrologic Modeling System. Interior Flood Modeling, World Environmental and Water Resources Congress, HEC-HMS, pp 1–3
Silva MMGTD, Weerakoon SB, Herath S (2014) Modeling of event and continuous flow hydrographs with HEC–HMS: case study in the Kelani River basin Sri Lanka. J Hydrol Eng (ASCE) 19(4):800–806
Singh V, Woolhiser D (2002) Mathematical modeling of watershed hydrology. J Hydrol Eng (ASCE) 7(4):270–292
SWAT (2017). http://swat.tamu.edu/media/115179/qswat-manual_v13.pdf; http://swat.tamu.edu/software/qswat/; http://swat.tamu.edu/media/114860/usermanual_swatcup.pdf; http://swat.tamu.edu/publications/peer-reviewed-publications/. Accessed 31 Jan 2017
SWMM (2017). http://www2.epa.gov/water-research/storm-water-management-model-swmm. Accessed 31 Jan 2017
VIC (2017). (http://vic.readthedocs.org/en/vic.4.2.c/and; http://www.hydro.washington.edu/Lettenmaier/Models/VIC/Overview/ModelOverview.shtml. Accessed 31 Jan 2017
XPSWMM (2017). http://xpsolutions.com/Software/XPSWMM/. Accessed 31 Jan 2017
Xu CY (2002) Text book of Hydrologic models. Uppsala University, Sweden

Suggested Further Reading

Borah DK (2011) Hydrologic Procedures of Storm Event Watershed Models: A Comprehensive Review and Comparison. Hydrol Process 25:3472–3489
Chen J, Wu Y (2012) Advancing Representation of Hydrologic Processes in the Soil and Water Assessment Tool (SWAT) Through Integration of the TOPographic MODEL (TOPMODEL) features. J Hydrol 420–421:319–328
Chen X, Yang T, Wang X, Xu CY, Yu Z (2013) Uncertainty Intercomparison of Different Hydrological Models in Simulating Extreme Flows. Water Resour Manage 27:1393–1409
Clark MP, Slater AG, Rupp DE, Woods RA, Vrugt JA, Gupta HV, Wagener T, Hay LE (2008) Framework for Understanding Structural Errors (FUSE): A Modular Framework to Diagnose Differences Between Hydrological Models, Water Resources Research, 44, W00B02
Clark MP, McMillan HK, Collins DBG, Kavetski D, Woods RA (2011) Hydrological Field Data from a Modeller's Perspective: Part 2: Process-Based Evaluation of Model Hypotheses. Hydrol Process 25:523–543

Clark MP, Nijssen B, Lundquist JD, Kavetski D, Rupp DE, Woods RA, Freer JE, Gutmann ED, Wood AW, Brekke LD, Arnold JR, Gochis DJ, Rasmussen RM (2015a) A unified approach for process-based hydrologic modeling: 1. modeling concept. Water Resour Res 51:2498–2514

Clark MP, Nijssen B, Lundquist JD, Kavetski D, Rupp DE, Woods RA, Freer JE, Gutmann ED, Wood AW, Gochis DJ, Rasmussen RM, Tarboton DG, Mahat V, Flerchinger GN, Marks DG (2015b) A unified approach for process-based hydrologic modeling: 2. model implementation and case studies. Water Resour Res 51:2515–2542

Case Studies

6

Abstract

This chapter presents various real-world global case studies in AR3 and AR5 perspective that are related to the evaluation of GCMs for maximum and minimum temperatures for India, intercomparison of statistical downscaling methods for projection of extreme precipitation in Europe, downscaling of climate variables using Support Vector Machine, Multiple Linear Regression for Malaprabha and Lower Godavari Basins, India and applicability of large-scale climate Teleconnections and Artificial Neural Networks for Regional Rainfall Forecasting for Orissa, India. In addition, the impact of climate change on semi-arid catchment water balance using an ensemble of GCMs for Malaprabha catchment, India; streamflow in four large African river basins; projection of rainfall–runoff for Murray–Hotham catchment of Western Australia; future changes in Mekong River hydrology are also parts of the chapter. The reader is expected to understand the impact studies through various case studies by studying this chapter.

Keywords

Africa · Artificial neural networks · Australia · Climate change · India · Precipitation · Teleconnections · Temperature

6.1 Introduction

In the present chapter, various case studies are presented to provide an overall view regarding the applicability of various techniques studied in previous chapters of the book. Eight case studies pertaining to India and Global context on various subjects

---

Electronic supplementary material The online version of this chapter (doi:10.1007/978-981-10-6110-3_6) contains supplementary material, which is available to authorized users.

K. Srinivasa Raju and D. Nagesh Kumar, *Impact of Climate Change on Water Resources*, Springer Climate, https://doi.org/10.1007/978-981-10-6110-3_6

Table 6.1 Description of various case studies

Case Study number	Title of the case study
6.2	Evaluation of global climate models for maximum (T_{max}) and minimum (T_{min}) temperatures
6.3	Downscaling of climate variables using support vector machine and multiple linear regression
6.4	Climate change impact on semi-arid catchment water balance using an ensemble of GCMs
6.5	Comparing impacts of climate change on streamflow in four large African river basins
6.6	Hydrologic impact of climate change on Murray–Hotham catchment of Western Australia: a projection of rainfall–runoff for future water resources planning
6.7	Intercomparison of statistical downscaling methods for projection of extreme precipitation in Europe
6.8	Future changes in Mekong river hydrology: impact of climate change and reservoir operation on discharge
6.9	Regional rainfall forecasting using large-scale climate teleconnections and Artificial Neural Networks

are presented in this chapter (Table 6.1). In the present study, Precipitation and rainfall are used interchangeably.

Table 6.2 presents various chapters and relevant topics covered in the case studies. Detailed case studies are presented in the next sections.

6.2 Evaluation of Global Climate Models for Maximum (T_{max}) and Minimum (T_{min}) Temperatures[1]

Summary: Thirty-six Coupled Model Intercomparison Project-5 (CMIP5)-based global climate models (GCMs) are explored to evaluate the performance of maximum (T_{max}) and minimum (T_{min}) temperature simulations for India covering 40 grid points. Two perspectives are considered. In the first perspective, three performance indicators are used for evaluating GCMs, namely, Correlation Coefficient

[1]Adopted from

Raju KS, Sonali P, Nagesh Kumar D (2017) Ranking of CMIP5-based Global Climate Models for India Using Compromise Programming, Theoretical and Applied Climatology, Springer, Vol. 128, No. 3, pp. 563–574, doi:10.1007/s00704-015-1721-6, https://link.springer.com/article/10.1007/s00704-015-1721-6

Raju KS, Nagesh Kumar D (2016) Selection of Global Climate Models for India using Cluster Analysis, Journal of Water and Climate Change, IWA, Vol. 7, No. 4, pp. 764–774, doi:10.2166/wcc.2016.112, http://jwcc.iwaponline.com/content/7/4/764?etoc

With permission from the publishers. More details are available in the original publication.

Table 6.2 Details of topics and relevant case studies

Chapter	Title of chapter	Topic wise remarks with relevance to Chapter	Relevant case study number
1	Introduction	SRES A1B, A2, B1 and COMMIT	6.3, 6.6, 6.8
		RCP 2.6, 4.5, 6.0, 8.5	6.3, 6.5
		El-Niño southern oscillation (ENSO)	6.9
2	Selection of global climate models	Performance indicators	6.2, 6.4, 6.5, 6.8, 6.9
		Entropy	6.2
		Compromise programming	6.2
		Group decision-making	6.2
		Ensemble of GCMs	6.2, 6.4
3	Downscaling techniques in climate modeling	Multiple linear regression	6.3
		Artificial neural networks	6.9
		Support vector machine	6.3
		Bias correction	6.5, 6.7
		Change factor approach	6.7
4	Statistical and optimisation techniques in climate modeling	Cluster analysis	6.2, 6.3
		Principal component analysis	6.3
		F-Statistic test	6.3
		Linear programming	6.8
5	Hydrological modeling	Arc GIS-Arc view extension of soil and water assessment tool (ArcSWAT)	6.4

(CC), Normalised Root Mean Square Error (NRMSE) and Skill Score (SS). Entropy technique is applied to compute weights of the three indicators. Compromise Programming (CP), a distance-based decision-making technique, is employed to rank the GCMs. Group decision-making technique is used to aggregate the ranking patterns obtained for individual grid points. In the second perspective, applicability of K-Means cluster analysis is explored for grouping GCMs. Skill score indicator is chosen for evaluation. Cluster validation techniques, namely, Davies–Bouldin Index (DBI) and F-Statistic test are used to obtain an optimal number of clusters of GCMs for India. Effective ensemble approach is also suggested.

6.2.1 Problem Description, Case Study and Data (Raju et al. 2017; Raju and Nagesh Kumar 2016)

Objectives chosen are as follows:

- Development of methodology for determining weights of the indicators
- To study the applicability of decision-making techniques for ranking GCMs
- To study the applicability of group decision-making for aggregating the grid-wise ranking of GCMs
- To study the applicability of cluster analysis and validation techniques for optimal grouping of GCMs
- Suitable ensemble of GCMs for T_{max}, T_{min} and combination of T_{max} and T_{min} (from now termed as T_{mm}) scenarios

Three indicators, Correlation Coefficient (CC), Normalized Root Mean Square Error/Deviation (NRMSE), and Skill Score (SS) are analyzed for evaluating the ability of 36 GCMs from CMIP5 (Taylor et al. 2012), namely, ACCESS1.0, ACCESS1.3, BCC-CSM1.1, BCC-CSM1.1-m, BNU-ESM, CCSM4, CESM1-BGC, CESM1-CAM5, CESM1-FAST CHEM, CESM1-WACCM, CNRM-CM5, CSIRO-Mk3.6, CanESM2, FGOALS-s2, FIO-ESM, GFDL-CM3, GFDL-ESM2G, GFDL-ESM2M, GISS-E2-H, GISS-E2-R-CC, GISS-E2-R, HadCM3, HadGEM2-AO, INM-CM4, IPSL-CM5A-LR, IPSL-CM5A-MR, IPSL-CM5B-LR, MIROC4h, MIROC5, MIROC-ESM-CHEM, MIROC-ESM, MPI-ESM-LR, MPI-ESM-MR, MPI-ESM-P, MRI-CGCM3, NorESM1-M for both T_{max} and T_{min} for India.

World Climate Research Programmes (WCRP's) CMIP5 multi-model data set for climate of the twentieth-century simulations have been used (Taylor et al. 2012). CMIP5-based GCMs providing monthly T_{max} and T_{min} data were considered. Outputs from 36 models from CMIP5 were used in this model evaluation. Forty-one years (1969–2009) of observed gridded T_{max} and T_{min} is available. India Meteorological Department (IMD) data is available at a 1° resolution (Srivastava et al. 2009), while most of the climate models from CMIP5 have historical data up to 2005 only (Taylor et al. 2012). Hence, 37 (1969–2005) years of the observed data from IMD was considered. This data may be sufficient for model assessment but could not be used for productive climate studies during first and second half of twentieth century. For a trend analysis, short length data may not be sufficient and may lead to erroneous conclusions. Due to this drawback, gridded T_{max} and T_{min} data set from Climate Research Unit (CRU) developed by University of East Anglia (Carter et al. 2004; Mitchell and Jones 2005) and reanalysis data from NCEP (National Centers for Environmental Prediction) were additionally used as substitutes (data record availability: IMD: 1969–2009; CRU2.1: 1901–2002; NCEP: 1948 till date). The three data sets were interpolated to common grid points of 2.5° × 2.5°. Using these re-gridded data sets, correlations and absolute gridded mean differences between annual values were compared for CRU2.1 and NCEP in relation to the IMD data set for a common period of 1969–2002. This exercise is repeated for all the months and all the seasons [January, February (JF), March, April, May (MAM), June, July, August, September (JJAS), October, November, and December (OND)] for both T_{max} and T_{min}. Significantly higher absolute gridded average differences were obtained between NCEP and IMD compared to CRU2.1 and IMD over most regions. CRU2.1 showed a better correlation with IMD than for NCEP reanalysis data. Hence, the observed T_{max} and T_{min} data set obtained from the CRU2.1 can be used as a proxy to IMD data set. Absolute gridded average differences between CRU2.1 and IMD were found to be higher, over Western Himalaya region and for a few grid points over Northeast region. A similar pattern has been observed for all the seasons and months. Therefore, the grid points over Western Himalaya region and the few over Northeast region were not considered (as the observational data sets themselves had significant differences). The period between 1961–1999 is considered as the base period for the present model evaluation study. Single ensemble realizations for each of the models have been used.

6.2.2 Results and Discussion

6.2.2.1 Ranking of Global Climate Models (Raju et al. 2017)

Gridwise (various latitude and longitude combinations resulting in 40 grid points) values of T_{max} and T_{min} are analyzed to assess how the individual GCMs can be ranked with reference to the three indicators using compromise programming. Weights are computed using entropy technique. The group decision-making technique is employed to aggregate the ranking patterns of GCMs for whole of India (Morais and Almeida 2012). Related results are presented in the following sections.

Maximum Temperature Scenario: Statistics of weights of indicators over 40 grid points obtained by entropy technique are presented in Table 6.3. It is observed

Table 6.3 Distribution of weights over 40 grid points obtained by entropy technique into various ranges for T_{max} and T_{min}

Weight range (in %)	T_{max}			T_{min}		
	CC	SS	NRMSE	CC	SS	NRMSE
≤ 10	29	29		39	34	
>10 and ≤ 20	9	5		1	5	
>20 and ≤ 30	2	6				
>30 and ≤ 40						
>40 and ≤ 50					1	1
>50 and ≤ 60						
>60 and ≤ 70			4			
>70 and ≤ 80			6			
>80 and ≤ 90			16			7
>90 and ≤ 100			14			32

that weights of performance indicators are varying for each grid point and are expected to effect the ranking pattern of GCMs gridwise.

A glimpse of ranking pattern obtained over 40 grid points by compromise programing is presented in Table 6.4. CESM1-CAM5, CNRM-CM5, FGOALS-s2 and MIROC5 are occupying first position, for 65% of grid points. Figure 6.1 shows the spatial distribution of GCMs which are occupying first position. In case of second position, BNU-ESM, CESM1-BGC, CESM1-CAM5, CESM1-WACCM, CNRM-CM5, FGOALS-s2, MIROC4h and MIROC5 are occupying 77.5% of grid points. However, there is a huge spread of GCMs in the third position. It is interesting to note that HadCM3 occupied last position 27 times and has never occupied first three positions.

Effort is also made to rank the GCMs for whole of India using group decision-making technique (Morais and Almeida 2012). It is observed from Fig. 6.2 that first three-ranked ranked GCMs, i.e., CNRM-CM5, MIROC5, and FGOALS-s2 have net strength of 493, 440, and 339, respectively, whereas CESM1-BGC, BNU-ESM, and CESM1-WACCM (4th, 5th and 6th ranks) have net strength of 253, 246, and 242, respectively, which are nowhere comparable to the first three GCMs. Last position (rank 36) is occupied by HadCM3 (net strength of −616) whereas, 35th position is occupied by ACCESS1.3 with a net strength of −477.

From the extensive studies (i.e., analyzing suitability of GCMs for 40 grid points individually and as a group), it is inferred that no single GCM can be suggested for India, so an ensemble of GCMs can be employed. As an initiation, the GCMs which occupied the first three positions in group decision-making perspective, i.e., CNRM-CM5, FGOALS-s2, and MIROC5 are suggested for ensemble. This aspect is also supported, as these GCMs occupied first position for 21 grid points among a total of 40. However, GCMs occupying fourth, fifth and sixth positions were not considered for ensemble due to their huge difference between the net strengths of these GCMs compared to the first three.

Minimum Temperature Scenario: Impact of climate variable T_{min} on ranking of GCMs is explored. Statistics of weights of indicators and compromise programing results are presented in Tables 6.3 and 6.4. It is observed that CESM1-CAM5, GFDL-CM3, MIROC4h, MIROC5, and Nor ESM-I occupied first position for 60% of grid points. Figure 6.3 presents spatial distribution of GCMs occupying the first position over 40 grid points for T_{min}. CESM1-BGC, CESM1-CAM5, CESM1-WACCM, MIROC4h, MIROC5 and NorESM1-M occupied 70% of grid points for second position whereas CCSM4, CESMI-FASTCHEM, CESM1-WACCM, FIO-ESM, MROC4h and NorESM1-M occupied 52.5% of grid points for third position. Quite interestingly, HadCM3 and INM-CM4 occupied the last position for 22 and 17 grid points, respectively.

In group decision-making perspective, four GCMs namely, MIROC4h, NorESM1-M, MIROC5, and CESM1-CAM5 occupied first four positions with net strengths of 430, 422, 338, 329, respectively, whereas the last position is occupied by INM-CM4 with a net strength of −661 (Fig. 6.2). This can also be supported by the fact that the above mentioned four GCMs occupied first position in several number of grids (in this case 20) and can be considered for ensemble purpose in case of T_{min}.

Table 6.4 Number of grid points the GCMs occupied the first three positions and last position for T_{max} and T_{min}

Model name	T_{max}				T_{min}			
	First	Second	Third	Last	First	Second	Third	Last
ACCESS1.0			3			1	1	
ACCESS1.3		1		4		1	1	
BCC-CSM1.1	2	1	1			1		
BCC-CSM1.1-m	1	1	2					
BNU-ESM	3	3	3		3	1		
CCSM4			2		2	2	4	
CESM1-BGC		4	2			3	2	
CESM1-CAM5	5	5	1	1	6	4	2	
CESM1-FAST CHEM		1	1		1	1	3	
CESM1-WACCM	2	3	4		1	4	3	
CNRM-CM5	7	7	4			1		
CSIRO-Mk3.6								
CanESM2	1		1		2			
FGOALS-s2	7	3	4		1	1	1	
FIO-ESM	2		4	1	1	1	3	
GFDL-CM3		1			4		1	
GFDL-ESM2G					1			
GFDL-ESM2M							1	
GISS-E2-H			1	3				
GISS-E2-R-CC				1		1		
GISS-E2-R			1				1	
HadCM3				27				22
HadGEM2-AO	1	1	1				1	
INM-CM4				1				17
IPSL-CM5A-LR								
IPSL-CM5A-MR					1			
IPSL-CM5B-LR	1			2				1
MIROC4h		3	1		4	5	5	
MIROC5	7	3	1		5	7	1	
MIROC-ESM-CHEM	1	1					1	
MIROC-ESM			1					
MPI-ESM-LR					1		2	
MPI-ESM-MR								
MPI-ESM-P						1	1	
MRI-CGCM3					2		2	
NorESM1-M		2	2		5	5	3	

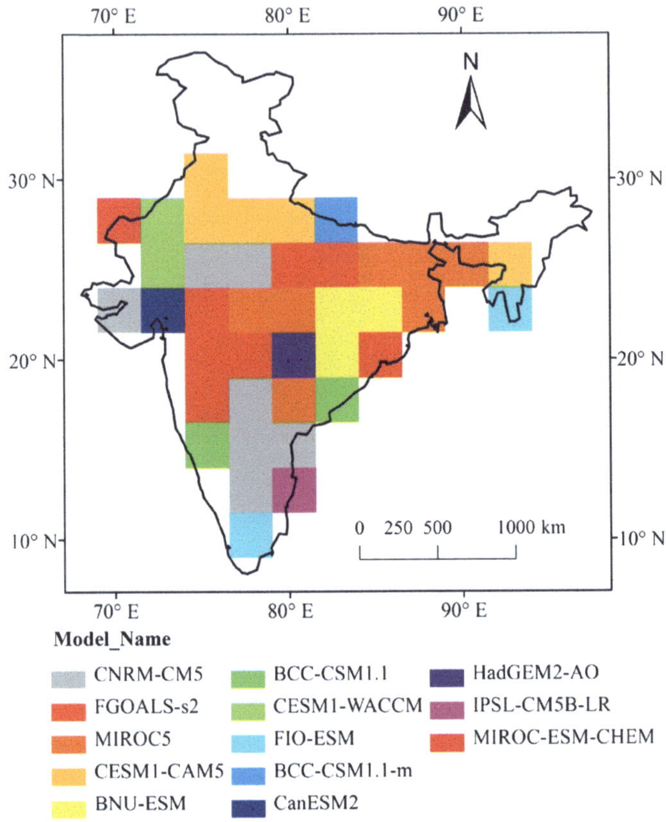

Fig. 6.1 Spatial distribution of GCMs occupying the first position over 40 grid points for T_{max}

6.2.2.2 Clustering of Global Climate Models (Raju and Nagesh Kumar 2016)

Data matrix of 36 GCMs and skill scores for each grid point are developed for three variables, i.e., T_{max}, T_{min} and T_{mm} (Raju and Nagesh Kumar 2016).

Cluster Validity Analysis Platform (CVAP) developed by Wang (2007) is used for K-Means cluster analysis. Cluster analysis is performed for 2–9 clusters and even beyond (Wang et al. 2009; Pennell and Reichler 2011; Raju and Nagesh Kumar 2014). CVAP is run for 2–30 clusters (five times for each cluster) to assess the occupancy of GCMs in each cluster for a total of 145 runs. It is observed that as the number of clusters increased beyond nine, most of the clusters are empty. Therefore, the analysis is restricted to nine clusters. For each chosen group, Davies–Bouldin Index (DBI) (Davies and Bouldin 1979) and F-Statistic test (Burn 1989) are evaluated and are used as basis for determining the optimal cluster.

Figure 6.4 presents the number of GCMs in clusters 2–9 and representative GCM in each subcluster for $T_{max.}$ Notation for cluster 2 in Fig. 6.4 is as follows: 1 represents the sub-cluster, GFDL-ESM2M is a representative GCM in the

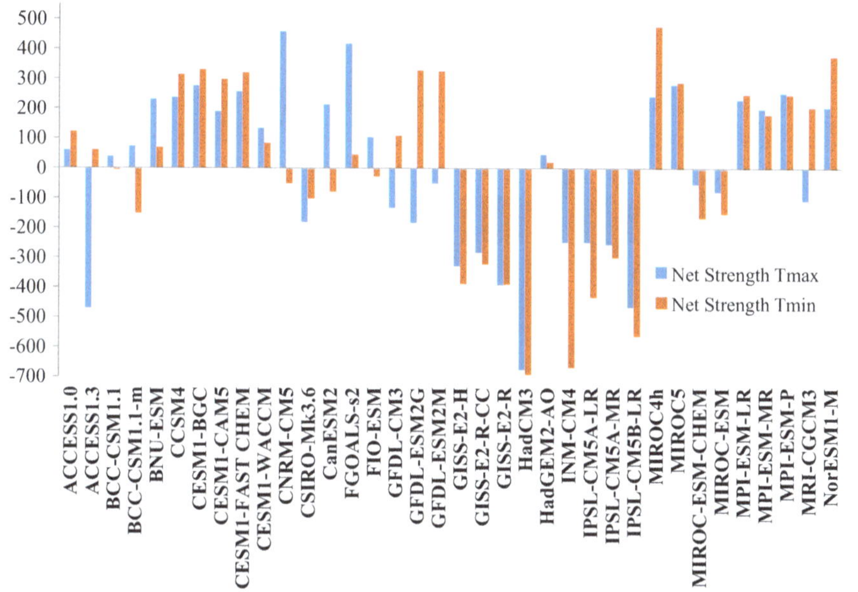

Fig. 6.2 Net strength of GCMs in Group Decision-Making for T_{max} and T_{min}

subcluster; 32 represents number of GCMs in that subcluster whereas 2 represents the subcluster, GISS-E2-H is a representative GCM in the subcluster; 4 represents number of GCMs in that subcluster. It is observed that as the number of clusters increase, there is a wide spread in the number of GCMs. It is also interesting to note that as the number of clusters increases to four and more, only one GCM (HadCM3) is observed in subclusters belonging to clusters 4–9 and only two GCMs are observed in few subclusters belonging to clusters 7–9. This outcome is on the expected lines as GCMs are only 36 and number of clusters are too many. It is to be noted that there is no specific trend while classifying GCMs from clusters 2–9. Most of the times GCMs are in the same subcluster irrespective of increase in size of cluster. This may be due to the similarities in their structure and a reason that they are developed by the same agency. Substantial similarities are noticed between GCMs from the same institution in the cluster analysis outcome. For example, GISS-E2-H, GISS-E2-R, GISS-E2-R-CC developed by NASA Goddard Institute for Space Studies are subclustering always jointly; similar to IPSL-CM5A-MR and IPSL-CM5A-LR developed by Institute Pierre–Simon Laplace whereas IPSL-CM5B-LR which is developed by the same institute is not part of the same subcluster. Variation of total squared error for 2–9 clusters is shown in Fig. 6.5. It is noticed from Fig. 6.5 that total squared error is reducing from 15.1453 (squared error value for subclusters 1 and 2 are 11.248 and 3.8973 totalling 15.1453) to 4.1211 for clusters 2–9 with a range of 11.0242. Average squared error per cluster is found to be 1.378. There is steep fall of squared error from cluster 2–3, i.e., 4.4097. Thereafter difference is approximately of 1 unit each from clusters 3–8

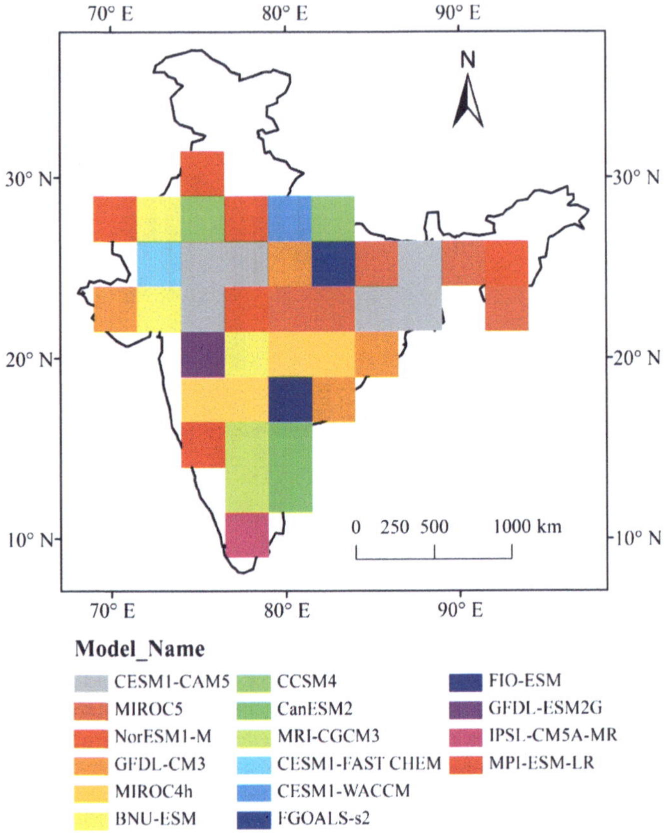

Fig. 6.3 Spatial distribution of GCMs occupying the first position over 40 grid points for T_{min}

whereas it is very nominal from 8 to 9. Squared error values between group average and the skill score values for each GCM in the subcluster are computed. The summation of the squared error values for all 40 grid points gives the total squared error value corresponding to each GCM in a subcluster. The GCM that gives the minimum squared error value in a subcluster is chosen as the representative GCM for that particular subcluster. The squared error value is one of the important input parameters for computation of DBI and F-Statistic test which provides the basis for optimal clusters. Figure 6.6 presents DBI values and these are varying from cluster 2 (1.2675) to cluster 7 (0.8261) but not in sequential order. According to DBI principles, optimum cluster size is 7. However, it is also interesting to note that, group 3, 7, and 9 have almost equal DBI values, i.e., 0.8829, 0.8261, and 0.8909, respectively, with a nominal difference of 0.0568 (with reference to clusters 7 and 3) and 0.0648 (with reference to clusters 7 and 9). This aspect makes it difficult to assume a reasonable margin of error to determine optimum clusters with

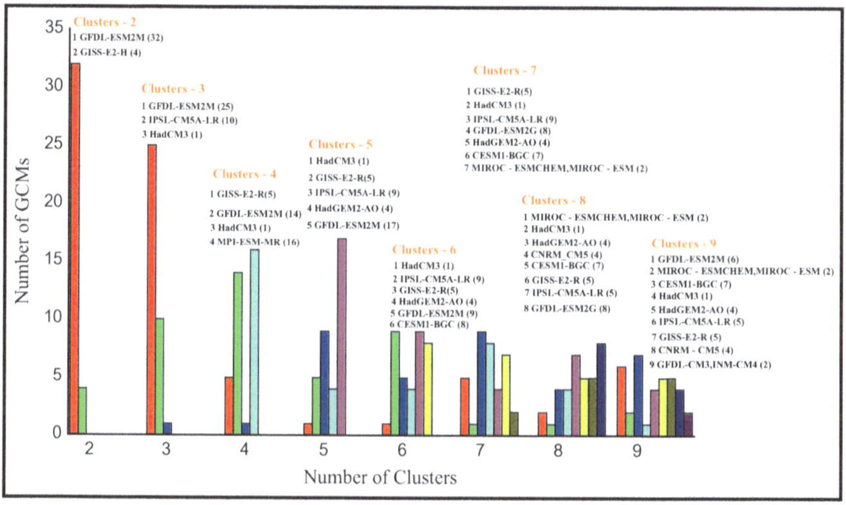

Fig. 6.4 Number of GCMs in clusters 2–9 for T_{max}

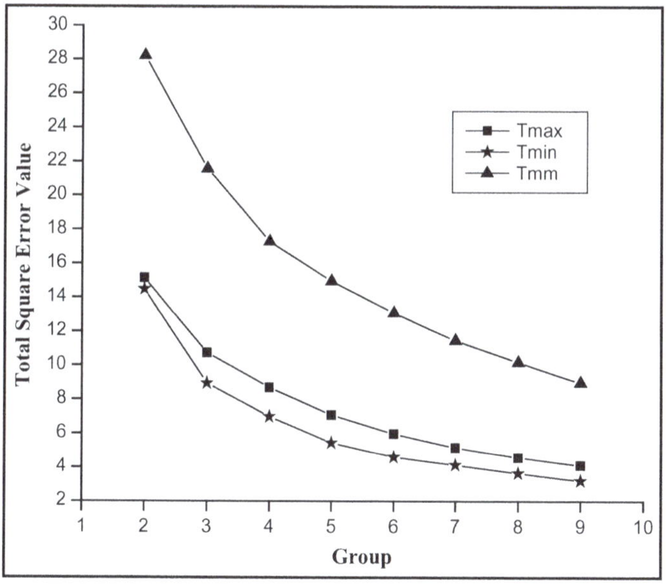

Fig. 6.5 Variation of total squared error values for T_{max}, T_{min} and T_{mm}

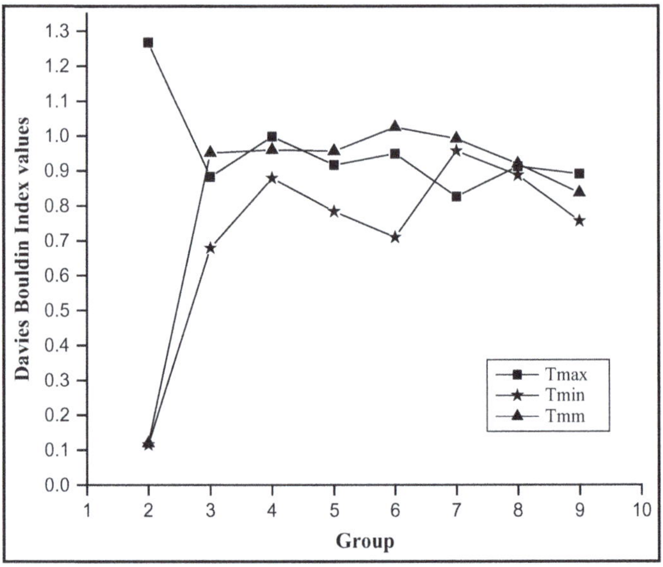

Fig. 6.6 Variation of davies-bouldin index values for T_{max}, T_{min} and T_{mm}

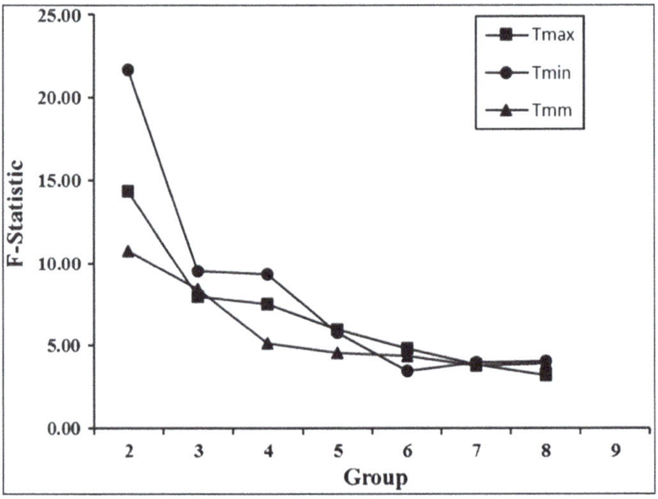

Fig. 6.7 Variation of F-Statistic values for $T_{max,}$ $T_{min,}$ and T_{mm}

confidence. This necessitates using complementarily the F-Statistic test (Burn 1989) along with DBI to obtain optimality more effectively.

Figure 6.7 presents F-Statistic values for clusters 2–8. It is observed that F-Statistic value is varying from 14.3764 (cluster 2) to 3.2060 (cluster 8) in a sequential order. Optimal clusters are two (based on the philosophy of preferring cluster whose F-statistic value is greater than 10). However, optimum clusters are fixed as 3 based on (a) preference of cluster 3 over 7 due to narrow marginal difference of DBI values, (b) compatibility with the output of F-Statistic and (c) steep error reduction from cluster 2–3. Accordingly, the ensemble of HadCM3, IPSL-CM5A-LR and GFDL, ESM2M is suggested.

Figures 6.5, 6.6 and 6.7 also present squared error, DBI and F-Statistic information for T_{min}, T_{mm} which are in similar lines of T_{max}.. Ensemble of ACCESS1.3, HadCM3 for T_{min} and ensemble of MPI-ESM-MR, HadCM3 can be chosen for T_{mm}.

6.2.3 Summary and Conclusions

The following main conclusions are emanated from the study as follows:

- Ensembles of CNRM-CM5, FGOALS-s2, MIROC5 for T_{max} and MIROC4h, NorESM1-M, MIROC5, CESM-CAM5 for T_{min} are suggested (from section "Ranking of Global Climate Models (Raju et al. 2017").
- Optimum cluster is two for both T_{min} and T_{mm} whereas it is three in case of T_{max} on the basis of F-Statistic test and DBI. Ensembles of (HadCM3, IPSL-CM5A-LR, GFDL-ESM2M), (ACCESS1.3, HadCM3), and (MPI-ESM-MR, HadCM3) are suggested for T_{max}, T_{min}, and T_{mm}, respectively, (from section "Clustering of Global Climate Models (Raju and Nagesh Kumar 2016)").

6.3 Downscaling of Climate Variables Using Support Vector Machine and Multiple Linear Regression[2]

Summary: Support Vector Machine (SVM)-based models are developed to downscale monthly sequences of hydrometeorological variables [Precipitation, maximum temperature (T_{max}) and minimum temperature (T_{min})] in Malaprabha River catchment (upstream of Malaprabha reservoir), India. The large-scale atmospheric variables simulated by CGCM3 for Special Report on Emissions Scenarios (SRES) A1B, A2, B1 and COMMIT scenarios were used to prepare inputs for the

[2]Adopted from

Anandhi A, Srinivas VV, Nagesh Kumar (2013) Impact of Climate Change on Hydro-meteorological Variables in a River Basin in India for IPCC SRES Scenarios, Chapter 12 in Climate Change Modelling, Mitigation and Adaptation, Rao YS, Zhang TC, Ojha CSP, Gurjar BR, Tyagi RD, Kao CM (Editors), American Society of Civil Engineers, pp. 327–356, doi:10.1061/9780784412718.ch12, http://ascelibrary.org/doi/10.1061/9780784412718.ch12

Akshara G (2015) Downscaling of Climate Variables Using Multiple Linear Regression-A Case Study Lower Godavari Basin, India, M.E. Dissertation, BITS Pilani.

With permission from the publishers. More details are available in the original publication.

SVM models. Precipitation, T_{max} and T_{min} are projected to increase in future for A1B, A2 and B1 scenarios, whereas no trend is discerned with the COMMIT.

Similarly, Multiple Linear Regression (MLR)-based downscaling technique for Lower Godavari basin, India is employed using GFDL-CM3 GCM. Two Representative Concentration Pathways (RCPs) 4.5 and 6.0 are analyzed. The implications of climate change on monthly basis of each of the hydrometeorological variables are assessed.

6.3.1 Problem Description, Case Study and Data (Anandhi et al. 2008, 2012, 2013; Akshara 2015; Akshara et al. 2017)

Objectives chosen are as follows:

- Development of a Support Vector Machine (SVM) and Multiple Linear Regression (MLR)-based downscaling models
- Exploring the developed models to obtain projections of Precipitation, T_{max} and T_{min}

Case Study 1: The study region is Malaprabha River catchment (upstream of Malaprabha reservoir), India. The region covers an area of 2093.46 km^2 and is situated between 15°30′ N and 15°56′ N latitudes, and 74°12′ E and 75°8′ E longitudes. It lies in the extreme western part of the Krishna River basin in India, and includes parts of Belgaum, Bagalkot and Dharwad districts of North Karnataka (Fig. 6.8). The average annual rainfall of the basin is 1051 mm. Malaprabha River originates in a region of high rainfall, and it is the main source of surface water for arid and semiarid regions downstream of Malaprabha reservoir.

The data consists of monthly mean atmospheric variables simulated by CGCM3. The data comprised of the twentieth-century simulations (20C3M) for the period 1971–2000, and future simulations forced by four SRES scenarios namely, A1B, A2, B1 and COMMIT for the period 2001–2100. Reanalysis data of the monthly mean atmospheric variables prepared by National Centers for Environmental Prediction (NCEP) for the period 1971–2000 were used. The data on observed precipitation were obtained from the Department of Economics and Statistics, Government of Karnataka, India, for 1971–2000. The data on observed temperature were obtained from IMD for 1978–2000. The GCM data were re-gridded to NCEP grid using Grid Analysis and Display System (GrADS) (Doty and Kinter 1993).

Case Study 2: Lower Godavari Basin, India (Godavari river basin 2017) lies between North latitudes 16°19′–19°03′ and East longitudes 80°01′–82°94′. GCM chosen is GFDL-CM3 with Representative Concentration Pathways (RCPs) scenarios of 4.5 and 6.0 (Chaturvedi et al. 2012). Figure 6.9 presents Google Earth image of Lower Godavari Basin. Three time periods 2020s (2020–2029), 2050s (2050–2059) and 2080s (2081–2089) were considered in order to observe the climatic changes occurring during these periods.

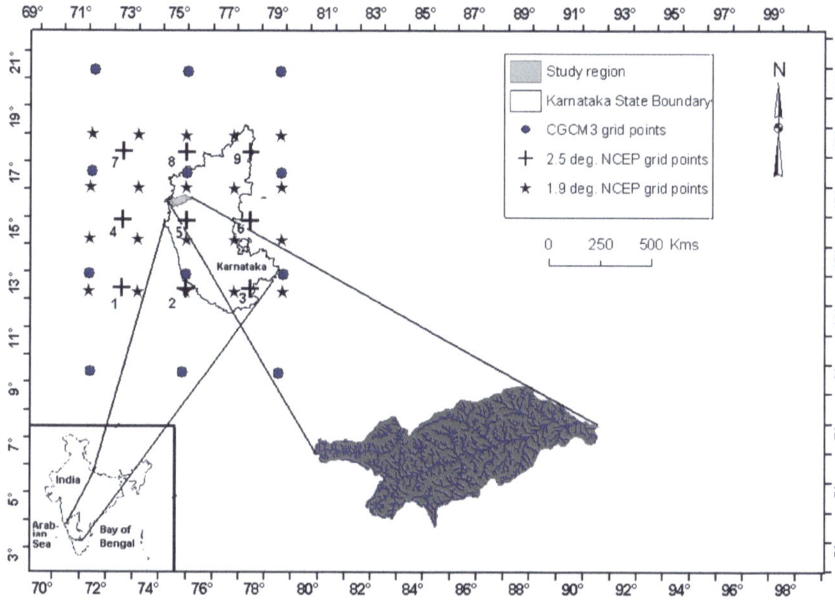

Fig. 6.8 Location of the study region in Karnataka state, India. the latitude, longitude and scale of the map refer to Karnataka State. The Data extracted at CGCM3 and 1.9° NCEP grid points are re-gridded to the nine 2.5° NCEP grid Points. Among the nine grid points 1, 4 and 7 are on Arabian sea, and the remaining points are on land

Fig. 6.9 Lower Godavari basin showing IMD grid points. (*Source* http://india-wris.nrsc.gov.in/wris.html—processed using Arc GIS and Google Earth)

6.3.2 Results and Discussion

6.3.2.1 Downscaling of Climate Variables Using Support Vector Machine (Anandhi 2007; Anandhi et al. 2013)

Selection of Probable Predictors: The selection of appropriate predictors is one of the most important step in a downscaling exercise (Fowler et al. 2007). The choice of predictors could vary from region to region, and it also depends on the characteristics of the large-scale atmospheric circulation and the predictand to be downscaled. The number of probable predictors is referred as m_1 in this chapter.

Stratification of Predictors: For stratification of predictors, the m_2 climate variables (potential predictors), which are realistically simulated by the GCM, were selected from the m_1 probable predictors, by specifying a threshold value (T_{ng1}) for correlation between the probable predictor variables in NCEP and GCM data sets. For the estimation of correlation, Product moment correlation (Pearson 1896), Spearman's rank correlation (Spearman 1904a, b) and Kendall's tau techniques (Kendall 1951) were considered.

Depending on the predictand variable to be downscaled, the stratification of the corresponding potential predictors was carried out in space (land and ocean) or in time (e.g., wet and dry seasons). When precipitation was considered as predictand, the stratification of the predictors was carried out in the time domain to form clusters corresponding to wet and dry seasons. When T_{max} and T_{min} were considered as predictands, the stratification of predictors was carried out in space domain.

Stratification of Predictors for Downscaling Precipitation: The climate of any region can be broadly classified into seasons for analysing precipitation. The predictor variables for downscaling a predictand could vary from season to season. Further, the relationship between the predictor variables and the predictand varies seasonally due to the seasonal variation of the atmospheric circulation (Karl et al. 1990). Hence, seasonal stratification has to be performed to select the appropriate predictor variables for each season and to facilitate the development of an individual downscaling model for each of the seasons. The seasonal stratification can be carried out by defining the seasons as either conventional (fixed) seasons or floating seasons. In fixed season stratification, the starting date and lengths of seasons remain the same for every year. In contrast, in floating season stratification, the date of onset and duration of each season is allowed to change from year to year. Past studies have shown that floating seasons are better than the fixed seasons, as they reflect "natural" scenario, especially under altered climate conditions (Winkler et al. 1997). Therefore, identification of the floating seasons under altered climate conditions helps to effectively model the relationships between predictor variables and predictands for each season, thereby enhancing the performance of the downscaling model. In the present scenario, the floating technique of seasonal stratification is considered to identify dry and wet seasons in a calendar year for both NCEP and GCM data sets. In the floating technique of seasonal stratification, the NCEP data are partitioned into two clusters depicting wet and dry seasons by using the K-means clustering technique (MacQueen 1967), whereas the GCM data are partitioned into two clusters by using the nearest neighbor rule.

From NCEP data of the m_2 variables, n Principal Components (PCs), which preserve more than 98% of the variance, are extracted using principal component analysis (PCA). The PCs corresponding to each month are used to form a feature vector for a month. The PCs are also extracted from GCM data, but along the principal directions obtained for the NCEP data. They are used to form feature vectors for GCM data. Each feature vector (representing a month) can be visualized as an object having a specific location in multidimensional space, whose dimensionality is defined by the number of PCs.

The feature vectors of the NCEP data are partitioned into two clusters (depicting wet and dry seasons) using the K-means cluster analysis. The clustering should be such that the feature vectors within each cluster should be close to each other as possible in space, and as far as possible in space from the feature vectors of the other clusters. The distance between each pair of feature vectors in space is estimated using Euclidian measure. Subsequently, each feature vector of the NCEP data is assigned a label that denotes the cluster (season) to which it belongs. Following this, the feature vectors prepared from GCM data (past and future) are labeled using the nearest neighbor rule to get the past and future projections for the seasons. As per this rule, each feature vector formed using the GCM data is assigned the label of its nearest neighbor among the feature vectors formed using the NCEP data. To determine the nearest neighbors for this purpose, the distance between each pair of NCEP and GCM feature vectors is computed using Euclidean measure. Comparison of the labels of contemporaneous feature vectors formed from NCEP and GCM past data is useful in checking if the GCM simulations could represent the regional climate fairly well, during the past period.

Optimal T_{ngl} is identified as a value for which the wet and dry seasons are formed for the study region using NCEP data, which are well correlated with the possible true seasons for the region. For this analysis, the plausible true wet and dry seasons in the study region are identified using a technique based on Truncation Level (TL). In this technique, the dry season is considered as consisting of months for which the estimated Theissen Weighted Precipitation (TWP) values for the region are below the specified TL, whereas the wet season is considered as consisting of months for which the estimated TWP values are above the TL. Herein, two options have been used to specify the TL. In the first option, the TLs are chosen as percentages of the observed Mean Monthly Precipitation (MMP) (70–100% of MMP at intervals of 5%). In the second option, the TL is chosen as the mean monthly value of the actual evapotranspiration in the river basin. The actual evapotranspiration is obtained for Krishna basin from Gosain et al. (2006). The potential predictors corresponding to optimal T_{ngl} are noted.

Stratification of Predictors for Downscaling Surface Temperature: The surface temperature in a region is dominated by local factors such as evaporation, sensible heat flux and vegetation in the region. Therefore the potential predictor variables influencing surface temperature in the study region are stratified based on the location of grid points (land and/or ocean) corresponding to the variables, to assess the impact of their use on downscaled temperature. Out of the nine 2.5° NCEP grid points considered in the study region, six are above land and the remaining three are over the sea. As there are no distinct seasons based on temperature, seasonal stratification as in the case of precipitation is not relevant.

Support Vector Machine (SVM) Downscaling Model: For downscaling the predictand, the m_1 probable predictors at each of the NCEP grid points are considered as probable predictors. Thus, there are m_3 (= m_1 × number of NCEP grid points) probable predictors. The potential predictors (m_4) are selected from the m_3 probable predictor variables. For this purpose, the cross-correlations are computed between the probable predictor variables in NCEP and GCM data sets, and the probable predictor variables in NCEP data set and the predictand. A pool of potential predictors is then identified for each season by specifying threshold values for the computed cross-correlations. The threshold value for cross-correlation for variables in NCEP and GCM data sets are denoted by T_{ng2}, whereas the same threshold for NCEP variables and predictand are depicted as T_{np}. The T_{np} should be reasonably high to ensure choice of appropriate predictors. Similarly, T_{ng2} should also be reasonably high to ensure that the predictor variables used in downscaling are realistically simulated by the GCM for the past so that the future projections of the predictand obtained using GCM data would be acceptable.

The downscaling model is calibrated to capture the relationship between NCEP potential predictors and the predictand. The data on potential predictors is first standardised for each season or location separately for a baseline period. Such standardization is widely used prior to statistical downscaling to reduce systemic bias (if any) in the mean and variance of the predictors in the GCM data, relative to the predictors in the NCEP reanalysis data (Wilby et al. 2004). This step typically involves subtraction of mean and division by the standard deviation of the predictor for the baseline period. The standardized NCEP predictor variables are then processed using PCA to extract PCs which are orthogonal and which preserve more than 98% of the variance originally present. A feature vector is formed for each month using the PCs. The feature vector forms the input to the SVM model, and the contemporaneous value of predictand is its output. The PCs account for most of the variance in the input and also remove the correlations, if any, among the input data. Hence, the use of PCs as input to a downscaling model makes it more stable and it also reduces the computational load.

To develop the SVM downscaling model, the feature vectors obtained are partitioned into a training set and a testing set. The partitioning was initially carried out using multifold cross-validation procedure in an earlier work (Haykin 2003; Tripathi et al. 2006). In this procedure, about 70% of the feature vectors are randomly selected for training the model, and the remaining 30% is used for validation. However, in this study the multifold cross-validation procedure is found to be ineffective as the time span considered for analysis is small and more extreme events occurred in the past decades than in the recent decade. Therefore, the feature vectors are formed from approximately first 70% of the available data, are chosen for calibrating the model and the remaining feature vectors are used for validation. The Normalized Mean Square Error (NMSE) is used as an index to assess the performance of the model. The training of SVM involves selection of the model parameters σ and C. The width of Radial Basis Function (RBF) kernel σ gives an idea about the smoothness of the derived function whereas C controls the trade-off between errors of the SVM on training data. Smola et al. (1998) explained the regularisation capability of RBF kernel and have shown that a large kernel width acts as a low-pass filter in the frequency domain. It attenuates the higher order frequencies, resulting in a smooth function. Alternately, RBF with a small kernel width retains most of the higher order frequencies leading to an approximation of a complex function by the learning machine. In this study, grid search procedure (Gestel et al. 2004) is used to find the optimum range for each of the parameters. Subsequently, the optimum values of the parameters are obtained from within the selected ranges, using the stochastic search technique of genetic algorithm (Haupt and Haupt 2004). The feature vectors obtained from GCM simulations are processed through the validated SVM downscaling model to obtain future projections of the predictand, for each of the four emission scenarios considered (i.e., SRES A1B, A2, B1 and COMMIT). Subsequently, for each scenario, the projected values of the predictand are chronologically divided into five parts (2001–2020, 2021–2040, 2041–2060, 2061–2080 and 2081–2100) to determine the trend in the projected values of the predictand. The procedure is illustrated as flowchart in Fig. 6.10. Relevant results and discussion are as follows:

Predictor Selection: For downscaling precipitation, the predictor variables are screened on the twin basis, i.e., monsoon rain is dependent on dynamics through advection of water from the surrounding seas and thermodynamics through effects of moisture and temperature, both of these can modify the local vertical static stability. In a changed climate scenario, both the thermodynamic and dynamic parameters may undergo changes. In the present study, probable predictor variables, which incorporate both the effects, are chosen. Wind during southwest monsoon season advect moisture into the region while temperature and humidity are associated with local thermodynamic stability and hence these parameters are considered as predictors. Meridional wind has more local effects, and together these winds are responsible for convergence of moisture and hence related to precipitation. Temperature affects the moisture holding capacity of the wind and the pressure at a location. The pressure gradient affects the circulation which in turn affects the moisture of the place and hence the precipitation. Higher precipitable water in the atmosphere means the presence of more moisture, which in turn causes statically unstable atmosphere leading to more vigorous overturning, resulting in more precipitation. Lower pressure leads to huge winds and more precipitation. At 925 mb pressure height, the boundary layer (near surface effect) is important. A pressure of 850 mb has a low-level response to regional precipitation. The 200 mb pressure level depicts the global scale effects. Temperatures at 700 and 500 mb represent the heating process of the atmosphere due to monsoonal precipitation. Monsoon precipitation is maximum at mid-troposphere for a constant pressure height. Geopotential height represents the pressure variation, which reflects the flow, and in turn moisture changes. Due to these reasons, fifteen probable predictors are extracted from the NCEP reanalysis and CGCM3 data sets.

For downscaling temperature, large-scale atmospheric variables, such as air temperature, zonal and meridional wind velocities at 925 mb are considered as

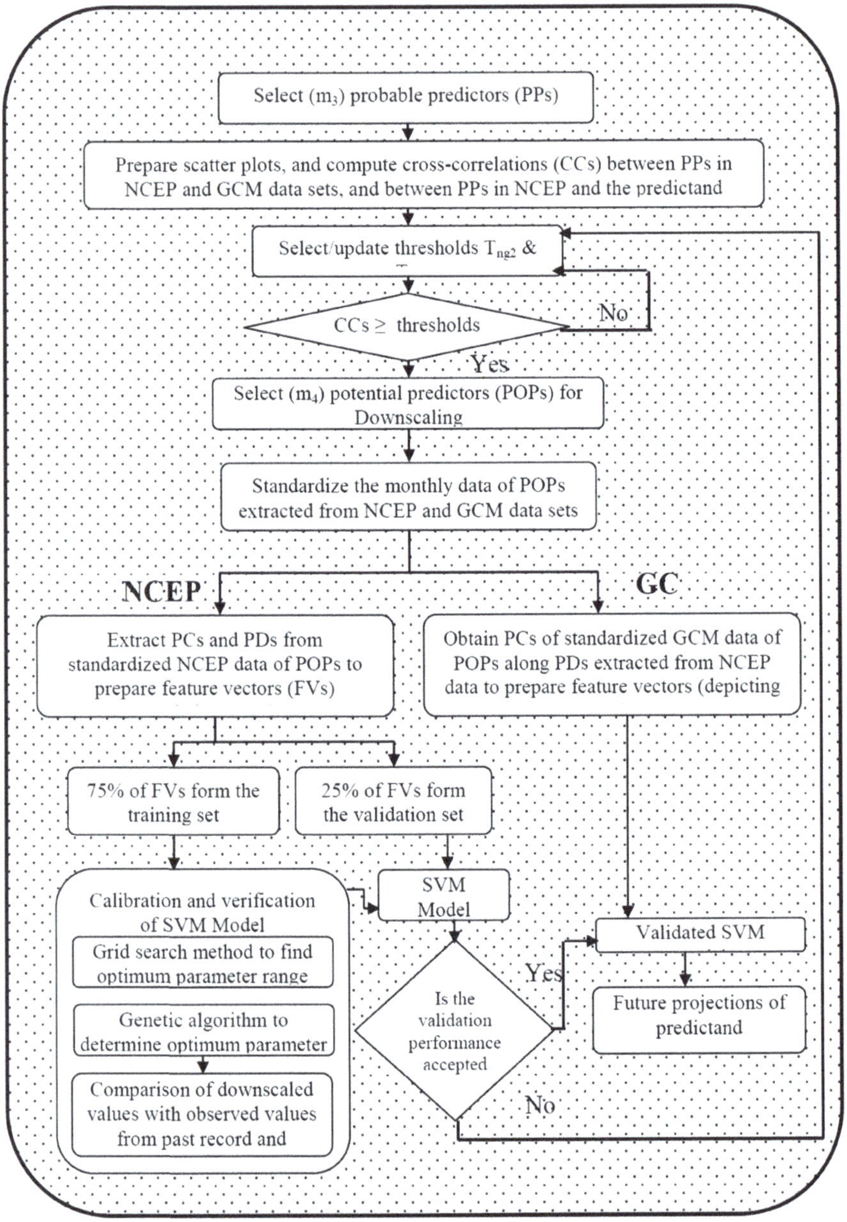

Fig. 6.10 Methodology followed for SVM downscaling. Notations in figure: PCs and PDs denote principal components and principal directions, respectively. T_{ng2} is the threshold between predictors in NCEP and GCM data sets. T_{np} denotes the threshold between predictors in NCEP data and the predictand (Ta 925), 700 mb (Ta 700), 500 mb (Ta 500) and 200 mb (Ta 200) pressure levels, geopotential height at 925 mb (Zg 925), 500 mb (Zg 500) and 200 mb (Zg 200) pressure levels, specific humidity at 925 mb (Hus 925) and 850 mb (Hus 850) pressure levels, zonal (Ua) and meridional wind velocities (Va) at 925mb (Ua 925, Va 925) and 200mb (Ua 200, Va 200) pressure levels, precipitable water (prw) and surface pressure (ps)

predictors. Surface flux variables, namely latent heat, sensible heat, shortwave radiation and longwave radiation fluxes are also considered for downscaling temperature, as these factors control the temperature of the Earth's surface. The incoming solar radiation heats the surface, while latent heat flux, sensible heat flux, and longwave radiation cool the surface. Due to these reasons, seven probable predictors are extracted from the NCEP reanalysis and CGCM3 data sets to downscale temperature. The chosen predictors are air temperature, zonal, and meridional wind velocities at 925 mb, and four fluxes: Latent Heat (LH), Sensible Heat (SH), Short Wave Radiation (SWR), and Long Wave Radiation (LWR).

SVM Downscaling Models: From the selected potential predictors for each season, PC are extracted to form feature vectors. These feature vectors are input to develop SVM downscaling model. For obtaining the optimal range of each of the SVM parameters (kernel width σ, and penalty term C), the grid search procedure is used. Typical results of the domain search to estimate the optimal ranges of the parameters for wet and dry seasons are shown in Fig. 6.11 and the range of σ and C having the least NMSE is selected as the optimum parameter range. The NMSE values are indicated in the bar code provided in Fig. 6.11. Using genetic algorithm, the optimum parameter is selected from the parameter range. Thus, the optimal values of SVM parameters C and σ obtained are 550 and 50 for the wet season, and 850 and 50 for the dry season, respectively. The optimal values of SVM parameters C and σ for T_{max}, are 2050 and 50 while for T_{min}, these are 1050 and 50, respectively. The results of downscaling are compared to observed variables as shown in Fig. 6.12.

Projected Future Scenarios: The future projections of three meteorological variables (Precipitation, T_{max} and T_{min}) were obtained for each of the four SRES scenarios (A1B, A2, B1 and COMMIT) using the developed SVM downscaling models. The projections were subsequently divided into five 20-year intervals (2001–2020, 2021–2040, 2041–2060, 2061–2080, and 2081–2100). The mean monthly values of observed and projected precipitation were estimated using the Theissen Polygon technique. For each of the four SRES scenarios, average of the

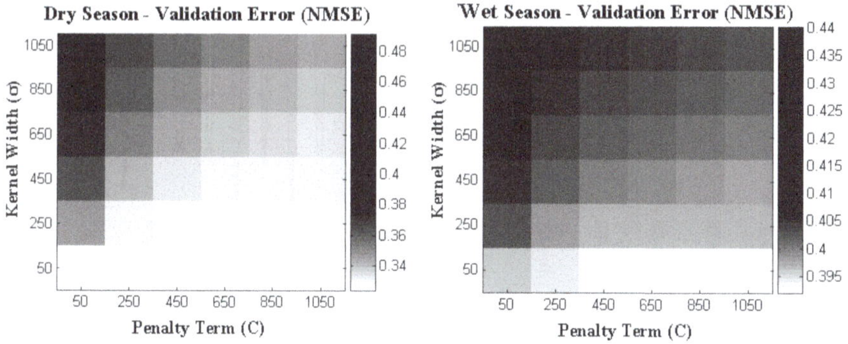

Fig. 6.11 Illustration of the domain search performed to estimate optimal values of kernel width (σ) and Penalty (C) for the SVM, for dry and wet seasons

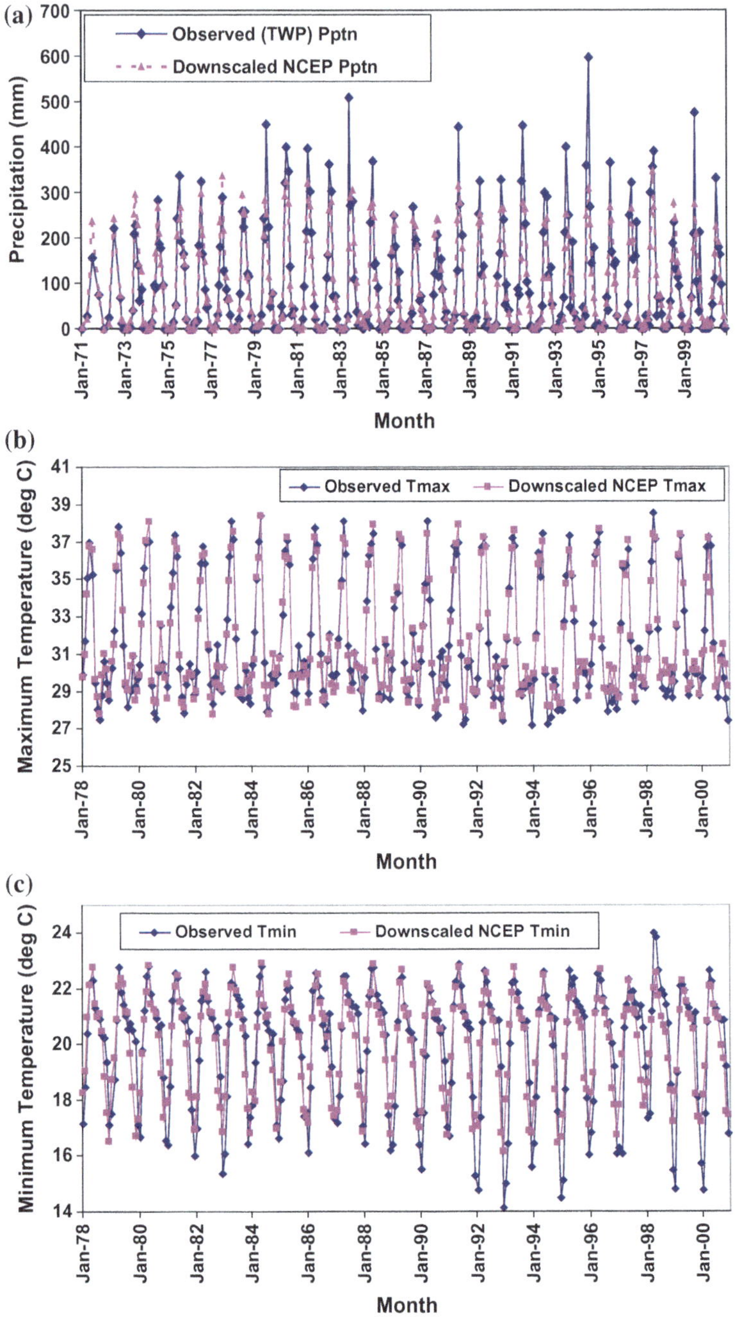

Fig. 6.12 Comparision of the monthly observed meteorological variable with the corresponding simulated variable using SVM downscaling model for NCEP data **a** Thiessen weighted precipitation (TWP) **b** T_{max} **c** T_{min}

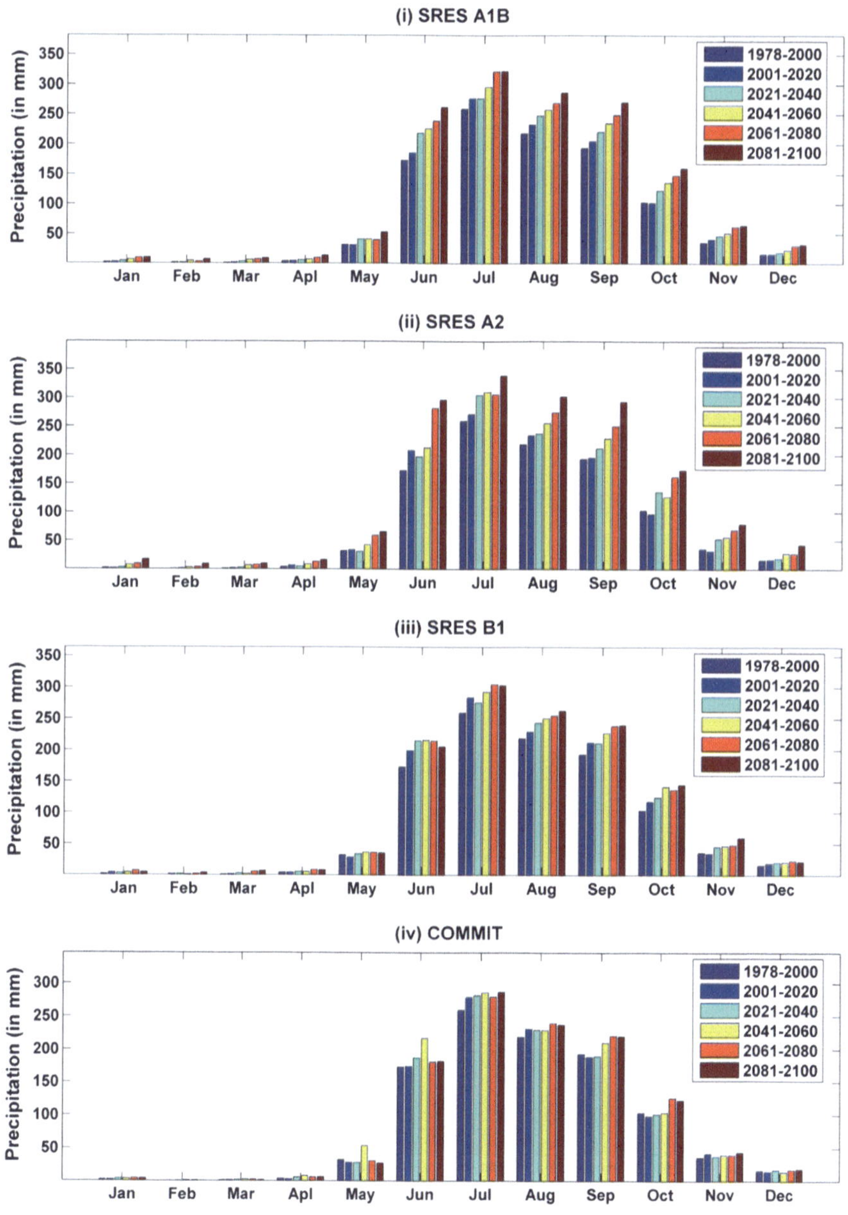

Fig. 6.13 Mean monthly precipitation for the period 1971–2100 for the four scenarios considered

mean monthly values of Theissen-weighted Precipitation, T_{max}, T_{min} are presented as bar plots, for all the five 20-year intervals in Figs. 6.13, 6.14 and 6.15, respectively. These plots facilitate in assessing the projected changes in each meteorological variable across a period of 2001–2100 at 20 years interval with

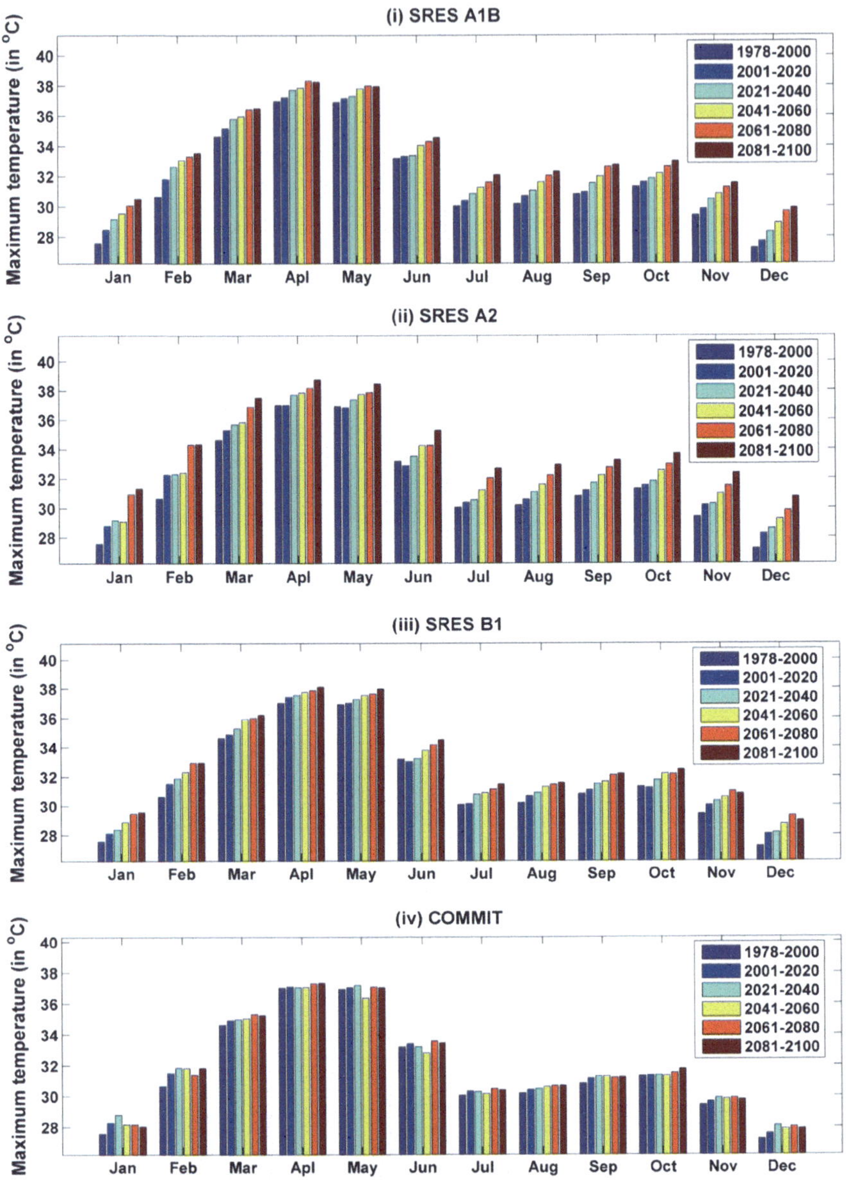

Fig. 6.14 Mean monthly T_{max} for the period 1978–2100 for the four scenarios considered

respect to the past (20C3M), for each SRES scenario. From the figures, it is observed that Precipitation, T_{max} and T_{min} tend to increase in future for A1B, A2 and B1 scenarios, whereas no trend is discerned with the COMMIT. The projected increase is high for A2 scenario, whereas least for B1 scenario. This is because

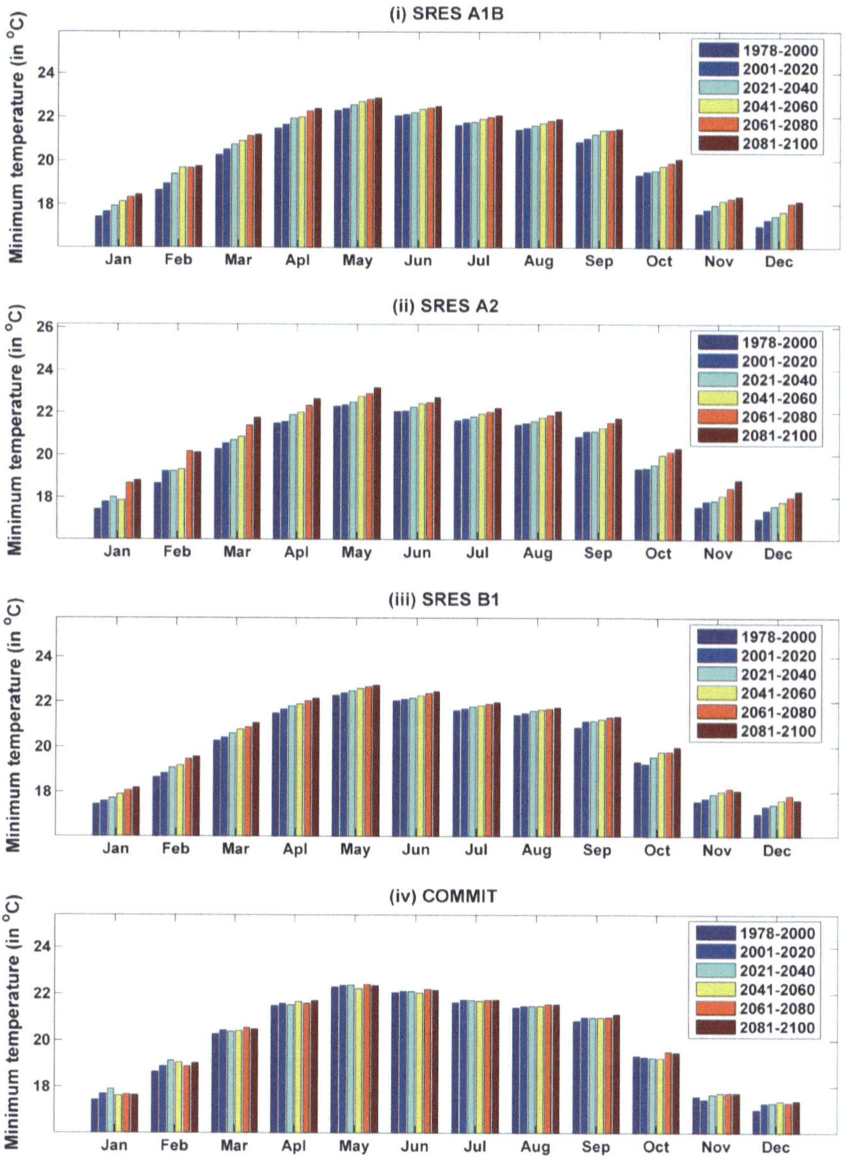

Fig. 6.15 Mean monthly T_{min} for the period 1978–2100 for the four scenarios considered

among the scenarios considered, the scenario A2 has the highest concentration of carbon dioxide (CO_2) equal to 850 ppm whereas the values for A1B, B2 and COMMIT scenarios are 720 ppm, 550 ppm and ≈370 ppm, respectively. Rise in the concentration of CO_2 in atmosphere causes the Earth's average temperature to increase, which in turn causes an increase in evaporation especially at lower latitudes. The evaporated water would eventually precipitate. In the COMMIT scenario, where the emissions are same as in the year 2000, no significant trend in the pattern of projected future precipitation could be discerned. It is observed that the change from past to future is gradual, and the change is high for A1B scenarios, while it is the least for B1 scenario. In A2 scenario, the change is much higher and different from A1B. In the case of COMMIT, no clear pattern change is visible. Change in the variables is minimum in the first 20-year interval (2001–2020) and maximum in the last 20-year interval (2081–2100).

6.3.2.2 Results and Discussion [Downscaling of Climate Variables Using Multiple Linear Regression (Akshara et al. 2017)]

Lower Godavari Basin was spread into five grid points based on the grid resolution available for the IMD data, i.e., 17.5 N-80.5 E, 17.5 N-81.5 E, 18.5 N-17.5 E, 18.5 N-80.5 E, and 18.5 N-82.5 E. T_{max}, T_{min} and precipitation are selected as predictands in downscaling methodology. Historical monthly data of T_{max}, T_{min} and precipitation for all the five grids were obtained from IMD, for the period 1969–2005 (baseline period). Accordingly, predictor variables are also chosen (Mujumdar and Nagesh Kumar 2012; Akshara et al. 2017). The predictor variables of NCEP/NCAR from 1969 to 2005 are used to develop the relationship with observed T_{max}, T_{min} and precipitation obtained from IMD. The relationship obtained by Multiple Linear Regression (MLR) is used to predict the monthly rainfall using the surface predictors obtained from GCM outputs. Fifteen regression equations are formulated with the three predictand variables, T_{max}, T_{min} and precipitation and five grids for the study area.

The standardized GCM predictors for the period 2006–2100 for both RCP 4.5 and RCP 6.0 are used in the regression equations for predicting future T_{max}, T_{min} and precipitation values. Three time periods 2020s (2020–2029), 2050s (2050–2059) and 2080s (2081–2089) were considered in order to observe the changes occurring during these periods. Related results for grid 1 are as follows:

RCP 4.5: Fig. 6.16 shows that there is a significant change in monthly temperatures for future decades compared to the baseline period (1969–2005); Trend followed in future time periods seems parallel and the change in temperature from 2020 to 2050 is high compared to changes during 2050–2080; The maximum T_{max} in the period 2020 is 44 °C, 2050 is 54 °C and 2080 is 57 °C.

RCP 6.0: Similar trend as that of RCP 4.5 is observed for RCP 6.0 as shown in Fig. 6.17. Change in temperature from 2050 to 2080 is larger than the change in temperature from baseline to 2020 as well as 2020–2050; T_{max} value for 2080 is nowhere comparable to either 2020 or 2050.

RCP 4.5: The projected T_{min} as observed from Fig. 6.18 implies that the temperatures tend to decrease further from February to August in 2020 compared to the baseline and shows a significant increase in 2050 and 2080; trend followed in 2050 and 2080 is almost similar.

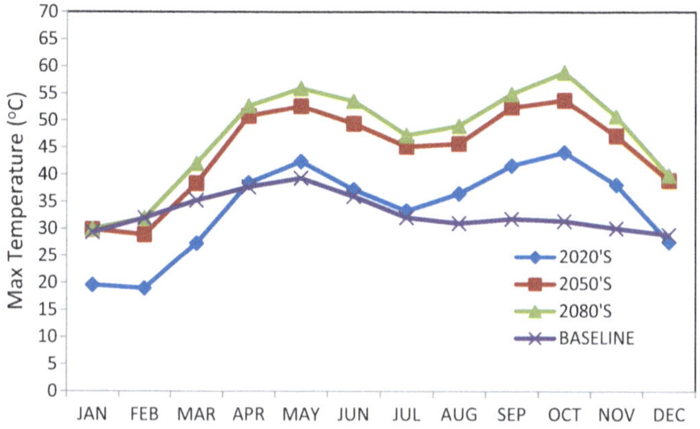

Fig. 6.16 T_{max} trend for RCP 4.5 for different time periods for grid 1 (17.5 N-80.5 E)

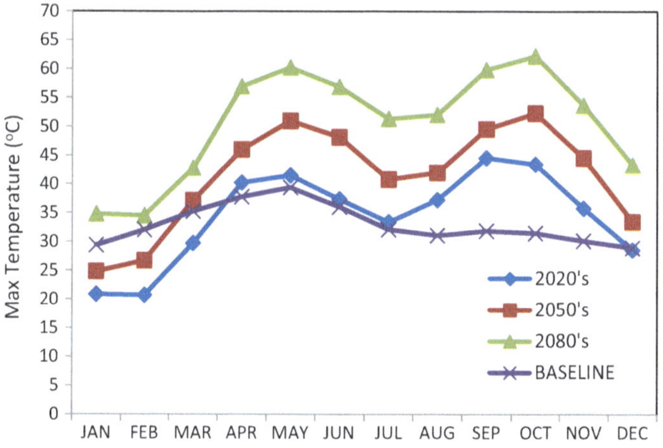

Fig. 6.17 T_{max} trend for RCP 6.0 for different time periods for grid 1 (17.5 N-80.5 E)

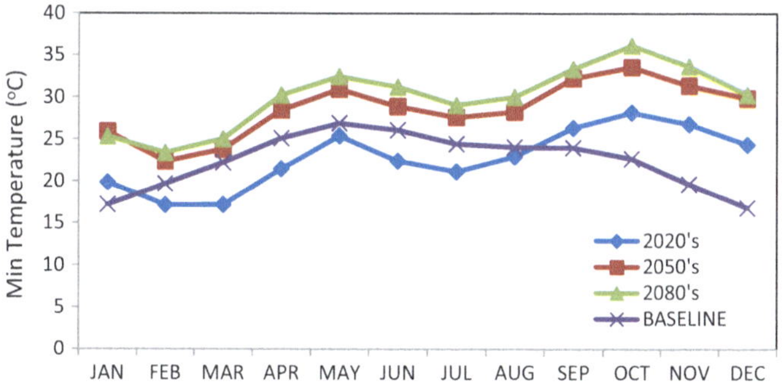

Fig. 6.18 T_{min} trend for RCP 4.5 for different time periods for grid 1 (17.5 N-80.5 E)

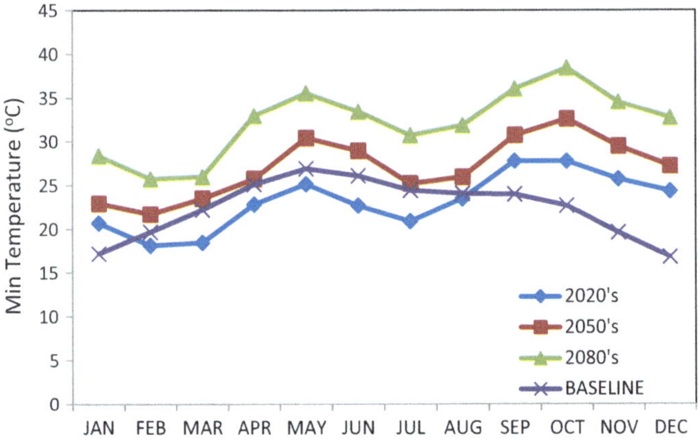

Fig. 6.19 T_{min} trend for RCP 6.0 for different time periods for grid 1 (17.5 N-80.5 E)

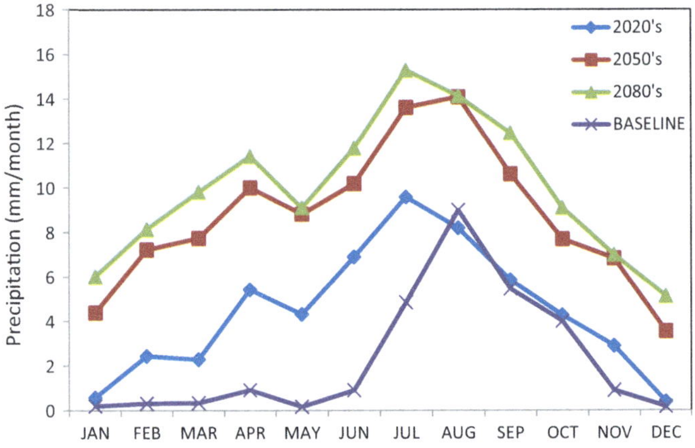

Fig. 6.20 Precipitation trend for RCP 4.5 for different time periods for grid 1 (17.5 N-80.5 E)

RCP 6.0: Fig. 6.19 shows the same trend for T_{min} for RCP 6.0 as that of RCP 4.5 and the maximum value reached in 2080 is around 38 °C compared to 26 °C in the baseline period which clearly shows the effect of global warming; Change in temperature from 2050 to 2080 is almost similar to the change from 2020 to 2050.

RCP 4.5: It is observed from Fig. 6.20 that there is minimum rainfall in January and December and maximum in July and August; no similarities are observed in the pattern of rainfall over years; the change in rainfall is significant from baseline period to 2020 as well as 2020–2050.

RCP 6.0: It is observed from Fig. 6.21 that precipitation is increasing in the future. Heavy rainfall may occur in July and August. Maximum rainfall observed in

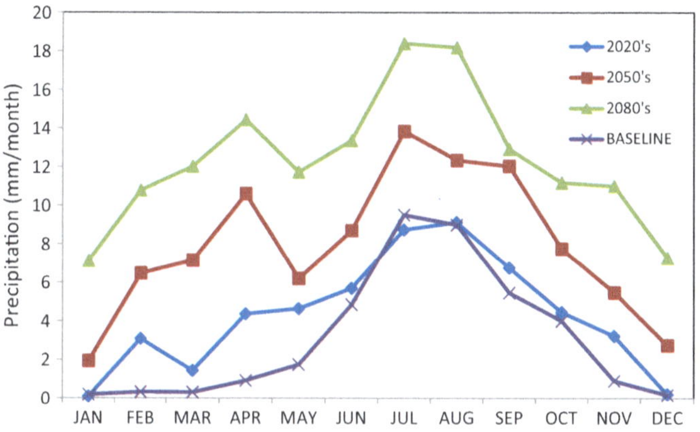

Fig. 6.21 Precipitation trend for RCP 6.0 for different time periods for grid 1 (17.5 N-80.5 E)

2020, 2050 and 2080 are 9 mm/month, 14 mm/month and 19 mm/month, respectively.

Similar observations were made for other grids also. Trend of the results are as follows:

- T_{max} for all the grids on an average has increased by 4–5 °C from 2020s to 2050s and 3–4 °C from 2050s to 2080s. Maximum temperatures are observed in May and June. A change in trend is observed with respect to the baseline period in pre monsoon and post monsoon temperatures. A slight decrease was observed in pre monsoon temperatures and a considerable increase in post monsoon temperatures resulting in an increase of warmer periods.
- Future T_{min} is in the range of 8–16 °C (minimum) and 36–40 °C (maximum) for all five grids considered.
- The region will become gradually warmer in the future periods and the warming up is generally higher during the post monsoon and summer. No noticeable difference between RCP 4.5 and RCP 6.0 is observed except for slight variations of temperatures in few months.
- Precipitation predictions show that there is a significant increase in the amount of precipitation. Precipitation for all the grids on an average has increased from 8–10 mm/month in the baseline period to 20–25 mm/month in 2080s. Considering the RCP's, there is no noticeable difference between RCP 4.5 and RCP 6.0 except for slight variations of precipitation in few months. It is observed that the amount of precipitation will be high in June, July and August compared to remaining months of the year and there is very less amount of precipitation in January and December.

6.3.3 Summary and Conclusions

The following main conclusions are emanated from the study:

- Precipitation, T_{max} and T_{min} are projected to increase in future for A1B, A2 and B1 scenarios, whereas no trend is discerned with the COMMIT for Malaprabha. The River catchment projected increase in predictands is high for A2 scenario and is less for B1 scenario.
- The region will become gradually warmer in the future and the warming up is generally higher during the post monsoon and summer for Lower Godavari Basin whereas precipitation predictions show that there will be a significant increase in the amount of precipitation in the future.

6.4 Climate Change Impact on Semi-arid Catchment Water Balance Using an Ensemble of GCMs[3]

Summary: Impact of climate change on the Water balance for a Semi-arid catchment in India is evaluated. Rainfall and hydro-meteorological variables for current (20C3M from 1981 to 2000) and two future time periods: mid of the twenty-first century (2046–2065) and end of the century (2081–2100) are simulated using Modified Markov Model-Kernel Density Estimation (MMM-KDE) and k-nearest neighbor downscaling models. An ensemble of five GCMs (MPI-ECHAM5, BCCR-BCM2.0, CSIRO-mk3.5, IPSL-CM4 and MRI-CGCM2) is used. Hydrologic simulations for current as well as future climates are carried out using ArcSWAT hydrologic model. The results showed a marginal reduction in runoff ratio and annual streamflow toward the end of the century. Increased temperature and evapotranspiration project an increase in the irrigation demand toward the end of the century. Rainfall projections for the future show marginal increase in the annual average rainfall. Short and moderate wet spells are projected to decrease, whereas short and moderate dry spells are projected to increase in the future. Projected reduction in streamflow and groundwater recharge along with the increase in irrigation demand is likely to aggravate the water stress in the region for the future scenario.

[3]Adopted from
Reshmidevi TV, Nagesh Kumar D, Mehrotra R, Sharma A (2017) Estimation of the climate change impact on a catchment water balance using an ensemble of GCMs, Journal of Hydrology, Elsevier, doi:10.1016/j.jhydrol.2017.02.016, http://www.sciencedirect.com/science/article/pii/S0022169417300963
With permission from the publishers. More details are available in the original publication.

6.4.1 Problem Description, Case Study and Data (Reshmidevi et al. 2017)

Objectives chosen are as follows:

- Exploring ensemble of five GCMs, Modified Markov Model-Kernel Density Estimation (MMM-KDE) and k-nearest neighbor downscaling models and ArcSWAT to evaluate the hydrologic impact of climate change for a semi-arid Malaprabha catchment in India.

Details of Malaprabha are presented in case study 6.3. Digital elevation model (DEM), digital soil map and land use/-land cover (LU/LC) map are used to represent the catchment heterogeneity for the hydrologic analysis. DEM of Malaprabha catchment at 30 m spatial resolution is obtained from Advanced Spaceborne Thermal Emission and Reflection Radiometer (ASTER) Global DEM (GDEM) data set released by the Japan's Ministry of Economy, Trade and Industry (METI) and NASA. Multi-season Landsat-7 ETM + imageries are used to extract LU/LC map of the catchment. In this study, a combination of visual and digital image interpretation technique is used to extract the LU/LC map from the satellite imagery (Reshmidevi and Nagesh Kumar 2014). Seven main LU/LC classes' viz., water, agricultural land, barren/fallow land, rocky area, forest, urban settlement and grass land are extracted in the first step. Appropriate bands that show unique band ratio are identified for each land cover class. A combination of visual interpretation and unsupervised classification using band ratio (Lillesand et al. 2004) is used to identify the seven major land cover classes. The second level classification, i.e., classification of the crop types is achieved using multi-temporal satellite images (Dutta et al. 1998) representing different cropping seasons. Depending upon the presence or absence of crop in each image, different crop types are classified. Field information and the district statistical information about the crop production are used to substantiate the classification.

Soil map of the area is obtained from the National Bureau of Soil Survey and Land Use Planning, Nagpur, India. Monthly inflow into the Malaprabha reservoir for the period 1973–2000 is obtained from the Water Resources Development Organisation, Bangalore, India and is used as the observed streamflow data to calibrate the hydrologic model.

Rainfall and meteorological variables, viz. T_{max}, T_{min}, relative humidity and wind speed at daily time steps are used for hydrologic simulation. Daily rainfall data at these nine stations in the catchment are available for the period 1971–2000, whereas observed meteorological data are available only for a short period 1993–2000. Mass curve analysis of the rainfall data was performed, from which the period 1993–2000 was found to be insufficient to represent the entire study period 1971–2000. Therefore, rainfall and meteorological variables are downscaled from the National Centers for Environmental Prediction (NCEP) reanalysis data for the period 1971–2000. In addition, for historic and future time periods, rainfall and hydro-meteorological variables are downscaled from multiple GCMs. Modified

Markov Model–Kernel Density Estimation (MMM–KDE) model (Mehrotra and Sharma 2010) is used to downscale rainfall from large-scale atmospheric variables and the k-nearest neighbor resampling technique is used to downscale the meteorological variables to the single location. Daily data of the atmospheric variables is obtained from five GCMs from the World Climate Research Programme's Coupled Model Intercomparison Project phase 3 (CMIP3), namely, BCCR-BCM2.0, MRI-CGCM2, CSIRO-mk3.5, MPI-ECHAM5, and IPSL-CM4. The twentieth-century climate experiment (20C3M) for the period 1981–2000 is selected to represent the historic time period. Hydrologic impact of climate change is also studied for two 20 year future time periods: 2046–2065 (referred hereafter as mid of the century) to represent the mid-twenty-first century, and 2081–2100 (referred hereafter as end of the century) to represent the end of twenty-first century.

For each time period, required atmospheric variables at grid nodes over the catchment are extracted from a single continuous (transient) run corresponding to SRES A2 emission scenario. Hydrologic model ArcSWAT is used to simulate the catchment hydrologic responses for historic and two future time periods under the A2 scenario.

6.4.2 Results and Discussion

Statistical Downscaling of Rainfall and Meteorological Variables: The variable convergence score (Johnson and Sharma 2009) is used to identify the GCM atmospheric variables for use in downscaling daily rainfall. These variables include Mean Sea Level Pressure (MSLP), North–South (N–S) gradient of MSLP, Temperature Depressions (TD) at 850, 700, 500 hPa, N–S gradient of TD at 850 hPa, U and V components of the wind velocities at 850 hPa, Equivalent potential temperature (EPT) at 850 hPa, N–S gradient of the Geopotential Height (GPH) at 700 hPa, Specific Humidity (SPH) at 500 hPa, N–S gradient of SPH at 500 hPa and EW gradient of SPH at 850 hPa (Mehrotra et al. 2013). The selected GCM atmospheric variables for 20C3M (1981–2000) and future time periods (2046–2065 and 2081–2100) are then bias-corrected by adopting a nested bias correction procedure (Johnson and Sharma 2012).

ArcSWAT Calibration and Validation: ArcSWAT is applied at daily time scale over the study area. The catchment is first divided into 12 sub-basins and each subbasin is further divided into Hydrological Response Units (HRUs) using the LU/LC, soil and slope information. Irrigated areas in the catchment are identified and the irrigation application is defined when plant water stress exceeds a threshold of 0.95. Daily values of rainfall at nine stations and hydrometeorological variables at the single location are downscaled from the reanalysis data and are used for model calibration and validation. Simulated discharge is compared with the observed monthly streamflow data. The period 1971–2000 is selected in this study, out of which the first two years are used as the warm-up period for the model and the period 1973–2000 is used for model calibration and validation. Multiple realizations of the downscaled rainfall and climate variables produced by MMM–KDE technique is used in the hydrologic modeling. In order to derive stable values of model parameters during calibration, concatenated data set is formed using multiple realizations. From the ensemble of 20 realizations, first nine realizations are used for calibration and the remaining 11 realizations are used for model validation. Each realization being of 28 years (excluding the warm-up period) total length of the calibration run is thus 252 years.

Model sensitivity analysis is performed using Latin Hypercube (LH) and One-factor-At-a-Time (OAT) techniques included in the ArcSWAT (Van Griensven, 2005) and the sensitive parameters are manually calibrated. Capability of the model to accurately produce Flow-Duration Curve (FDC) for the annual and monsoon (June–September) streamflows is considered as the evaluation criteria during calibration. Using monthly streamflow simulations from the model, annual and monsoon streamflows are calculated for each realization and the average FDCs for the annual and monsoon periods are generated. Deviation of the simulated FDC from the observed FDC is measured using Mean Absolute Relative Error (MARE).

Multi-model Ensemble Projection under the Future Scenario: Ensemble of simulations from the hydrologic model, obtained by using rainfall and meteorological data from the five GCMs is used to evaluate the hydrologic impact of climate change in the catchment by comparing hydrologic responses in the future time periods with that obtained under 20C3M scenario. When the temporal periods do not overlap, FDCs are commonly used for comparing the flow regimes in the hydrologic analyses (Sugawara 1979; Yu and Yang 2000; Westerberg et al. 2011). Since the temporal periods of the 20C3M and the future scenarios do not overlap, FDCs of annual and monsoon flows are used here for the comparison.

A weighted ensemble average technique is used to derive ensemble average streamflow simulation from the five GCMs. ArcSWAT is run using the downscaled rainfall and meteorological data from each GCM, and FDCs for the annual and monsoon flows are generated. Simulated FDCs for 20C3M scenario are then compared against FDCs of the observed streamflow data for the same period. Deviation of the simulated FDCs from the observed FDC, expressed in terms of Mean Absolute Relative Error (MARE) is used as the evaluation criterion. For each hydrologic simulation (using input from different GCMs), MARE of the annual and monsoon FDCs are estimated and the mean of these two values is used to derive the weight for each simulation.

GCMs simulating the historic scenario satisfactorily are expected to be capable of simulating the future scenarios reasonably well (Reichler and Kim 2008; Errasti et al. 2010). Therefore, the weights derived for the hydrologic simulations under 20C3M scenario are adopted for the future time periods as well. Hydrologic responses of the catchment under the future time periods (mid of the century and end of the century), are simulated by using the downscaled rainfall and meteorological variables from each GCM, and the FDCs of both annual and monsoon flows are generated. Further, using the set of weights derived from the 20C3M scenario, weighted average FDCs for the future scenarios are generated. Future projections are compared with 20C3M simulations to quantify the streamflow variation under future scenarios. A flowchart of the methodology is shown in Fig. 6.22.

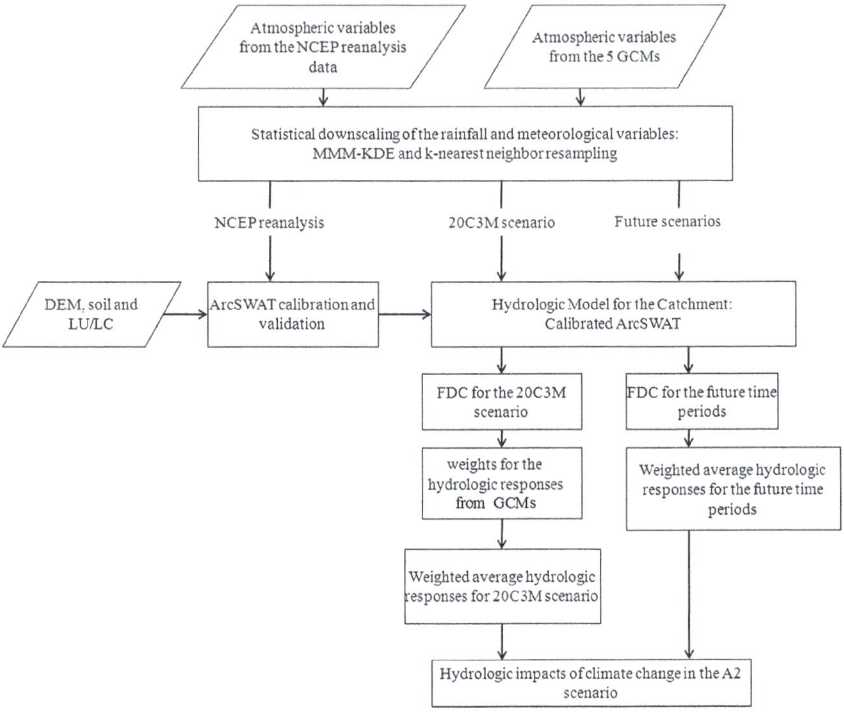

Fig. 6.22 Schematic representation of the methodology for estimating hydrologic impacts of climate change

Hydrologic simulations from the five GCMs are aggregated to find variations in other water budget components, viz. Potential Evapotranspiration (PET), Actual Evapotranspiration (ET), irrigation demand and groundwater recharge in the future time periods.

Test for Statistical Significance: Statistical significance of projected changes in the mean annual and monsoon streamflows is evaluated using non-parametric rank-based, Mann–Whitney test (Wilcoxon 1945; Mann and Whitney 1947). In this study, the weighted average annual and monsoon streamflow series of the future time periods are compared against the 20C3M scenario and the test statistic is derived. This test statistic is used to find the corresponding a, which indicates the statistical significance of difference between the medians of the two streamflow series compared.

Hydrologic Responses in the Future: DEM, soil map and LU/LC map of the Malaprabha catchment are used as spatially referenced input to ArcSWAT to simulate the catchment hydrological processes. The model is run at daily time scale using downscaled rainfall and hydrometeorological data from the reanalysis data set. Streamflow simulations at daily time scale are aggregated to monthly scale, and are compared with the observed data. MARE is found to be less than 0.1 both for

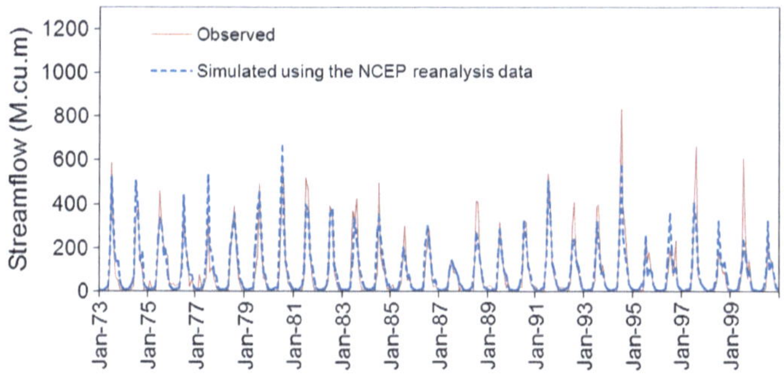

Fig. 6.23 Comparison of the monthly streamflow hydrograph simulated using the rainfall and meteorological variables downscaled from the NCEP reanalysis data with the observed data

calibration and validation phases for annual and monsoon flows. Further, monthly streamflow data series is generated taking average of 20 realizations (used for calibration and validation), and the same is compared with the observed monthly streamflow data in Fig. 6.23. Simulated monthly streamflow series matches well with the observed data, albeit slight underestimation of the peak flow values. Nash–Sutcliffe Efficiency (NSE) coefficient is found to be 0.82, which is considered to be excellent according to the general performance rating for monthly streamflow recommended by Moriasi et al. (2007). Since, a simple averaged monthly flow is used as the benchmark model in NSE, higher values of NSE may be due to the high seasonality of the monthly data. Therefore, as recommended by Schaefli and Gupta (2007), an additional index NSEB is also used in this study to evaluate the model performance. Long-term average streamflow for each month is used as the benchmark model to calculate NSEB. During the calibration period, NSEB is obtained as 0.74 for the monthly streamflow simulations. Higher values of NSE and NSEB indicate better performance of the current hydrologic model compared to the selected benchmark models.

The calibrated hydrologic model is used to simulate hydrological responses under 20C3M scenario. Simulated FDCs of annual and monsoon flows obtained using the downscaled rainfall and meteorological data from the five GCMs are compared with the FDCs of the observed data in Fig. 6.24. Table 6.5 shows the MARE of the annual and monsoon flows for the five GCMs and the corresponding weights. Using the set of weights, weighted average FDC for the 20C3M scenario is derived, which is also plotted in Fig. 6.24. The weighted average annual FDC matches well with the observed data with the exception of some underestimation of moderate and low monsoon flows. This may be mainly due to the small differences between the observed and the downscaled rainfall data in terms of annual wet days and the amount of rainfall per wet day (Mehrotra et al. 2013).

Since bias-corrected GCM outputs are used in the downscaling, such variations may be due to systemic errors from the downscaling model and therefore, may persist for the future scenarios as well. Hence, to quantify the future changes in the

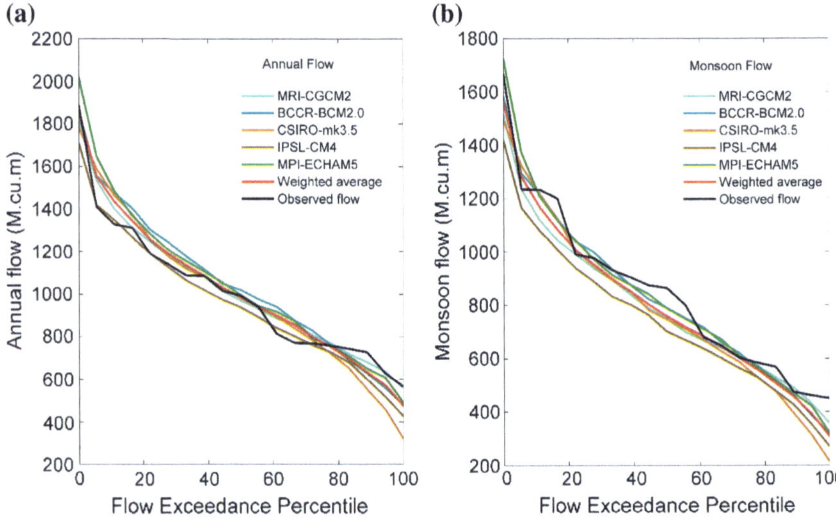

Fig. 6.24 Simulated flow-duration curves of the **a** Annual and **b** Monsoon flows for the 20C3M scenario

Table 6.5 MARE values of the GCMs for the 20C3M scenario and the corresponding weights

	MRI-CGCM2	BCCR-BCM2.0	CSIRO-mk3.5	IPSL-CM4	MPI-ECHAM5
Annual MARE	0.039	0.081	0.095	0.063	0.068
Monsoon MARE	0.066	0.064	0.109	0.132	0.058
Weight	0.28	0.20	0.14	0.15	0.23

streamflow, hydrologic responses in the future time periods are compared with the weighted ensemble average streamflow for the 20C3M scenario.

Figure 6.25 presents the weighted ensemble average FDCs of the annual and monsoon flows for the 20C3M, mid of the century and end of the century scenarios, from which some interesting observations can be drawn. Moderate (corresponding to 40–60% exceedance probability) and high (corresponding to 10% exceedance probability) streamflows at both annual and monsoon timescales show nominal increases for the mid of the century scenario, whereas low flows (corresponding to 90% exceedance probability) show some reduction in the future (Fig. 6.25a, b). Streamflow projections for the end of the century show an overall reduction in monsoon flows as seen from Fig. 6.25b. This may be attributed to the changes in the rainfall pattern projected toward the end of the century and an increase in the evaporation demand. Study conducted by Mehrotra et al. (2013) shows a plausible reduction in the number of short (2–4 days duration) and moderate (5–7 days duration) wet spells, whereas an increase in the rainfall amounts from short, moderate and long (more than 7 days duration) wet spells over the study area by the

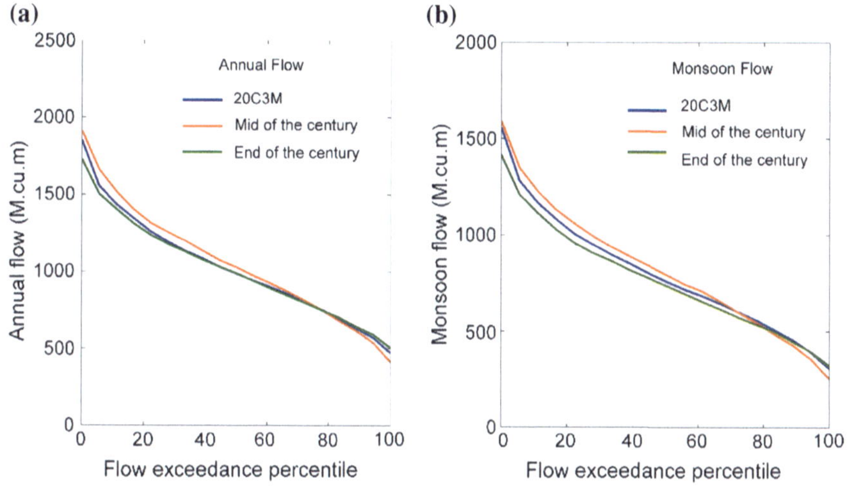

Fig. 6.25 Weighted ensemble average projections of the **a** Annual and **b** Monsoon streamflows for the 20C3M, mid and end of the century scenarios

mid of the century. On the other hand, short (2–9 days duration) and moderate (10–18 days duration) dry spells are projected to increase under the future scenarios (Mehrotra et al. 2013). Projected increase in the dry spells is likely to cause a reduction in the annual and monsoon low flows.

Using the weighted ensemble simulation, mean annual and monsoon flows are calculated for mid of the century and end of the century scenarios. In addition, annual and monsoon streamflow values corresponding to 10, 90 and 95% exceedance probability are also estimated from the FDCs, which are given in Table 6.6. Since streamflow values with less than 10% exceedance probability indicate peak flows, years corresponding to the peak flows are called wet years. On the other hand, 90 and 95% dependable flows indicate lower annual flows, and the years in which annual average flow corresponds to 90% or 95% dependable flows are considered as the low flow years or dry years. The results show an increase in the average annual and monsoon flows in the mid of the century scenario. The study also shows an increase in peak flows as indicated by an increase in the annual and monsoon streamflow values corresponding to 10% exceedance probability. Accordingly, in the mid of the century, more number of years are projected to have annual streamflow in excess of the current peak flow corresponding to 10% exceedance probability. In other words, more number of wet years are projected under the mid of the century scenario. Further, a reduction in the 90 and 95% dependable flows indicate an increase in the number of dry years. In other words, frequencies of both wet and dry years are projected to increase toward the mid of the century. Some of the previous studies analysing the historic rainfall in the twentieth century have also reported an increase in the wet and dry years in the changing climate conditions (Changnon 1987; Sousa et al. 2009).

Table 6.6 Annual and monsoon flow statistics simulated using the five GCMs for the 20C3M scenario and the future time periods for the A2 scenario

GCM	Scenario	Flow statistics (M.cu.m)[a]							
		Annual				Monsoon			
		1	2	3	4	5	6	7	8
MRI-CGCM2	20C3M	1028.5	1427.2	673.1	630.1	796.6	1143.9	480.4	425.8
	Mid of the century	1038.1	1441.8	668.3	605.8	793.1	1156.3	467.0	404.8
	End of the century	1064.3	1459.1	736.7	685.1	768.5	1096.9	494.9	444.6
BCCR-BCM2.0	20C3M	1050.9	1492.1	616.3	551.4	825.5	1233.9	446.0	381.2
	Mid of the century	1145.0	1809.5	508.2	432.8	871.4	1434.5	307.9	246.5
	End of the century	1145.5	1560.6	759.3	703.6	924.1	1313.4	574.8	522.6
CSIRO-mk3.5	20C3M	1002.2	1490.0	534.7	445.8	786.0	1239.9	379.8	305.9
	Mid of the century	991.1	1449.0	533.7	457.5	791.8	1203.5	383.4	314.6
	End of the century	795.3	1251.7	373.2	332.2	610.8	1031.4	242.3	205.4
IPSL-CM4	20C3M	964.0	1363.5	584.2	509.6	744.4	1098.0	412.8	346.6
	Mid of the century	994.5	1438.7	582.2	535.1	777.6	1170.2	388.4	334.4
	End of the century	986.4	1389.2	593.6	537.5	731.7	1097.3	401.0	343.8
MPI-ECHAM5	20C3M	1057.3	1517.8	643.8	598.0	838.2	1242.2	461.4	414.3
	Mid of the century	1087.0	1574.9	657.4	553.9	864.3	1292.1	489.3	401.9
	End of the century	987.7	1392.0	587.9	546.7	749.7	1110.8	394.9	350.9
Weighted average	20C3M	1026.3	1460.5	622.1	562.8	802.7	1191.4	444.8	385.4
	Mid of the century	1057.6	1546.6	601.8	527.6	822.7	1252.1	416.7	349.2
	End of the century	1013.0	1424.0	633.9	584.7	767.4	1134.2	437.9	389.6

[a](1) Average annual flow (2) Annual streamflow corresponding to 10% exceedance probability (3) 90% dependable annual flow (4) 95% dependable annual flow (5) Average monsoon flow (6) Monsoon streamflow corresponding to 10% exceedance probability (7) 90% dependable monsoon flow (8) 95% dependable monsoon flow

Table 6.7 Mann–Whitney Test Statistics for the Percentage Change in the Mean Annual and Monsoon Flows in the Future

	Mid of the century		End of the century	
	Annual flow	Monsoon flow	Annual flow	Monsoon flow
Test statistic	0.19	0.10	0.10	0.10
Significance level (α)	0.849	0.92	0.92	0.92

Streamflow simulations for the end of the century scenario show some reduction in mean annual and monsoon flows corresponding to 10% exceedance probability, whereas no significant changes in the lower flows (90 and 95% dependability). This implies that frequency of wet years is likely to decrease toward the end of the century. These changes, however, are insignificant as indicated by the Mann–Whitney test. The test statistics and significance levels of the percentage changes in the mean annual and monsoon flows are presented in Table 6.7.

Although the projected changes in mean annual and monsoon flows are not significant, the intra-annual variation of flows needs to be analyzed. Box-whisker plots of the weighted ensemble average monthly flows for the 20C3M and the two future scenarios are shown in Fig. 6.26. These plots show some variations in the monthly flows during the monsoon period. Median and the 75 percentile flows are projected to increase in June while some reduction is noted during July and August under the future scenario. Close to 25%, reduction in the streamflow is observed in July toward the end of the century. Being the peak sowing period for the Kharif crops in India, any drop in the water availability in July in the future scenario may have an adverse impact on the agriculture.

Climate Change Impact on Water Budget Components: Monthly hydrologic simulations of the catchment under the 20C3M and future scenarios are used to analyze the changes in other catchment water budget components as well.

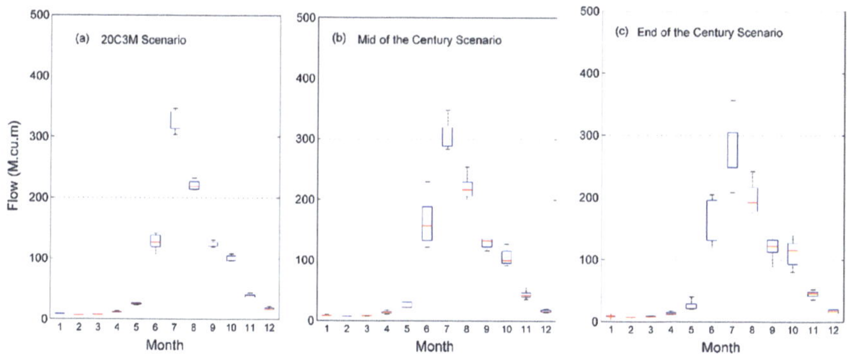

Fig. 6.26 Comparison of the weighted average monthly flows for the 20C3M scenario and the future time periods. In the box plot, upper and lower hinges represent the 75 and 25 percentiles, respectively. The whiskers show the other data points except the outliers. The line within the box shows the median

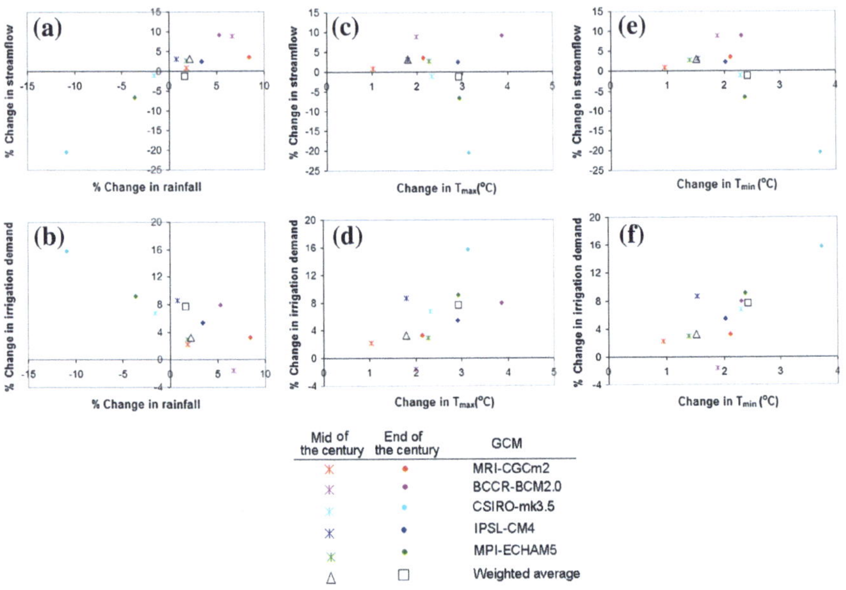

Fig. 6.27 Relationship between the projected changes in streamflow and irrigation demand with projected changes in the climate from the five GCMs

Variations in ET, irrigation demand and groundwater recharge are estimated under the projected climate conditions. Figure 6.27 shows the variations in streamflow and irrigation demand with the changes in rainfall and temperature. In general, an increase in rainfall causes an increase in streamflow. Rainfall projections from various GCMs vary from −10.9 to 8.4% with respect to the 20C3M scenario, whereas the streamflow projections vary from −20.6 to 9.0% toward the end of the twenty-first century (Fig. 6.27a). Likewise, the irrigation demand is projected to increase by 3.2–15.7% across various GCMs by the end of the twenty-first century (Fig. 6.27b). From Fig. 6.27a, b, it can be observed that changes in the streamflow and irrigation demand are largely related to the variation in rainfall and the GCM. Nevertheless, small deviations may be observed, which may be attributed to the changes in T_{max} and T_{min} as shown in Figs. 6.27c–f. For example, using BCCR-BCM2.0 larger increases in irrigation demand toward the end of the century is projected, which may be attributed to the highest increase in T_{max} as shown in Fig. 6.27d. Likewise, results from CSIRO-mk3.5 project significant reduction in streamflow and increase in irrigation demand, which may be attributed to the combined effect of large increases in T_{max} and T_{min}, and significant reduction in rainfall.

Figure 6.28 shows the catchment water budget components for the 20C3M scenario at the annual time scale as well as for the monsoon period. ET is the major abstraction from the rainfall amounting close to 70% annually and close to 45% during the monsoon period. Irrigation supplements the crop water demand to a large

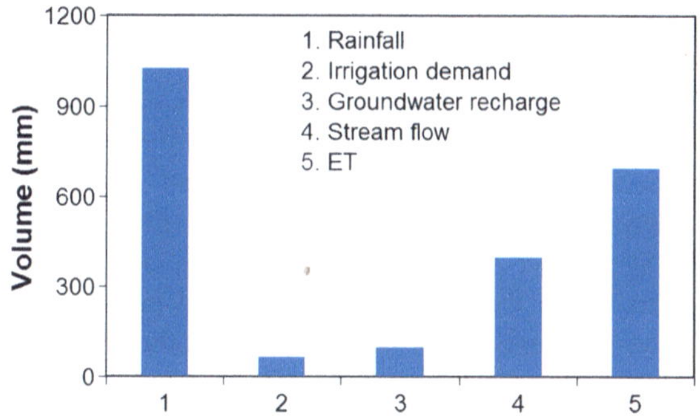

Fig. 6.28 Water budget components in the Malaprabha catchment for the 20C3M scenario

extent. Groundwater recharge is the amount of water that reaches the shallow aquifer. A part of it appears as groundwater flow and contributes to the streamflow at the watershed outlet.

Percentage changes in the water budget components projected under the future scenario with respect to 20C3M scenario at the annual time scale are also analyzed. Weighted ensemble average simulations show marginal increases in annual average rainfall under the future scenarios (2.2% toward the mid of the century and 1.6% toward the end of the century). Though the changes in the annual average rainfall are nominal, the variations in the rainfall pattern in terms of number and durations of wet and dry spells are of major concern. T_{max} is projected to increase by 0.51 and 0.84 °C (1.8 and 2.9%) toward the mid of the century and end of the century, respectively. Similarly, T_{min} is projected to increase by 0.29 and 0.46 °C (1.5 and 2.4%) toward the mid of the century and end of the century, respectively. With changes in temperature and rainfall pattern, evapotranspiration rates are projected to increase by 2.3% and 4.1%, respectively, for the mid of the century and end of the century scenarios. Runoff ratio (ratio of average annual runoff to average annual rainfall) of the catchment is found to be 0.4184, 0.4178 and 0.4083 for 20C3M, mid of the century and end of the century scenarios, respectively. The 2.5% reduction in the runoff ratio shows 1.2% reduction in the annual average streamflow toward the end of the century. In addition, groundwater recharge rates are also projected to decline (by 7.3%) toward the end of the century.

6.4.3 Summary and Conclusions

The following main conclusions are emanated from the study:

- The results show only marginal changes in annual average rainfall in the catchment under the future scenarios. Corresponding changes in the hydrologic components are also found to be statistically insignificant both for the annual and monsoon periods.
- Even though the changes in the streamflow and irrigation demand are strongly related to the variation in rainfall, they are not directly proportional to each

other. Such deviations may be attributed to the GCMs used, as well as the changes in rainfall pattern and atmospheric temperature.

- With 0.84 and 0.46 °C increases in the daily maximum and minimum temperatures toward the end of the century, evapotranspiration rate is projected to increase by 4.1%, irrigation demand is projected to increase by 7.7% and groundwater recharge is projected to decline by 7.3%.
- Streamflow projections for the end of the century scenario show nominal reduction in the average annual and monsoon flows. Changes in rainfall and temperatures are projected to reduce the runoff ratio by 2.5% by the end of the century.
- Projected increase in the evapotranspiration and irrigation demand, associated with the decrease in the groundwater recharge and streamflow is an indication of possible aggravation of the water stress in the catchment in future.

6.5 Comparing Impacts of Climate Change on Streamflow in Four Large African River Basins[4]

Summary: In this study, impacts of climate change on streamflow are studied in four large representative African river basins: the Niger, the Upper Blue Nile, the Oubangui and the Limpopo. Eco-hydrological model SWIM (Soil and Water Integrated Model) was set up for all four basins individually. The validation of the models for four basins shows results from adequate to very good, depending on the quality and availability of input and calibration data. For the climate impact assessment, the model was driven with outputs of five bias-corrected Earth system models of Coupled Model Intercomparison Project 5 (CMIP5) for the representative concentration pathways (RCPs) 2.6 and 8.5. This climate input is put into the context of climate trends of the whole African continent and compared to a CMIP5 ensemble of 19 models in order to test their representativeness. Subsequently, the trends in mean discharges, seasonality and hydrological extremes in the twenty-first century were compared. The uncertainty of results for all basins is found to be high. Still, climate change impact is clearly visible for mean discharges of the projections and is the lowest in the Upper Blue Nile, where an increase in streamflow is most likely. Attention to the possible risks of increasing high flows in the face of the dominant water scarcity in Africa is felt. In conclusion, the study shows the impact

[4]Adopted from open access publication under the Creative Commons Attribution 3.0 License V. Aich, S. Liersch, T. Vetter, S. Huang, J. Tecklenburg, P. Hoffmann, H. Koch, S. Fournet, V. Krysanova, E. N. Müller, and F.F. Hattermann (2014) Comparing impacts of climate change on streamflow in four large African river basins, Hydrol. Earth Syst. Sci., Copernicus, 18, 1305–1321, doi: 10.5194/hess-18-1305-2014, http://www.hydrol-earth-syst-sci.net/18/1305/2014/ with permission from authors. More details are available in the original publication.

that intercomparisons have added value to the adaptation discussion and may be used for setting up adaptation plans in the context of a holistic approach.

6.5.1 Problem Description, Case Study and Data (Aich et al. 2014)

Objectives chosen are as follows:

- To investigate differences in the sensitivity of modeled annual discharge to climate parameters among the basins
- To study climate impacts on river discharge for four basins in terms of quantity and seasonality
- To explore changes in hydrological extremes (high flow, low flow) for the four basins
- To analyze the uncertainties of the projections
- To identify and discuss the implications for adaptation.

The selected basins Niger, Upper Blue Nile, Oubangui and Limpopo are distributed all over sub-Saharan Africa, in the West, East centre and South (Fig. 6.29). In addition, they cover all climate groups of sub-Saharan Africa according to the Köppen (1900) classification after Strahler (2013). Besides the tropical humid climates (A), dry climates (B), sub-tropical climates (C) and highland climates (H) they also cover most of the climatic types and subtypes of the continent. All four African basins were modeled using the eco-hydrological model SWIM (Krysanova et al. 1998). SWIM is a process-based model and it simulates the

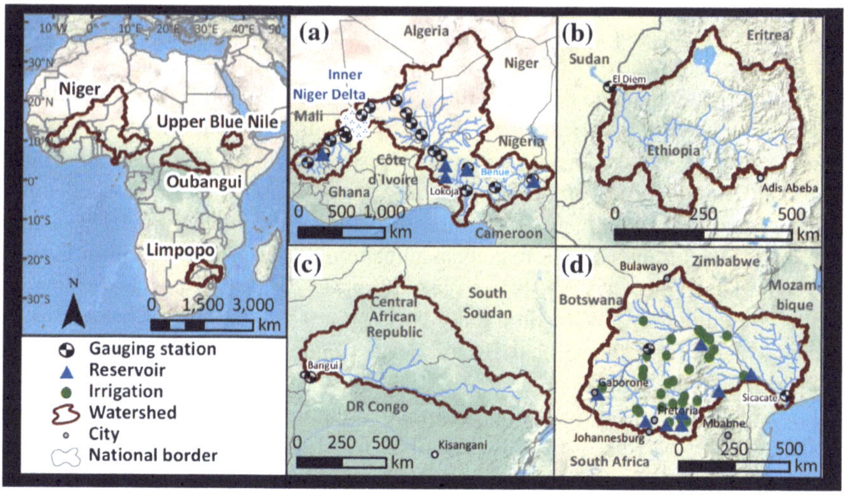

Fig. 6.29 Map of the four modeled basins **a** Niger. **b** Upper Blue Nile. **c** Oubangui. **d** Limpopo

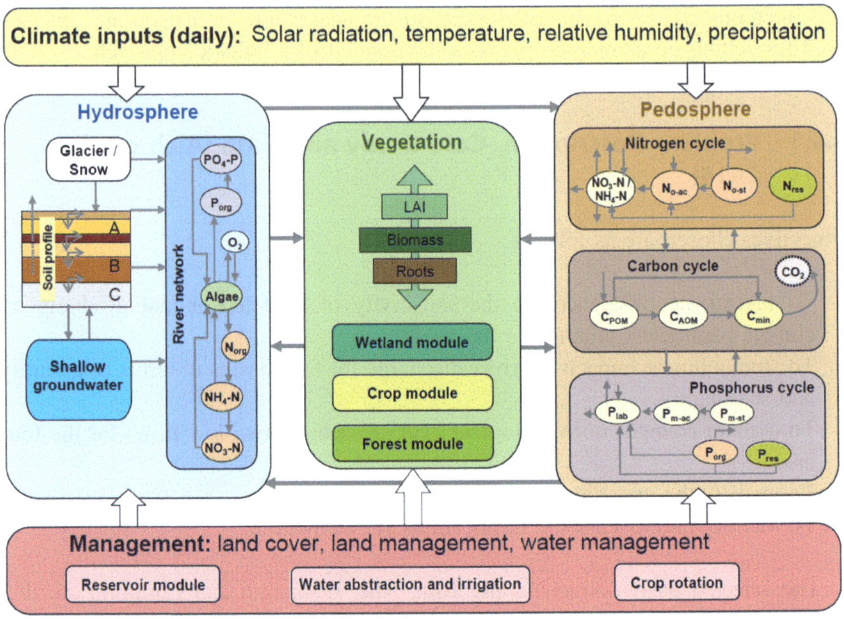

Fig. 6.30 Structure of the Eco-hydrological model SWIM

dominant eco-hydrological processes such as evapotranspiration, vegetation growth, runoff generation and river discharge, and also considers feedbacks among these processes (Krysanova et al. 2005) (Fig. 6.30). For all four regions, DEM was derived from the Shuttle Radar Topography Missions (SRTM) with 90 m resolution (Jarvis et al. 2008). Soil parameters were derived from the Digital Soil Map of the World (FAO et al. 2012). Relevant soil data for SWIM include its depth, clay, silt and sand content, bulk density, porosity, available water capacity, field capacity, and saturated conductivity for each of the soil layers. Land use data were reclassified from the global land cover (Bartholomé and Belward 2005). Land use classes of SWIM include water, settlement, industry, road, cropland, meadow, pasture, mixed forest, evergreen forest, deciduous forest, wetland, savannah (heather) and bare soil. Climatic observations are generally sparse in Africa and homogeneously distributed over the continent. Therefore, to compare the results, model was calibrated for four basins using a reanalysis climate data set produced within the EU FP6 WATCH project (WFD 2011; Weedon et al. 2011). This data contains all variables required for SWIM on a daily basis on a 0.5×0.5 grid. Observed river discharge data from the Global Runoff Data Centre was used to calibrate and validate the model (Fekete et al. 1999).

For analysing climate trends, output of an ensemble of 19 CMIP5 Earth System Models (ESMs) was used. Of this ensemble, five ESMs' (HadGEM2-ES, IPSL-5 CM5A-LR, MIROC-ESM-CHEM, GFDL-ESM2M, and NorESM1-M) outputs were used for driving the hydrological model. The five chosen ESMs have been downscaled using a trend-preserving bias correction technique with the WFD reanalysis data, and have been resampled on a 0.5 × 0.5 grid for the time period 1950–2099 (Hempel et al. 2013). Representative Concentration Pathways (RCPs) covers different emission concentrations, and in this study, the RCP 2.6 and 8.5 scenarios were employed for all five ESMs to cover the low and high ends of possible future climatic projections.

6.5.2 Results and Discussion

Calibration and Validation: Table 6.8 summarises the SWIM model setup and calibration information as well as the results of the validation. The SWIM model was validated for the gauging stations at the outlets of the four basins; the results are presented in Fig. 6.31. Nash and Sutcliffe (1970), and Percent Bias (PBIAS) were used for evaluating the model error, i.e., quantifying the efficiency of the model. The SWIM model was basically able to reproduce the hydrological characteristics of each basin reasonably well, with NSE of the monthly runoff rate ranging between 0.63 and 0.9 and the daily runoff rate ranging from 0.55 to 0.89.

Table 6.8 Characteristics of basin models and validation results

	Niger	Upper Blue Nile	Oubangui	Limpopo
Number of subbasins	1,923	558	377	2,020
Number of hydrotopes	13,883	1,700	1,734	13,085
Number of included reservoirs	5	0	0	8
Number of included irrigation schemes	0	0	0	31
Number of gauging stations used for calibration	18	1	1	2
Gauging station(s) used for calibration/validation	Lokoja[a]	El Diem	Bangui	Sicacate, Oxenham Ranch
Calibration period	1972–1982[a]	1961–1970	1981–1990	1971–1978
NSE[b] (daily)	0.92	0.81	0.66	0.72, 0.73
PBIAS[c]	8.6	20.9	19.1	11.5, −6.7
Validation period	1983–1992[a]	1971–1980	1971–1980	1980–1987[d]
NSE[b] (daily)	0.89	0.63	0.6	0.55
NSE[b] (monthly)	0.9	0.73	0.63	0.8
PBIAS[c]	2.1	39	15.7	3.4

[a]In the Niger basin 18 gauging stations have been used for the calibration
[b]Nash–Sutcliffe Efficiency
[c]Percent bias of monthly average
[d]The gauging station Oxenham Ranch was only used for calibration and not validated

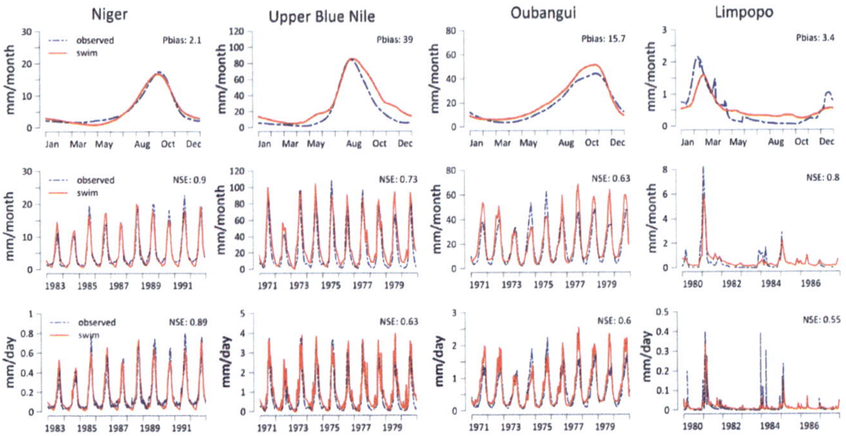

Fig. 6.31 Validation of SWIM at the outlets of the four basins. In the top row the seasonality of monthly runoff rate in validation period and PBIAS, in the middle row the monthly runoff rate and in the bottom row the daily runoff rate in the validation period, both with Nash–Sutcliffe efficiency

Climate Trends: Precipitation and temperature are the key drivers for the hydrological regime of rivers, and climate change has its main impact on these two variables. Figure 6.32 presents mean trends for these two parameters from 2006 until 2100 projected by 19 CMIP5 models for the whole African continent. Shown are the results for RCP 8.5 in order to illustrate the most pronounced trends under extreme scenario conditions. All models agree on a distinct temperature rise over

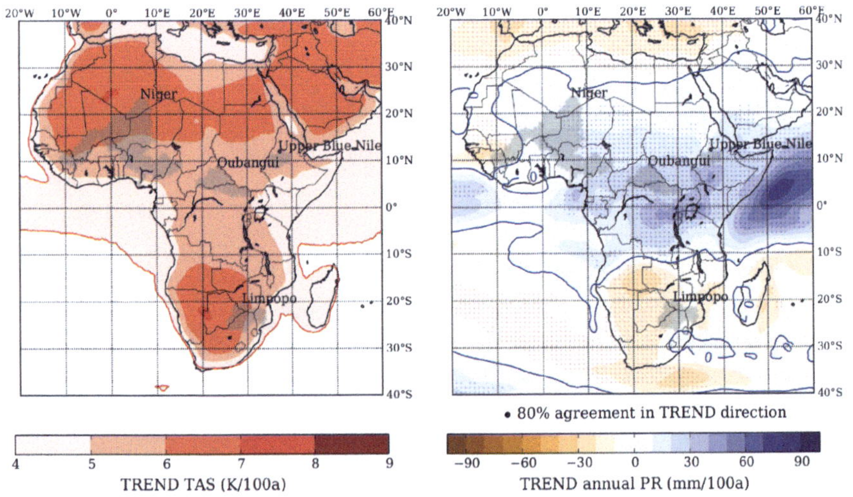

Fig. 6.32 Mean temperature (left) and precipitation (right) trends over the African continent for 19 CMIP5 Models from 2006–2100 for RCP 8.5. For precipitation, an agreement in trend direction of 80% or more of the models is marked with a dot

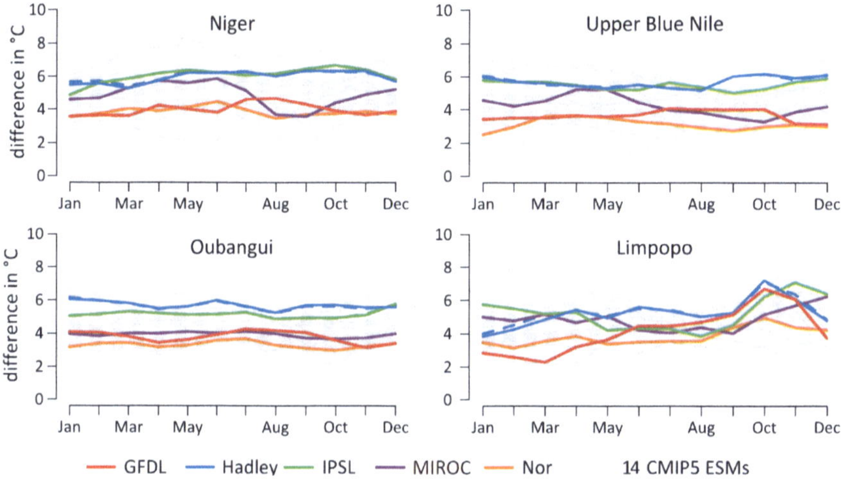

Fig. 6.33 Difference in monthly mean temperature in the far projection period (2070–2099) relative to the base period (1970–1999) for RCP 8.5 for five bias-corrected model projections (coloured lines), the uncorrected ESMs (coloured dashed lines) and 14 ENSEMBLE ESMs (grey dashed lines)

the whole African continent, while in the tropics much of the additional energy input is converted to latent heat. The highest temperature increase of 6–7 °C and in some parts even up to 8 °C, is projected over the already driest and hottest areas in the Sahara and Southern African savannahs and deserts. For the projection of streamflow, bias-corrected model output of five ESMs (HadGEM2-ES, IPSL-5 CM5ALR, MIROC-ESM-CHEM, GFDL-ESM2M and NorESM1-M) were used.

In Figs. 6.33 and 6.34, temperature and precipitation of these climate runs were compared to the uncorrected runs and 14 other CMIP5 models in order to display the influence of the bias correction and where the respective models lie in a larger ensemble (i.e., if the model is especially dry or wet, warm or cold or in the middle of the whole ensemble). In Fig. 6.33, the seasonal changes between mean monthly temperatures show a distinctly homogeneous pattern. In all four basins, the temperature rises between 3 and 6 °C. The bias correction hardly influenced the temperature. The five selected model outputs cover the temperature range of the CMIP5 ensemble in all four basins well. Figure 6.34 presents monthly precipitation in the same periods for the RCP 8.5. Figure 6.35 illustrates the sensitivity of river discharge to climate variability and change in the four basins. Changes in precipitation are shown in the range from −50 to 100%, and for discharge the range is from −100 to 200%. Values outside this range are not shown but are included in the calculation

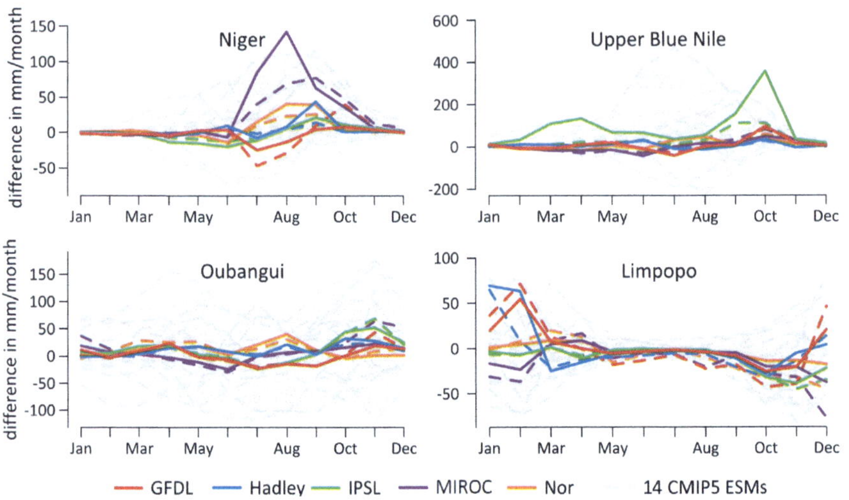

Fig. 6.34 Difference in monthly precipitation in the far projection period (2070–2099) relative to the base period (1970–1999) for RCP 8.5 for five bias-corrected model projections (coloured lines), the uncorrected ESMs (coloured dashed lines) and 14 ENSEMBLE ESMs (grey dashed lines)

of the fitted local regression, plotted as a black line. They also studied the impact of climate change on discharge and seasonality, changes in extremes, differences in climate change sensitivity among the basins, changes of streamflow under climate change, changes in hydrological extremes, sources of uncertainties and implications for adaptation.

6.5.3 Summary and Conclusions

The following main conclusions are emanated from the study:

- Large differences between the sensitivities of streamflow regimes to climate variability among the four basins are observed.
- Most uncertainty exists in regional impact studies derived from climate models, even after the application of bias correction to inputs.

It is also mentioned by Aich et al. (2014) that detailed future studies for planning adaptation strategies are required to develop flood protection measures.

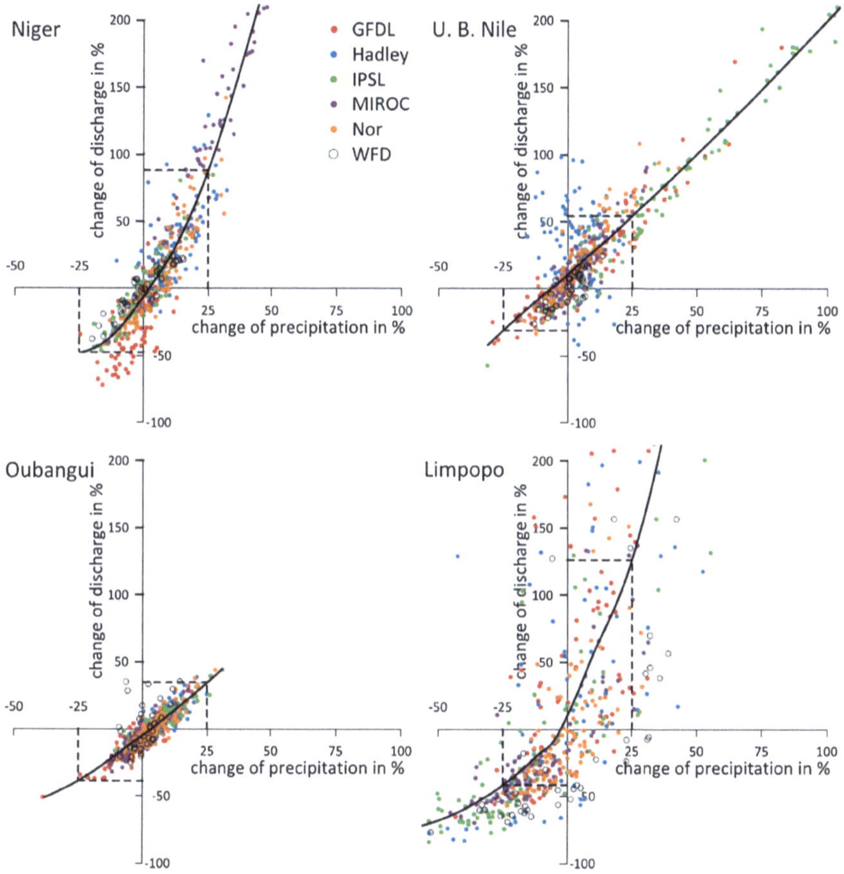

Fig. 6.35 Climate sensitivity in the four basins. Change in modeled annual discharge [percent] per change of precipitation [percent] for 2006–2099 compared to the mean of base period 1970–1999 for five climate models in RCP 8.5 and WFD. Curve shows fitted local regression over all values

6.6 Hydrologic Impact of Climate Change on Murray–Hotham Catchment of Western Australia: A Projection of Rainfall–Runoff for Future Water Resources Planning[5]

Summary: The hydrologic impact of climate change on the Murray–Hotham catchment in Southwest Western Australia (SWWA) has been investigated using a multi-model ensemble approach through projection of rainfall and runoff for the periods mid (2046–2065) and late (2081–2100) of the century. The Land Use Change Incorporated Catchment (LUCICAT) model was used for hydrologic modeling. Model calibration was performed using (5 km) grid rainfall data from the Australian Water Availability Project (AWAP). Downscaled and bias-corrected rainfall data from 11 general circulation models (GCMs) for Intergovernmental Panel on Climate Change (IPCC) emission scenarios A2 and B1 were used in LUCICAT model to derive rainfall and runoff scenarios for 2046–2065 (mid this century) and 2081–2100 (late this century). The results of the climate scenarios were compared with observed past (1961–1980) climate. The mean annual rainfall averaged over the catchment during recent time (1981–2000) was reduced by 2.3% with respect to the observed past data (1961–1980) and the resulting runoff reduction was found to be 14%. Compared to the past, the mean annual rainfall reductions, averaged over 11 ensembles and over the period for the catchment for A2 scenario are 13.6 and 23.6% for mid and late this century, respectively, while the corresponding runoff reductions are 36 and 74%.

For B1 scenario, the rainfall reductions were 11.9 and 11.6% for mid and late this century and the corresponding runoff reductions were 31 and 38%. Spatial distribution of rainfall and runoff changes showed that the rate of changes were higher in high-intensity rainfall areas compared to low-intensity rainfall areas. Temporal distribution of rainfall and runoff indicates that high-rainfall events in the catchment reduced significantly and further reductions are projected, resulting in significant runoff reduction. A catchment scenario map has been developed by plotting decadal runoff reduction against corresponding rainfall reduction at four gauging stations for the observed and projected periods. This could be useful for future water resources planning in the catchment. Projection of rainfall and runoff result based on the GCMs varied significantly for the time periods and emission scenarios. Hence, there is considerable uncertainty involved in this study even though ensemble mean was used to explain the findings.

[5]Adopted from open access publication under the Creative Commons Attribution 3.0 License
S A. Islam, M. A. Bari, and A.H.M.F. Anwar (2014)Hydrologic impact of climate change on Murray–Hotham catchment of Western Australia: a projection of rainfall–runoff for future water resources planning, Hydrol. Earth Syst. Sci., Copernicus, 18, 3591–3614, doi: 10.5194/hess-18-3591-2014, http://www.hydrol-earth-syst-sci.net/18/3591/2014/
With permission from authors. More details are available in the original publication.

6.6.1 Problem Description, Case Study and Data (Islam et al. 2014)

Objectives chosen are as follows:

- To analyze spatial and temporal variability along with probability of exceedance of observed and projected rainfall and runoff data.
- To develop catchment scenario map plotting decadal rainfall and runoff changes for observed and projected periods.

The Murray River catchment, with an area of 6,736 km^2, lies within the Murray river basin and the Peel–Harvey sub-region, around 110 km southwest of Perth in Western Australia (Fig. 6.36). To distinguish this study area from the well-known Murray–Darling catchment in Eastern Australia, it is referred to as the Murray–Hotham catchment in reference to the two major rivers in the catchment. Geologically, the catchment is located in the Darling Plateau, the surface of Yilgarn Block. The climate of the catchment is temperate based on the Köppen classification system (Stern et al. 2000) with hot dry summers and cool winters with most of the rainfall

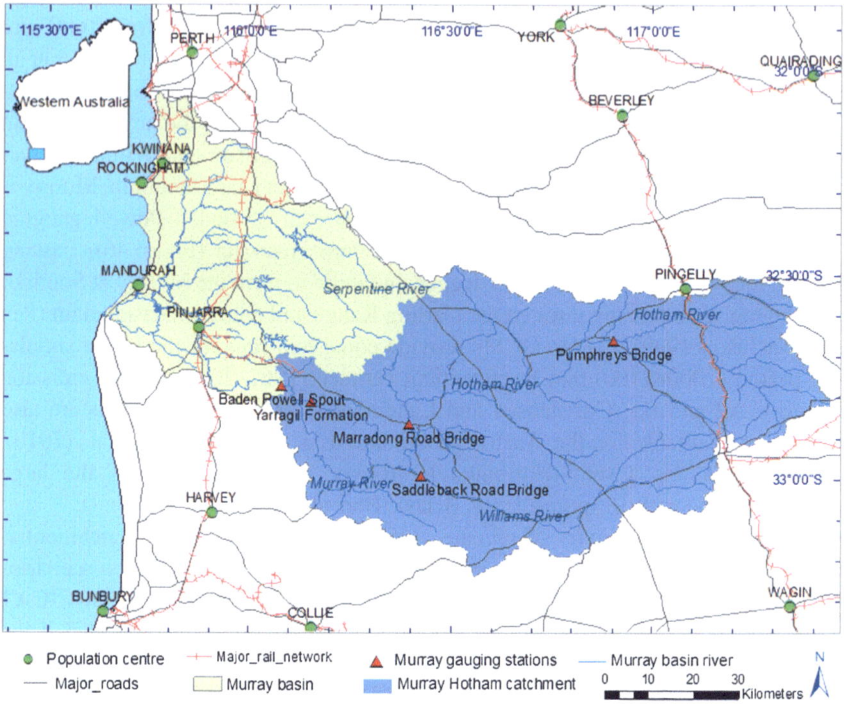

Fig. 6.36 Murray–Hotham catchment of Western Australia with major rivers and gauging stations in the study area

Table 6.9 Goodness of fit for daily streamflow simulations

Gauging station	Measure of fit	Nash–Sutcliffe efficiency (E^2)	Correlation coefficient (CC)	Overall water balance (E)	Flow-period error index (EI)
Baden powell water spout	Overall	0.70	0.84	0.07	1.00
	Calibration	0.70	0.84	0.07	1.01
	Validation	0.80	0.91	−0.03	0.98
Marradong road bridge	Overall	0.48	0.80	−0.03	0.99
	Calibration	0.47	0.79	−0.03	0.99
	Validation	0.81	0.94	−0.03	0.99
Saddleback road bridge	Overall	0.49	0.76	−0.04	1.02
	Calibration	0.48	0.75	−0.03	1.02
	Validation	0.84	0.92	−0.12	1.00
Yarragil formation	Overall	0.56	0.75	−0.01	0.86
	Calibration	0.56	0.75	−0.01	0.90
	Validation	0.68	0.80	0.08	1.05

(around 75%) occurring during winter between May and September. Observed mean annual rainfall varied across the catchment from East to West, low (400 mm) to high (1100 mm) with a gradual increase. Mean annual evaporation ranges from 1600 mm toward the Southwest to 1800 mm in the North East corner of the basin (Mayer et al. 2005). Mean annual rainfall and runoff from contributing catchments at the gauging stations are presented in Table 6.9. The Murray River is one of the largest rivers in terms of flow volume in SWWA which begins as the Hotham and Williams River systems and drains into the Indian Ocean via the Peel Inlet near Mandurah. Murray is the only free-flowing river (devoid of dam upstream of Baden Powell gauging station) in the Northern Jarrah Forest in Western Australia (Fig. 6.36). Passing through the hilly country, the rivers deepen and unite to form the Murray at South of Boddington, then passing through the Darling Range and onto the coastal plain (Pen and Hutchison 1999). Eleven GCMs provide consistent runs for the future simulation period (2000–2100) and the twentieth century (1961–2000) for the emission scenarios A2 and B1 (Christensen and Lattenmaier 2007). These models are also found to be suitable for the Australian climate as studied by Bari et al. (2010). The LUCICAT is a semi-distributed hydrologic model that divides the large catchment into small Response Units (RUs) (Bari and Smettem 2003).

The hydrological impact of climate change on Murray–Hotham catchment is assessed through projection of rainfall–runoff for the two IPCC emission scenarios A2 and B1 for the periods 2046–2065 and 2081–2100, respectively. The LUCICAT hydrologic model is applied to simulate future rainfall–runoff using downscaled and bias-corrected rainfall data of GCMs. A conceptual diagram of the hydrologic modeling using the LUCICAT model is shown in Fig. 6.37. At first, input files with attribute of catchment, channels, nodes and rainfall stations were prepared through processing of a DEM of the catchment using ArcGIS. The attribute files were

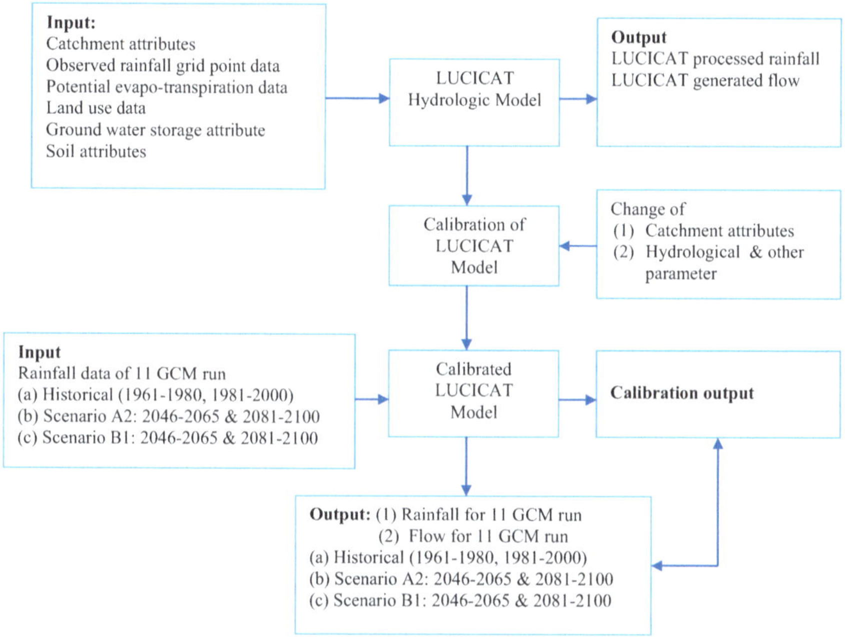

Fig. 6.37 Conceptual diagram of the LUCICAT modeling process with climate change scenarios (modified from Islam et al. 2011)

developed by dividing the catchment into 135 RUs. Land use history and pan evaporation data were considered as input for model calibration. The model was calibrated at five gauging stations (Fig. 6.36) for 1960–2004 and validated for 2005–2009 with recently developed 5 km grid rainfall produced by the Bureau of Meteorology, Australia (Jones et al. 2009). Next, downscaled GCM rainfall data were processed for hindcast (1961–2000) and different climate scenarios (A2 and B1) for 2046–2065 and 2081–2100. Downscaling of GCM data to a 5 km resolution (compatible to hydrologic modeling) was carried out by the Bureau of Meteorology Statistical Downscaling Model (BoMSDM) which works with an analogue approach (Timbal et al. 2009). The downscaled rainfall data was subsequently used as input to the calibrated model for generating various rainfall and runoff scenarios. The annual rainfall data processed for the hindcast period using downscaled GCMs data was compared with observed annual rainfall data. A scale factor was developed for each of the GCMs to match the hindcast annual rainfall with the observed annual rainfall. The corresponding scaling factors are applied to downscaled daily rainfall data (2046–2065, 2081–2100) for the emission scenarios of A2 and B1. Processed rainfall and runoff scenarios along with historical data were then analyzed, compared and presented. To address uncertainties involved with the GCM data, a multi-model ensemble approach (with 11 GCMs' data) was adopted and the ensemble mean was derived.

6.6.2 Results and Discussion

Catchment Hydrology: The runoff rate (runoff divided by rainfall) has changed across the catchment significantly in the last several decades. During the last three decades, overall, total runoff declined due to the absence of high rainfall in last three decades and this was found to be dominant in the high-rainfall receiving part of the catchment. Similar changes in the total runoff were also observed in other studies (CSIRO 2009; DoW 2010).

Calibration and Validation: The LUCICAT model has 29 model parameters which are grouped as follows: (i) an estimated a priori group which do not need calibration and (ii) a variable set of eight physically meaningful parameters, which need calibration. The calibration criteria are (i) joint plot of observed and simulated daily flow series, (ii) scatter plot of monthly and annual flow, (iii) flow period Error Index (EI), (iv) Nash–Sutcliffe efficiency (E^2), (v) explained variance, (vi) Correlation Coefficient (CC), (vii) overall water balance (E) and (viii) flow-duration curves.

The model was calibrated for daily flow. The model was found to be well fitted for depicting daily flow (e.g., high, medium and low flow). Results also revealed that the model is capable of describing peaks, duration of flow and recession for all types of flow conditions (e.g., high, medium and low flow). The daily flow model was validated with hydrographs at all gauging stations and it was found that it can predict future daily streamflow for the catchment effectively. Table 6.9 presents a summary of model performance based on observed and simulated daily flow. Results also revealed that the model can predict future annual flow successfully based on catchment rainfall.

Observed annual rainfall for the historical period (1960–2000) and projected annual rainfall for the mid (2046–2065) and late (2081–2100) century at four gauging stations of the catchment are presented in Fig. 6.38. Table 6.10 presents observed and projected rainfall scenarios which are self-explanatory. In addition, they studied spatial and temporal variation of rainfall and runoff. Table 6.11 presents observed and predicted runoff scenarios. Future projection for water resources planning along with uncertainty and its implication are also discussed in detail. According to the Islam et al. (2014) major causes of uncertainty that affect climate change impact studies are GCMs, choice of downscaling techniques, selection of the hydrologic model, appropriate model parameterization, assessing the assumptions and limitations of the model and estimation of uncertainty associated with the modeling approach.

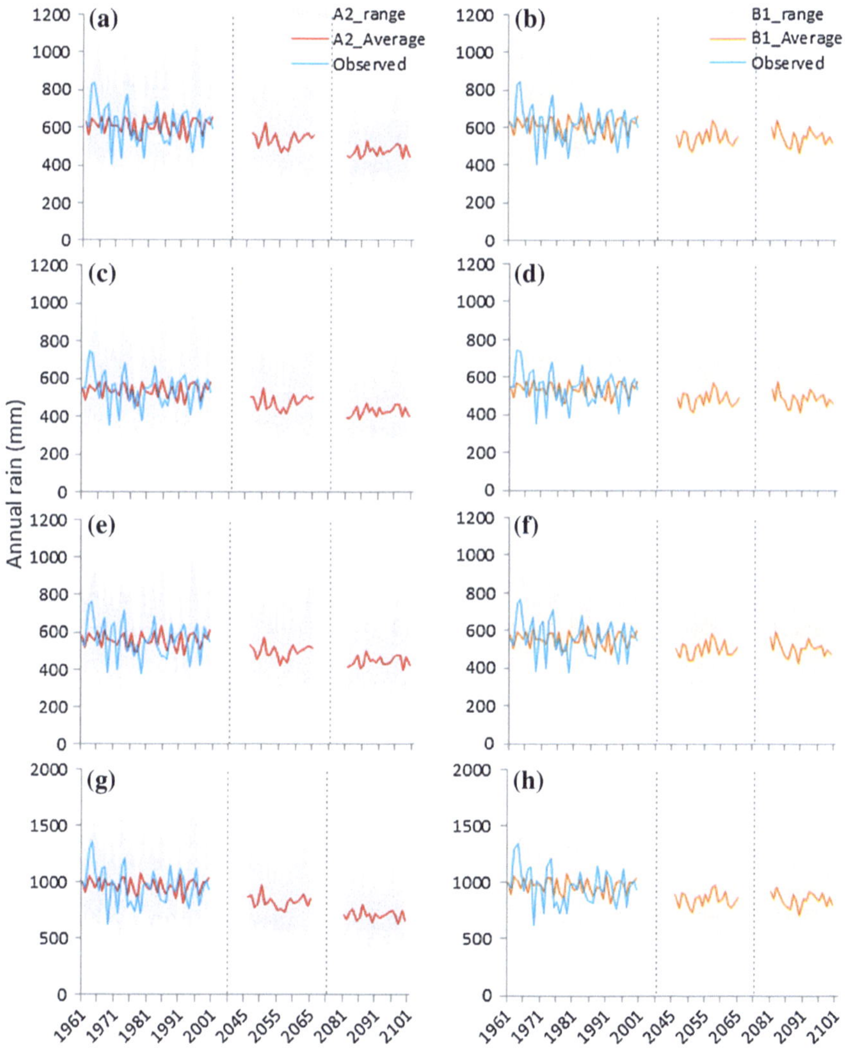

Fig. 6.38 Observed and projected annual rainfall under scenarios A2 and B1 for the four gauging stations: **a** and **b** at Baden Powell, **c** and **d** at Marradong Road Bridge, **e** and **f** at Saddleback Road Bridge, **g** and **h** at Yarragil Formation. The average for projected rainfall is the ensemble mean of 11 GCMs, and A2 and B1 range represent the maximum and minimum of all the GCMs

Table 6.10 Observed and projected rainfall scenarios

Gauging stations	Percentile	Observed rainfall (mm)				Change in average rainfall with respect to the past (%)[a]			
		Historical (1961–2000)	Past (1961–1980)	Recent (1981–2000)	Change (%)	2046–2065		2081–2100	
						A2	B1	A2	B1
Baden Powell	Q90	726	779	696	−11	−13	−11	−24	−12
	Q50	622	622	623	0	−15	−16	−24	−12
	Q10	489	437	508	16	−9	−4	−15	0
	Mean	616	623	609	−2	−13	−12	−24	−12
Marradong Road Bridge	Q90	646	690	607	−12	−15	−10	−29	−11
	Q50	550	549	556	1	−21	−15	−30	−12
	Q10	439	381	445	17	−17	−5	−22	0
	Mean	547	552	542	−2	−13	−12	−23	−12
Saddleback Road Bridge	Q90	677	717	645	−10	−13	−10	−22	−12
	Q50	566	585	566	−3	−16	−16	−25	−13
	Q10	423	398	451	13	−8	−4	−12	0
	Mean	564	573	555	−3	−13	−12	−22	−11
Yarragil Formation	Q90	1140	1217	1114	−8	−15	−13	−27	−15
	Q50	949	963	947	−2	−15	−14	−28	−11
	Q10	765	729	815	12	−15	−10	−25	−8
	Mean	964	975	953	−2	−15	−12	−27	−12

Table 6.11 Observed and projected runoff scenarios

Gauging stations	Percentile	Observed runoff (GL)				Change in average runoff with respect to the past (%)[a]			
		Historical (1961–2000)	Past (1961–1980)	Recent (1981–2000)	Change (%)	2046–2065		2081–2100	
						A2	B1	A2	B1
Baden Powell	Q90	537	692	389	−44	−40	−35	−77	−50
	Q50	220	233	210	−10	−43	−41	−79	−34
	Q10	92	85	114	34	−54	−34	−80	−34
	Mean	285	307	264	−14	−36	−31	−74	−38
Marradong Road Bridge	Q90	280	334	167	−50	−44	−44	−79	−59
	Q50	105	108	92	−16	−52	−53	−82	−49
	Q10	34	23	49	109	−39	−21	−72	−12
	Mean	129	136	121	−11	−41	−39	−76	−45
Saddleback Road Bridge	Q90	163	173	105	−39	−39	−35	−72	−47
	Q50	68	71	66	−7	−48	−45	−76	−44
	Q10	24	23	30	30	−55	−39	−72	−27
	Mean	76	80	72	−10	−36	−33	−69	−36
Yarragil Formation	Q90	6.7	8.3	3.5	−58	−57	−50	−92	−66
	Q50	1.9	3.4	1.6	−52	−81	−77	−98	−76
	Q10	0.6	0.6	0.6	−1	−86	−81	−99	−80
	Mean	3.0	4.1	1.9	−54	−64	−60	−93	−67

6.6.3 Summary and Conclusions

The following main conclusions are emanated from the study:

- Declining trend of rainfall and runoff in SWWA is likely to continue during mid and late this century, resulting in lower flow to the dams and a subsequent lower availability of surface water. Hence, water resources managers and policy makers will have to rely more on groundwater, desalination or other sources of water (e.g., recycling) for Perth water supply.
- GCM-derived annual rainfall and runoff show extreme high or low value (which is much higher or lower) beyond the observed range during a particular time period. Hence, considerable bias remains in the findings of this study.

6.7 Intercomparison of Statistical Downscaling Methods for Projection of Extreme Precipitation in Europe[6]

Summary: Present study compared eight Statistical Downscaling Methods (SDMs) used in climate change impact studies. Four techniques are based on Change Factors (CFs), three are Bias Correction (BC) techniques, and one is a perfect prognosis technique. The eight techniques are used to downscale precipitation output from 15 Regional Climate Models (RCMs) from the ENSEMBLES project for 11 catchments in Europe. The overall results shown an increase in extreme precipitation in most catchments in both winter and summer. For individual catchments, the downscaled time series tend to agree on the direction of the change but differ in the magnitude. Differences between the SDMs vary between the catchments and depend on the season analyzed. General conclusions cannot be drawn regarding the differences between CFs and BC techniques. The performance of the BC techniques during the control period also depends on the catchment, but in most of the cases, these techniques represent an improvement compared to RCM outputs. Analysis of the variance in the ensemble of RCMs and SDMs indicates that at least 30% and up to approximately half of the total variance is derived from the SDMs. Recommendations are provided for the selection of the most suitable SDMs to include in the analysis.

[6]Adopted from open access publication under the Creative Commons Attribution 3.0 License
M. A. Sunyer, Y. Hundecha, D. Lawrence, H. Madsen, P. Willems, M. Martinkova, K. Vormoor, G. Bürger, M. Hanel, J. Kriauciuniene, A. Loukas, M. Osuch, and I. Yücel (2015) Intercomparison of statistical downscaling methods for projectionof extreme precipitation in Europe, Hydrol. Earth Syst. Sci., Copernicus, 19, 1827–1847, doi:10.5194/hess-19-1827-2015, http://www.hydrol-earth-syst-sci.net/19/1827/2015/
With permission from authors. More details are available in the original publication.

6.7.1 Problem Description, Case Study and Data (Sunyer et al. 2015)

Objectives chosen are as follows:

- To assess and compare the changes in extreme precipitation obtained using a range of SDMs and RCMs in 11 European catchments
- To assess the possibility of identifying general advantages and deficiencies of different SDMs when applied to various catchments
- To assess whether there are common trends in projected changes in extreme precipitation over Europe and main source of variation in the extreme precipitation.

Figure 6.39 shows the location of the 11 catchments studied and the main properties of each catchment are summarized in Table 6.12. The observational data used are daily catchment precipitation. The climate model data used in this study is an ensemble of 15 RCMs from the ENSEMBLES project (van der Linden and Mitchell 2009). These 15 simulations are based on 11 RCMs driven by six different GCMs. The spatial resolution of all the models is 0.22° (approximately 25 km). For all the models, daily precipitation time series are available for the time period 1951–2100. In this study, they considered 1961–1990 and 2071–2100 as the control and future time periods, respectively. It must be noted that six RCMs do not have data available for the year 2100. The future period used for these models is 2071–2099; this is not expected to influence the results of this study. For each catchment, daily

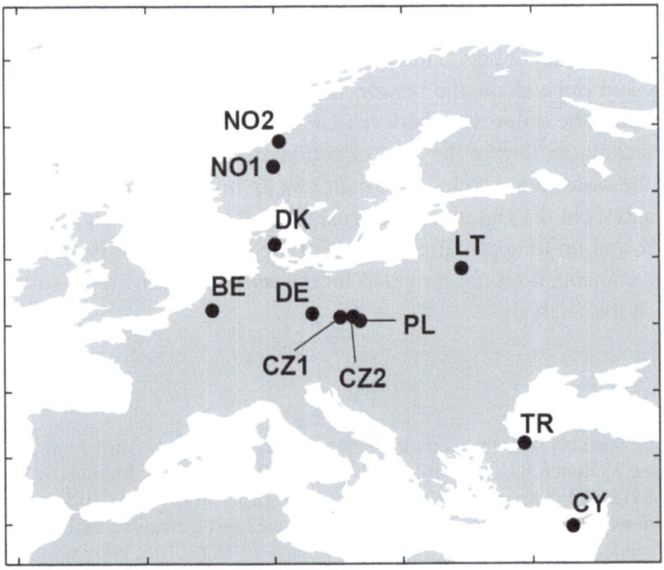

Fig. 6.39 Location of the 11 catchments studied

Table 6.12 Summary of the main characteristics of the catchments. The column with the label extremes indicates the season where most precipitation extremes occurred. The catchments are sorted from north to south, with the most northern catchment in the top row

Name	River, Country	Area (km^2)	Median altitude (m)	Data used for calculation of catchment precipitation	Mean annual precipitation ($mm\ yr^{-1}$)	Extremes	Observation period
NO2	Nordelva, Norway	207	349	1 × 1 km grid (Tveito et al. 2005)	2437	Winter	1957–2010
NO1	Atna, Norway	463	1204	1 × 1 km grid (Tveito et al. 2005)	852	Summer	1957–2010
DK	Aarhus A, Denmark	119	65	10 × 10 km grid (DMI, 2012)	868	Summer	1989–2010
LT	Merkys, Lithuania	4416	109	1 station	658	Summer	1961–1990
BE	Grote Nete, Belgium	383	32	6 stations	828	Summer	1986–2003
DE	Mulde, Germany	6171	414	43 stations	937	Summer	1951–2003
CZ2	Upper Metuje, Czech Republic	67	588	1 × 1 km grid (Sercl 2008)	788	Summer	1980–2007
CZ1	Jizera, Czech Republic	2180	365	10 stations	860	Summer	1951–2003
PL	Nysa Klodzka, Poland	1083	316	2 stations	589	Summer	1965–2000
TR	Gocbeylidere, Turkey	609	153	1 station	850	Autumn	1960–1990
CY	Yermasoyia, Cyprus	157	575	2 stations	640	Winter	1986–1997

precipitation has been extracted from the 15 RCMs for both the periods using the nearest neighbor interpolation centroid.

6.7.2 Results and Discussion

Eight SDMs are used to obtain downscaled RCM projections at the catchment scale. These techniques are based on the idea to define a relationship between the large-scale variables (RCM outputs) and local-scale variables (catchment precipitation). Eight statistical downscaling techniques employed are: Bias correction of mean, Bias correction of mean and variance, Bias correction by quantile mapping, Expanded downscaling (XDS), Change factor of mean, Change factor of mean and variance, Change factor quantile mapping, and Change factor quantile perturbation.

The outputs from all the SDMs are analyzed using an Extreme Precipitation Index (EPI). This is defined as the average change in extreme precipitation higher than a defined return period. In this study, the return period is set equal to 1 and 5 years. EPI is estimated separately for each SDM, RCM, catchment, threshold return period, season, and temporal aggregation. Four seasons are considered: winter (December to February), spring (March to May), summer (June to August), and autumn (September to November). Additionally, the index is estimated considering the whole year, i.e., without dividing into seasons. The temporal aggregations considered are 1, 2, 5, 10, and 30 days. These are estimated using a moving average from the daily time series. The variability in the EPI values was found when comparing the downscaled time series for control and future arises mainly from three sources: GCMs, RCMs, and SDMs. A variance decomposition approach is used to address the influence of each of these sources on the total variance for each catchment, return level, season, and temporal aggregation. The approach described in Déqué et al. (2007, 2012) is employed here.

The total variance of EPI, V, can be split into the different contributions as $V = R + G + S + RG + RS + GS + RGS$, where R, G, and S are the individual parts of the variance explained by the RCMs, GCMs, and SDMs, respectively; RG, RS and GS are the variance due to the interaction of RCM–GCM, RCM–SDM, and GCM–SDM, respectively, and RGS is the variance due to the interaction of all three sources.

Extreme Precipitation Index and Variance Decomposition from all Catchments to Three Selected Catchments: Fig. 6.40 summarizes the results of all the SDMs and RCMs for all the catchments for winter and summer for a temporal aggregation of first day. Additionally, it compares the results of the SDMs with the changes between the control and future periods projected by the RCMs. Overall results show that, in general, the SDMs do not modify the change projected by the uncorrected RCMs significantly. Nonetheless, in some cases, the use of few downscaling techniques might modify the magnitude of the change projected by the uncorrected RCMs. Figure 6.40 does not differentiate between the variability due to the use of different SDMs and different RCM–GCM simulations. The variance decomposition approach is used to assess variance of each of the sources

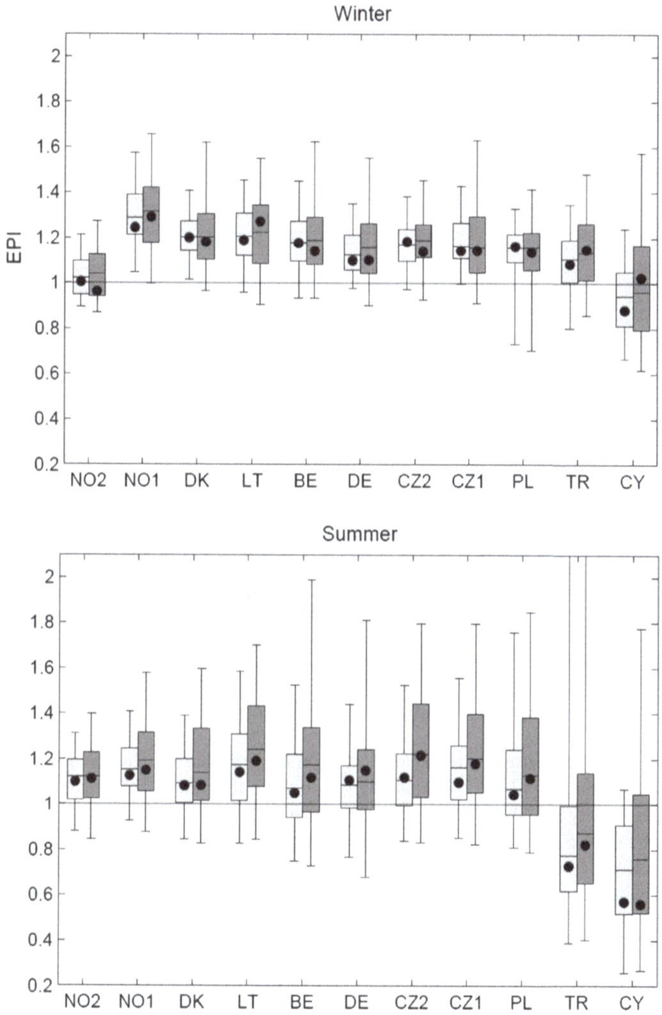

Fig. 6.40 EPI estimated from the comparison of the downscaled time series for control and future periods for 1-year (light grey boxes) and 5-year levels (dark grey boxes). The boxes indicate the 25, 50, and 75th percentiles and the whiskers the 5 and 95th percentiles. The circles show the median of all the values of EPI estimated from the comparison of the RCM outputs for the control and future periods. All the results represent a temporal aggregation of 1 day

individually. Figure 6.41 shows the total variance decomposed arising from the GCMs, RCMs, SDMs, and the interaction terms for all catchments for the 1- and 5-year levels and temporal aggregation of 1 day. The variance for the 5-year level is higher for all catchments and seasons than the variance for the 1-year level. In summer, the variance tends to increase from North to South for the 5-year level, and to some extent also for the 1-year level. This trend is not observed in winter.

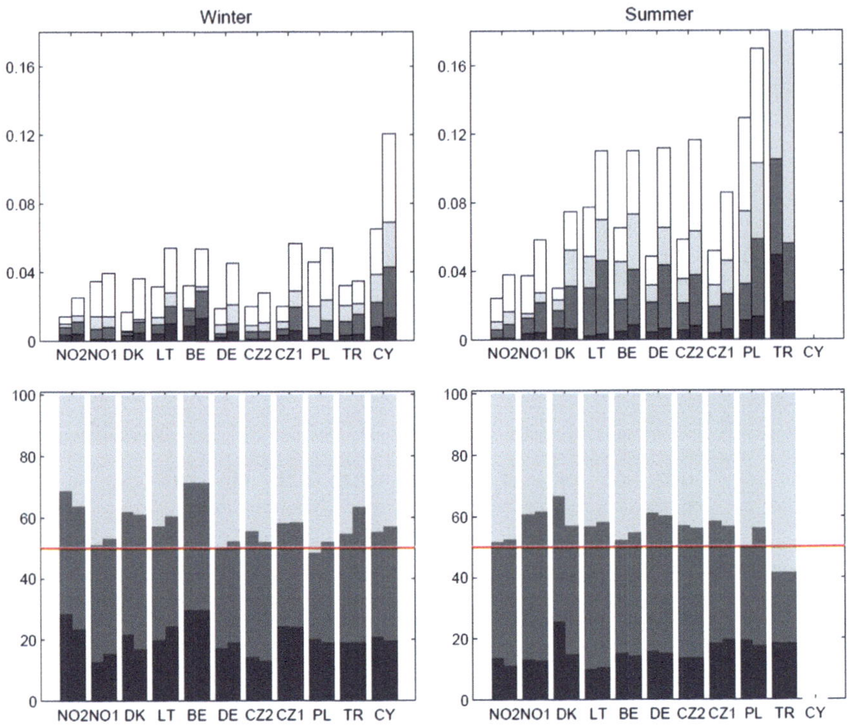

Fig. 6.41 In the top row, total variance decomposed in variance from GCMs, RCMs, SDMs, and all the interaction terms (darkest to lighter grey colours). In the bottom row, percentage of the total variance explained by GCMs, RCMs, and SDMs (darkest to lighter grey colours). All the results are shown for 1- and 5-year levels in the left and right column of each catchment, respectively. All the results are for a temporal aggregation of 1 day. In summary, the expected changes in extreme precipitation when considering all the RCMs and SDMs are analyzed

The larger variance in the Southern catchments for the 5-year level may be partially caused due to larger sampling variance (smaller number of extreme events).

Figure 6.41 shows that in most cases the variance due to the RCM–GCM simulations is larger than the variance due to SDMs. However, the interaction term is in both seasons and in most catchments similar or larger than the individual sources of variance. In all cases, the percentage of the variance obtained by the RCMs is larger than the percentage obtained by the GCMs. For both return levels, in winter the average percentage explained by the GCMs is approximately 20%, while in summer it is approximately 15%. The smaller percentage for the GCMs in the summer is due to the larger relative influence of both the RCMs and SDMs. This is likely due to the fact that in Europe, extreme precipitation from convective storms occurs more frequently during summer (e.g., Lenderink 2010; Hofstra et al. 2009), and this has a larger influence on the outputs from the RCMs and SDMs due to their higher spatial resolution. Several studies have shown that the errors of the RCMs are even larger for daily extreme precipitation in summer over Europe (e.g.,

Frei et al. 2006; Fowler and Ekström 2009). The results of the variance decomposition obtained for aggregation levels larger than 1 day (not shown) obtains a smaller total variance. For these temporal aggregations, the main source of variation is the RCM–GCMs, although the percentage explained by SDMs is slightly larger than for the 1-day aggregation. The decrease in total variance and in the percentage explained by RCM–GCMs mainly reflects that the model outputs are more similar for larger temporal aggregations. The results from the variance decomposition highlight the need for considering both ranges of SDMs and an ensemble of RCMs driven by different GCMs for assessing the uncertainty in the projection of changes in extreme precipitation.

Other Aspects: Authors analyzed expected changes in extreme precipitation, comparison of observations and bias-corrected RCMs for the control period and extreme precipitation for all the catchments, etc.

6.7.3 Summary and Conclusions

The outputs from all the statistical downscaling techniques are analyzed using an extreme precipitation index. The following main conclusions are emanated from the study:

- Extreme precipitation is expected to increase in most catchments in both winter and summer. A decrease in extreme precipitation is only expected for both winter and summer in CY and for summer in TR.
- In most catchments, larger changes are expected in winter than in summer.
- In most catchments for both winter and summer, the RCM–GCM projections are the main source of variability in the results compared to the differences between SDMs, although variability due to the SDMs explains at least 30% of the total variance in all cases.
- In all cases, the RCMs represent a larger percentage of the total variability than the GCMs, especially in summer. For this season, the total variance tends to be higher for most of the Southern catchments.
- The eight statistical downscaling techniques agree on the direction of the change but not the magnitude.
- There is a large variability in the changes estimated from different statistical downscaling techniques and RCMs.

It is not possible to draw general conclusions regarding differences between the downscaling techniques, as the differences depend on the physical geographical characteristics of the catchment and the season analyzed. However, they recommend the use of a set of statistical downscaling techniques as well as an ensemble of climate model projections. According to them, the selection of statistical downscaling techniques should include: techniques that are able to project changes in extreme precipitation, if they are expected to be different from other precipitation properties; techniques based on different underlying assumptions, for example, BC and CF techniques; and techniques that use different outputs from the RCMs as, for example, XDS, CF, or BC techniques include mean and variance of precipitation, and range of quantiles.

6.8 Future Changes in Mekong River Hydrology: Impact of Climate Change and Reservoir Operation on Discharge[7]

Summary: The transboundary Mekong River is facing two ongoing changes that are expected to significantly impact its hydrology and the characteristics of its exceptional flood pulse. The rapid economic development of the riparian countries has led to massive plans for hydropower construction, and projected climate change is expected to alter the monsoon patterns and increase temperature in the basin. The aim of the study is to assess the cumulative impact of these factors on the hydrology of the Mekong River within next 20–30 years. They downscaled the output of five General Circulation Models (GCMs) that were found to perform well in the Mekong region. For the simulation of reservoir operation, they used an optimization approach to estimate the operation of multiple reservoirs, including both existing and planned hydropower reservoirs. For the hydrological assessment, they used a distributed hydrological model, VMod, with a grid resolution of 5 km × 5 km. They found a high variation in the discharge, results depending on which of the GCMs is used as input. The simulated change in discharge at Kratie (Cambodia) between the baseline (1982–1992) and projected time period (2032–2042) ranges from −11 to +15% for the wet season and −10 to +13% for the dry season. Their analysis also shows that the changes in discharge due to planned reservoir operations are clearly larger than those simulated due to climate change: 25–160% higher dry season flows and 5–24% lower flood peaks at Kratie. The projected cumulative impacts follow rather closely the reservoir operation impacts, with an envelope around them induced by the different GCMs. Their results indicate that within the coming 20–30 years, the operation of planned hydropower reservoirs are likely to have a larger impact on the Mekong hydrograph than the impacts of climate change, particularly during the dry season. On the other hand, climate change will increase the uncertainty of the estimated reservoir operation impacts. Their results indicate that even the direction of the flow-related changes induced by climate change is partly unclear. Consequently, both dam planners and dam operators should pay closer attention to the cumulative impacts of climate change, and reservoir operation on aquatic ecosystems, including the multibillion-dollar Mekong fisheries.

[7]Adopted from open access publication under the Creative Commons Attribution 3.0 License
H. Lauri, H. de Moel, P.J. Ward, T.A. Rasanen, M. Keskinen, and M. Kummu (2012) Future changes in Mekong River hydrology: impact of climate change and reservoir operation on discharge, Hydrol. Earth Syst. Sci., Copernicus, 16, 4603–4619, doi:10.5194/hess-16-4603-2012, http://www.hydrol-earth-syst-sci.net/16/4603/2012/
With permission from authors. More details are available in the original publication.

6.8.1 Problem Description, Case Study and Data (Lauri et al. 2012)

Objectives chosen are as follows as follows:

- To assess in detail the individual and cumulative impacts of climate change (using multiple GCMs) and reservoir operation on the hydrology of the Mekong River

The Mekong River extends from the Tibetan Plateau in China to the Mekong Delta in Vietnam. The river basin is located between latitudes 8 and 34° N, containing uplands with mountains over 5000 m and alpine climate in the Northern part of the basin, and large tropical floodplains in the southern part of the basin. The Mekong river basin covers an area of 795 000 km^2, and has an average outflow of 15 000 $m^3 \cdot s^{-1}$ (475 $km^3 \cdot yr^{-1}$) (Mekong River Commission 2005). The basin is usually divided geographically into the upper and lower parts, with the division point at Chiang Saen, Thailand, which is the closest discharge measurement station at the border with China (Fig. 6.42). The upper basin, from the headwaters up to approximately Chiang Saen, is steep, and falls from elevations above 4500 m to about 500 m over a distance of 2000 km, with an average slope of 2 m km^{-1}. In the lower basin, from Chiang Saen to Kratie, the river has a moderately steep slope, with an elevation drop from 500 m to a few tens of meters over a course of 2000 km, or a slope of about 0.25 m km^{-1} on average. Downstream from Kratie, on the Mekong

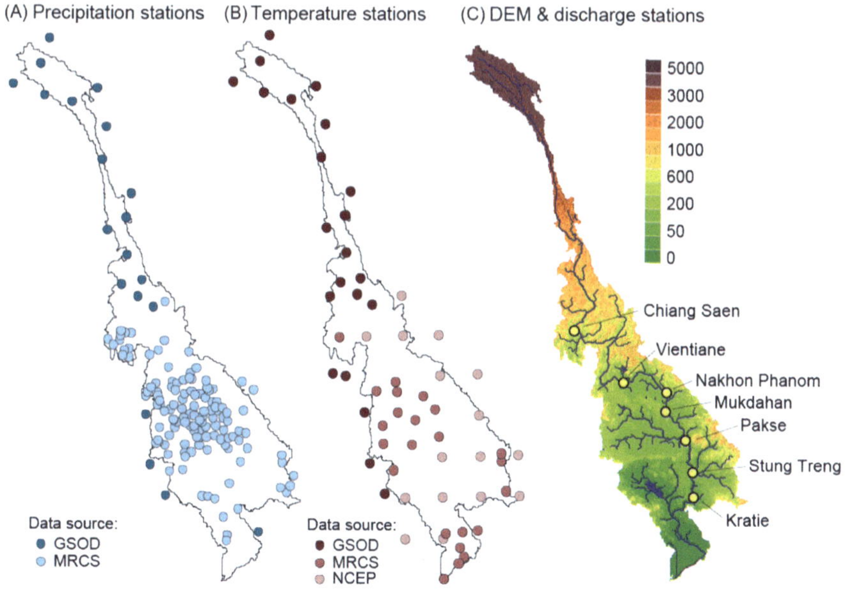

Fig. 6.42 Location of the Hydrometeorological stations used in the study. **A** precipitation stations; **B** temperature stations; **C** main river discharge gauging stations over the DEM (digital elevation model). GSOD stands for Global Surface Summary of Day data (NCDC, 2010); MRCS stands for Mekong River Commission Hydrometeorological database (Mekong River Commission, 2011); and NCEP for NCEP-DOE Reanalysis 2 data (NOAA, 2011)

floodplains and delta, the river bed is more or less flat, reaching the South China Sea after a distance of 500 km with a fall in elevation of 15 m, giving this section of the river an average slope of 0.03 m km^{-1} (Mekong River commission 2005). The lower part of the basin belongs mostly to tropical savannah and monsoon climate zones. The year is divided into dry and wet seasons. The wet season lasts approximately from early May to October and the dry season from November to April. The wet season climate is dominated by the summer monsoon, arriving partly from the Southwest and partly from the Southeast. In addition to the monsoon, the climate is affected by tropical cyclones coming from the East. These cyclones contribute to precipitation mainly during August, September, and October months of the year (Mekong River Commission 2005). The uppermost part of the basin is located in the Tibetan plateau, where the precipitation distribution is similar to that in the lower part of the basin, with most of the precipitation occurring during summer. The form of precipitation in winter is mainly snowfall due to lower temperatures caused by high elevation. In the upstream basin areas with highest altitudes, there are also several glaciers with a combined surface area of ca. 320 km^2 (Armstrong et al. 2005). Due to the monsoonal climate and the steepness of the riverbed in the upper and lower basins, the hydrograph of the Mekong River is single-peaked, with large differences between high and low flow values. At Stung Treng, where the River enters the Cambodian plains from Lao PDR, the average annual flow is about 13 000 $m^3\ s^{-1}$, while the average annual maximum is 51 500 $m^3\ s^{-1}$ and the minimum is 1700 $m^3\ s^{-1}$ (computed from the years 1970–2002 observed data). Simulated annual runoff in the catchment varies from less than 100 mm yr^{-1} in the Eastern part of Thailand to over 2000 mm yr^{-1} in the Central part of Laos (computed from years 1982–1992 simulated data). Average annual runoff for the whole basin is about 600 mm yr^{-1} (Mekong River Commission 2005). Data requirement consists of Meteorological, Discharge, Reservoirs related, and Climate change related. They modeled the hydrology of the Mekong Basin using VMod, which is a distributed hydrological model-based on a gridded representation of the modeled watershed. A detailed description of the model computation techniques and model equations can be found in the VMod model manual (Koponen et al. 2010). Delta technique is used for downscaling precipitation and temperature data.

6.8.2 Results and Discussion

Reservoir Operation Rules: Linear programming (LP) (e.g., Dantzig and Thapa 1997) was used to estimate monthly outflows for each reservoir separately. The aim of the LP objective function used is to maximise annual outflow from a reservoir through hydropower turbines, using the reservoir active storage, estimated monthly inflows, minimum outflow, and optimal outflow from the reservoir as parameters. An additional term was included into the objective function to force the filling of the reservoir during the wet season and emptying of the reservoir during the dry season. Constraints were also required to keep the reservoir outflow constant during the dry season. The monthly inflows for each reservoir, which are required in the optimization, were estimated from computed 24-year time series (April 1981–April 2005). The resulting operation rules aim to overestimate the reservoir usage and find an upper limit to the possible impact of reservoirs on Mekong discharges. Normal reservoir operation rules are often more careful and aim to make reservoir is fill to full capacity each year. The optimization of all upstream reservoirs were performed so that before optimising a given reservoir. The inflows to the reservoir to be optimized were then computed with the upstream reservoirs being active. They first performed the reservoir optimization for the baseline conditions. To ensure correct operation of the reservoirs under the climate change scenarios, the reservoir operation was optimized separately for each climate change scenario setup (i.e., model run). Authors selected 1982–1992 as the baseline period and 2032–2042 as the future time period so that both periods were of equal duration. The hydrological model runs, with their associated GCM, emission scenario, and reservoir configuration, are listed in Table 6.13.

Impact of Climate Change on Temperature, Precipitation, and Runoff: The temperature, precipitation, and runoff of different model runs for the years 2032–2042 were compared to the baseline data (1982–1992). Daily average temperature for the whole catchment, computed as the mean of minimum and maximum temperature, increased by 0.8–1.4° C in the model runs using the A1b emission scenario, and 0.6–1.3° C in the runs using the B1 scenario. The spatial distribution of annual average temperature increase is similar for all runs using the A1b emission scenario. The increases are greater in the Southern and Northern parts of the basin when compared to the middle part, and the largest temperature increases are found in the southeastern part and in the narrow mid-north part of the catchment. For the runs using the B1 emission scenario, the temperature changes show a similar pattern compared to the runs using the A1b scenario, but the magnitude of change is smaller in the former. Compared to temperature change, the spatial distribution of precipitation change differs much more between the model runs. The modeled

Table 6.13 Hydrological model runs and their settings used in this study. BL stands for baseline simulation, +rv stands for reservoirs (i.e., reservoir operation included in the simulations)

Group	Model run	GCM	Emission scenario	Reservoirs included
Baseline	BL	None	None	No
	BL + rv	None	None	Yes
A1b	ccA (+rv)	CCCMA_CGCM3.1	A1b	No (yes)
	cnA (+rv)	CNRM_CM3	A1b	No (yes)
	giA (+rv)	GISS_AOM	A1b	No (yes)
	mpA (+rv)	MPI_ECHAM5	A1b	No (yes)
	ncA (+rv)	NCAR_CCSM3	A1b	No (yes)
B1	ccB (+rv)	CCCMA_CGCM3.1	B1	No (yes)
	cnB (+rv)	CNRM_CM3	B1	No (yes)
	giB (+rv)	GISS_AOM	B1	No (yes)
	mpB (+rv)	MPI_ECHAM5	B1	No (yes)
	ncB (+rv)	NCAR_CCSM3	B1	No (yes)

runoff for the whole catchment increases in six model runs and decreases in four runs. The spatial pattern of runoff change in the lower part of the catchment is similar for all hydrological model runs, but varies in the middle and upper parts of the catchment. In the lower part, there is a decrease in runoff in the West, and varying amounts of increase in runoff in the East. Under emission scenario A1b, in the middle part of the catchment three model runs show increasing runoff while two model runs show decreasing runoff. Also, in the uppermost part of the catchment the model runs disagrees on the direction of change.

Other Perspectives: Authors also studied the impact of climate change and of reservoir operations on main river discharge. The cumulative impact of climate change and reservoir operations on main river discharge, Interannual variation of the cumulative impacts of climate change and reservoir operation, impact of climate change, and reservoir operation on selected flood pulse parameters are also studied.

6.8.3 Summary and Conclusions

The study assessed the impact of climate change and reservoir operation on the hydrology of the Mekong River within the next 20–30 year. The following main conclusions are emanated from the study:

- Climate change is likely to increase the precipitation and average temperature in the basin. The range between GCMs is, however, relatively large for both variables.
- Large variation in discharge results between the hydrological model using different GCMs is observed under the two emission scenarios, A1b and B1.
- Direction of climate change impacts on Mekong discharges remains uncertain.

Authors suggested the use of multiple GCMs for estimating the possible climate change impacts on Mekong discharge.

6.9 Regional Rainfall Forecasting Using Large-Scale Climate Teleconnections and Artificial Neural Networks[8]

Summary: Artificial Neural Networks technique for regional rainfall forecasting for Orissa state, India on monthly and seasonal time scales is proposed. The possible relation between regional rainfall and the large-scale climate indices like

[8]Adopted from
Nagesh Kumar D, Janga Reddy M, Maity R (2007) Regional Rainfall Forecasting using Large-Scale Climate Teleconnections and Artificial Intelligence Techniques, Journal of Intelligent Systems, De Grutyer, 16, 307–322, doi10.1515/JISYS.2007.16.4.307, https://www.degruyter.com/view/j/jisys.2007.16.4/jisys.2007.16.4.307/jisys.2007.16.4.307.xml
With permission from the publishers. More details are available in the original publication.

El-Niño Southern Oscillation (ENSO), Equitorial Indian Ocean Oscillation (EQUINOO), and a local climate index of Ocean–Land Temperature Contrast (OLTC) are used to forecast monsoon rainfall. The results showed reasonably good accuracy for monthly and seasonal rainfall forecasting and emphasises the value of using large-scale climate teleconnections for regional rainfall forecasting.

6.9.1 Problem Description, Case Study and Data (Nagesh Kumar et al. 2007)

Objectives chosen are as follows:

- Explore Artificial Neural Networks (ANNs) for regional rainfall forecasting for Orissa state, India on monthly and seasonal time scales using ENSO, EQUINOO, and OLTC indices.

The study uses data of various parameters, viz. monthly NINO3.4, Sea Surface Temperature Anomaly (SSTA), EQWIN index, OLTC index and monthly rainfall anomaly over Orissa subdivision. Monthly Nino3.4 SSTA and SSTA for region (10° S–10° N, 60° E–85° E) (1958–1990) data have been collected from the web site of Climate Analysis Center, National Centers for Environmental Prediction (NCEP 2017); Wind data (1958–1990) have also been collected from NCEP (2017) to obtain EQWIN index for EQUINOO. Monthly rainfall and temperature data (1901–1990) have been collected from the web site of Indian Institute of Tropical Meteorology (2017), Maity and Nagesh Kumar (2006), Sahai et al. (2000, 2003).

6.9.2 Results and Discussion

Analysis of Data and Modeling Aspects: Genetic Optimizer (Nagesh Kumar et al. 2007) is used to optimize the ANNs architecture. In brief, genetic optimizer consists of following processes: Initialize parameters, generate initial population, train the network and evaluate fitness, propagation of networks, check for termination criteria, output is the best solution obtained so far during the evolution.

The input and output patterns are scaled to 0–1 range through mapping with the help of minimum and maximum values of the patterns, whereby it can be modeled using Fermi function. Termination criteria is used to stop the learning process, when either the epoch counter reaches 1000 or the maximum value of squared deviation of neural networks output from the observed value among all training patterns is less than 0.001. To avoid over fitting, the model is first trained with Back Propagation ANN (BPANN) using training data set, then the model is cross validated, by testing its performance with a different validation data set. In this process the model makes better generalisation of the new data set. To identify the months that can be used as input to the ANNs model, cross correlation analysis is carried out for predictor variables. The inputs driven to the ANNs are monthly Nino 3.4

Table 6.14 Climate indices of predictor variables considered for rainfall prediction of June, July, August, September, and for summer monsoon (JJAS)

Rainfall	Predictor variables considered
June	Nino 3.4 (February, March, April), EQWIN (May, June), OLTC (May, June)
July	Nino 3.4 (February, March, April), EQWIN (June, July), OLTC (June, July)
August	Nino 3.4 (February, March, April), EQWIN (July, August), OLTC (July, August)
September	Nino 3.4 (February, March, April), EQWIN (August, September), OLTC (August, September)
JJAS	Nino 3.4 (February, March, April), EQWIN (May, June), OLTC (May, June)

SSTA, EQWIN, and the OLTC indices. The output variable is rainfall. Table 6.14 gives a complete list of inputs that have been used for forecasting rainfall for June, July, August, September (JJAS), and for the monsoon season.

The predictor variable data set is available for 33 years (1958–1990). To train the ANNs, 23 years data set is selected and then to test the performance of the trained model, the remaining 10 years data set is used. The models are trained with different combinations of network architectures by using genetic optimizer. The parameters considered for genetic optimizer are: population size (number of networks created per generation) = 50; maximum number of generations = 100; probability of crossover (probability for a child network to be crossed over with another child network), $P_c = 0.6$, and probability of mutation (probability for a network to be modified during rollover to a new generation), $P_m = 0.04$. For BPANNs, the parameters adopted are, learning parameter = 0.2; momentum parameter = 0.1; and maximum number of epochs = 1000. The best suited network architectures obtained for different monthly and seasonal models are given in Table 6.15. By using the Genetic Optimiser-ANNs methodology, the regional rainfall for Orissa state has been forecasted. For the month of June ANN model, the CC values obtained are 0.9941 and 0.8349 for training and testing periods, respectively. Figure 6.43a shows the comparison of ANN forecasted rainfall with observed rainfall for June. It can be seen that except for one year (1986), the model results are within reasonable accuracy and they well predict the low rainfall (1981, 1982, 1983 and 1987) and high rainfall (1984, 1989) values during the testing period. For the month of July, the CC values obtained are 0.9994 and 0.8002 for

Table 6.15 Architecture of ANNs selected and the performance of the models for rainfall forecasting in the monsoon season for Orissa sub-division

Month/season	Network architecture	Correlation coefficient (CC)	
		Training	Testing
June	7, 7, 7, 1	0.9941	0.8349
July	7, 8, 9, 1	0.9994	0.8002
August	7, 10, 1	0.9969	0.8102
September	7, 8, 1	0.9998	0.5775
JJAS	7, 8, 1	0.9975	0.8951

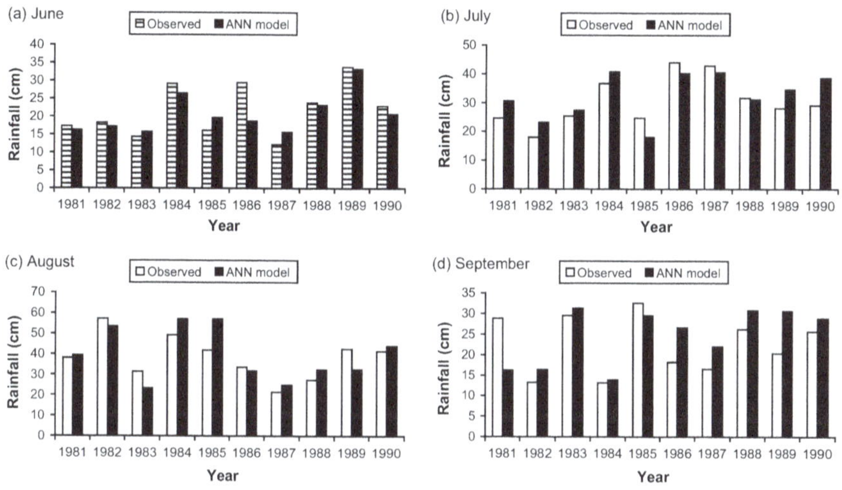

Fig. 6.43 Comparison of rainfall predicted for (1981–1990) using ANN model and observed rainfall for **a** June; **b** July; **c** August; **d** September

training and testing periods, respectively. Figure 6.43b compares ANN model forecasted rainfall with observed rainfall for July during the testing period.

It is observed that, even though there are a few deviations from observed rainfall (1981, 1982, 1985, 1989, and 1990), the model shows reasonable accuracy. For the month of August, the CC values obtained are 0.9969 and 0.8102 for training and testing periods, respectively. Figure 6.43c compares ANN model-forecasted rainfall with observed rainfall for August during the testing period. The model results show that for two years, ANN over forecasted (1984, 1985), and for other two years it under forecasted (1983, 1989), and for the remaining years, the forecasts are in reasonable agreement. For September, the CC values obtained are 0.9998 and 0.5775 for training and testing periods, respectively. Figure 6.43d shows the comparison of ANN-forecasted rainfall with observed rainfall for September during the testing period. It can be observed that the model has under forecasted for one year (1981), over forecasted for four years (1986, 1987, 1988, and 1989) and for the remaining years the results are in reasonable accuracy. Similarly, the ANN model is trained for summer monsoon season, June to September rainfall forecasting. Figure 6.44 compares observed rainfall with ANN model forecasted rainfall for monsoon season (JJAS) during testing period. The CC values obtained for JJAS seasonal model are 0.9975 and 0.8951 for training and testing periods, respectively. The model results are within reasonable accuracy with observed rainfall for most of the seasonal rainfall values, except for few minor deviations for one year (1982). It can be noticed that, when monthly rainfall forecasting is compared to seasonal rainfall forecasting, seasonal rainfall model is performing better to monthly models. This may be due to the dynamic nature of climate variables which lead to

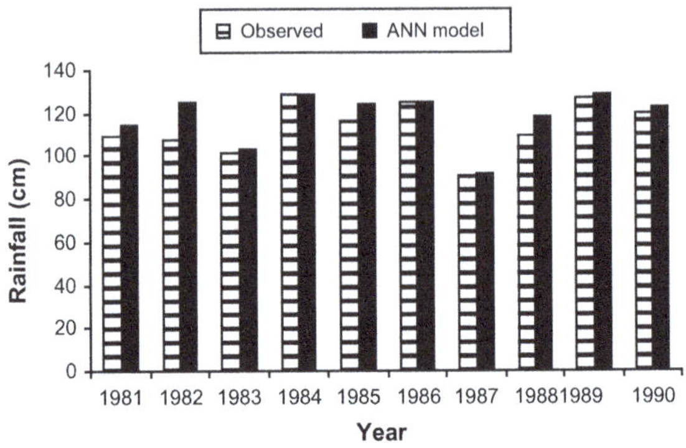

Fig. 6.44 Comparison of rainfall predicted for (1981–1990) using ANN model and observed rainfall for summer monsoon (JJAS)

more uncertainty in monthly time scale than seasonal scale, thus larger variability can be observed in intra-seasonal rainfall prediction.

6.9.3 Summary and Conclusions

The climate indices of ENSO (Nino 3.4 SSTA), EQUINOO (EQWIN), and Ocean–Land Temperature Contrast (OLTC) have been used as predictor variables to predict the monthly and seasonal rainfall. The following main conclusions are emanated from the study:

- The obtained results are encouraging and show improvement in rainfall forecasting.
- Incorporation of global climate information in rainfall prediction is proved useful.
- ANNs is found to be a suitable technique for this purpose.

In this chapter, various case studies related to ranking of GCMs, downscaling techniques, and adaptation studies are presented.

References

Aich V, Liersch S, Vetter T, Huang S, Tecklenburg J, Hoffmann P, Koch H, Fournet S, Krysanova V, Müller EN, Hattermann FF (2014) Comparing impacts of climate change on streamflow in four large African river basins. Hydrol Earth Syst Sci 18: 1305–1321

Akshara G (2015) downscaling of climate variables using multiple linear regression-a case study lower godavari basin, India, M.E. dissertation, BITS Pilani

Akshara G, Raju KS, Singh AP, Vasan A (2017) Application of multiple linear regression as downscaling methodology for lower godavari basin.In: Proceedings of International conference on water, environment, energy and society (ICWEES-2016), Springer

Anandhi, A (2007) Impact assessment of climate change on hydrometeorology of Indian River Basin for IPCC SRES scenaries. Ph.D thesis, Indian Institute of Science, Bangalore.

Anandhi A, Srinivas VV, Nanjundiah RS, Nagesh Kumar D (2008) downscaling precipitation to river basin in India for IPCC SRES scenarios using support vector machine. Int J Climatol Springer 28(3):401–420

Anandhi A, Srinivas VV, Nagesh Kumar D, Nanjundiah RS (2012) Daily relative humidity projections in an Indian river basin for IPCC SRES scenarios. Theor Appl Climatol Springer 108(1–2):85–104

Anandhi A, Srinivas VV, Nagesh Kumar (2013) Impact of climate change on hydrometeorological variables in a river basin in India for IPCC SRES scenarios. In: Rao YS, Zhang TC, Ojha CSP, Gurjar BR, Tyagi RD, Kao CM (eds) Chapter 12 in climate change modeling, mitigation, and adaptation. American Society of Civil Engineers, pp 327–356

Armstrong R, Raup B, Khalsa SJS, Barry R, Kargel J, Helm C, Kieffer H (2005) GLIMS glacier database. National Snow and Ice Data Center, Digital Media, Boulder, Colorado USA

Bari MA, Smettem KRJ (2003) Development of a salt and water balance model for a large partially cleared catchment. Aust J Water Resour 7:83–99

Bari MA, Amirthanathan GE, Timbal B (2010) Climate change and long term water availability in south-western Australia—an experimental projection. In: Practical responses to climate change national conference 2010, Hilton on the Park, Melbourne, Australia, 29 September–1 October 2010

Bartholomé E, Belward AS (2005) GLC2000: a new approach to global land cover mapping from earth observation data. Int J Remote Sens 29:1959–1977

Burn DH (1989) Cluster analysis as applied to regional flood frequency. J Water Resour Plan Manag ASCE 115:567–582

Carter TR, Jones PD, Hulme M, New M (2004) A comprehensive set of high521 resolution grids of monthly climate for Europe and the globe: the observed record (1901– 2000) and 16 scenarios (2001–2100). Tyndall Working Paper 55, Tyndall Centre, University of East Anglia, Norwich, United Kingdom

Changnon SA Jr (1987) Climate fluctuations and record-high levels of lake Michigan. Bull Am Meteoro Soc 68(11):1394–1402

Chaturvedi RJ, Joshi J, Jayaraman M, Bala G, Ravindranath NH (2012) multi-model climate change projections for India under representative concentration pathways. Curr Sci 103:7–10

Christensen NS, Lettenmaier DP (2007) A multimodel ensemble approach to assessment of climate change impacts on the hydrology and water resources of the Colorado River Basin. Hydrol Earth Syst Sci 11:1417–1434. doi:10.5194/hess-11-1417-2007

Commonwealth Scientific and Industrial Research Organisation (CSIRO) (2009) Surface water yields in south-west Western Australia, a report to the Australian government from the CSIRO southwest Western Australia sustainable yields project. CSIRO water for a healthy country flagship, Commonwealth Scientific and Industrial Research Organisation, Australia, 171 p

Danish Meteorological Institute (DMI) (2012) Climate Grid Denmark. Dataset for use in research and education. Daily and monthly values1989–2010 10 × 10 km observed precipitation 20 × 20 km temperature, potential evaporation (Makkink), wind speed, global radiation, Technical Report 12–10. http://beta.dmi.dk/fileadmin/Rapporter/TR/tr12-10.pdf. Accessed 3 June 2014

Dantzig GB, Thapa MN (1997) Linear programming 1: Introduction. Springer

Davies DL, Bouldin DW (1979) A Cluster Separation Measure. IEEE Trans Pattern Anal Mach Intell 1(2):224–227

Department of Water (DoW) (2008) Water solutions, winter'08, Perth, Western Australia, 9 p

Déqué M, Rowell DP, Lüthi D, Giorgi F, Christensen JH, Rockel B, Jacob D, Kjellström E, Castro M, van den Hurk B (2007) An intercomparison of regional climate simulations for Europe: assessing uncertainties in model projections. Clim Change 81:53–70. doi:10.1007/s10584-006-9228-x

Déqué M, Somot S, Sanchez-Gomez E, Goodess CM, Jacob D, Lenderink G, Christensen OB (2012) The spread amongst ENSEMBLES regional scenarios: regional climate models, driving general circulation models and interannual variability. Clim Dynam 38:951–964. doi:10.1007/s00382-011-1053-x

Doty B, Kinter JI (1993) The grid analysis and display system (GrADS): a desktop tool for earth science visualization. In: Proceedings of American Geophysical Union 1993 Fall Meeting, San Fransico, CA, 6–10 December, 1993

Dutta S, Patel NK, Medhavy TT, Srivastava SK, Mishra N, Singh KRP (1998) Wheat crop classification using multidate IRS LISS-I data. J Indian Soc Remote Sens 26(1–2):7–14

Errasti I, Ezcurra A, Sáenz J, Ibarra-Berastegi G (2010) Validation of IPCC AR4 models over the Iberian peninsula. Theor Appl Climatol 103:61–79

FAO, IIASA, ISRIC, ISSC and JRC (2012) Harmonized World Soil Database v 1.2. http://webarchive.iiasa.ac.at/Research/LUC/External-World-soil-database/HTML/. Accessed 26 June 2013

Fekete BM, Vorosmarty CJ, Grabs W (1999) Global, composite runoff fields based on observed river discharge and simulated water balances, GRDC Report 22. Global Runoff Data Center, Koblenz, Germany

Fowler HJ, Ekström M (2009) Multi-model ensemble estimates of climate change impacts on UK seasonal precipitation extremes. Int J Climatol 29:385–416. doi:10.1002/joc.1827

Fowler HJ, Blenkinsop S, Tebaldi C (2007) Linking climate change modelling to impacts studies: recent advances in downscaling techniques for hydrological modelling. Int J Climatol 27 (12):1547–1578

Frei C, Schöll R, Fukutome S, Schmidli J, Vidale PL (2006) Future change of precipitation extremes in Europe: intercomparison of scenarios from regional climate models. J Geophys Res 111:D06105. doi:10.1029/2005JD005965

Gestel T, Suykens J, Baesens B, Viaene S, Vanthienen J, Dedene G, Moor B, Vandewalle J (2004) Benchmarking least squares support vector machine classifiers. Mach Learn 54(1):5–32

Godavari River basin (2017) India WRIS—water resources information system of India. http://india-wris.nrsc.gov.in/wris.html. Last Accessed 31 Jan 2017

Gosain A, Sandhya R, Debajit B (2006) Climate change impact assessment on hydrology of indian river basins, in special issue on climate change and India. Curr Sci 90(3):346–353

Haupt R, Haupt S (2004) Practical genetic algorithms. Wiley, New Jersey, p 253

Haykin S (2003) Neural networks: a comprehensive foundation. Pearson Education, Singapore

Hempel S, Frieler K,Warszawski L, Schewe J, Piontek F (2013) A trend-preserving bias correction—the ISI-MIP approach. Earth Syst Dynam 4, 219–236, doi:10.5194/esd-4–219-5-2013.

Hofstra N, Haylock M, New M, Jones PD (2009) Testing EOBS European high-resolution gridded data set of daily precipitation and surface temperature. J Geophys Res 114:D21101. doi:10.1029/2009JD011799

Indian Institute of Tropical Meteorology (2017). http://www.tropmet.res.in/. Accessed 31 Jan 2017)

Islam SA, Bari M, Anwar AHMF (2011) Assessment of hydrologic impact of climate change on Ord river catchment of Western Australia for water resources planning: a multi-model ensemble approach. In: Proceedings of the 19th international congress on modelling and simulation, Perth, Western Australia, 12–16 December 2011, pp. 3587–3593

Islam SA, Bari MA, Anwar AHMF (2014) Hydrologic impact of climate change on Murray-Hotham catchment of Western Australia: a projection of rainfall–runoff for future water resources planning. Hydrol Earth Syst Sci 18:3591–3614

Jarvis A, ReuterHI, Nelson A, Guevara E (2008) Hole-filled SRTM for the globe Version 4. http://srtm.csi.cgiar.org/ Accessed 2 Mar 2013

Johnson F, Sharma A (2009) Measurement of GCM Skill in predicting variables relevant for hydroclimatological assessments. J Clim 22:4373–4382

Johnson F, Sharma A (2012) A nesting model for bias correction of variability at multiple time scales in general circulation model precipitation simulations. Water Resour Res 48:W01504

Jones DA, Wang W, Fawcett R (2009) High-quality spatial climate data-sets for Australia. Aust Meteorol Oceanogr J 58:233–248

Karl T, Wang W, Schlesinger M, Knight R, Portman D (1990) A method of relating general circulation model simulated climate to the observed local climate part i: seasonal statistics. J Clim 3:1053–1079

Kendall MG (1951) Regression, structure and functional relationship. Part I, Biometrika 38:11–25

Koponen J, Lauri H, Veijalainen N, Sarkkula J (2010) HBV and IWRM watershed modelling user guide, MRC information and knowledge management programme, DMS—Detailed modelling support for the MRC project. http://www.eia.fi/index.php/support/download.

Köppen W (1900) Versuch einer Klassifikation der Klimate, vorzugsweise nach ihren Beziehungen zur Pflanzenwelt, Geogr. Zeitschrift, 6, 593–611

Krysanova V, Müller-Wohlfeil D-I, Becker A (1998) Development and test of a spatially distributed hydrological/water quality model for mesoscale watersheds. Ecol Model 106:261–289

Krysanova V, Hattermann F, Wechsung F (2005) Development of the ecohydrological model SWIM for regional impact studies and vulnerability assessment. Hydrol Process 19:763–783

Lauri H, de Moel H, Ward PJ, Rasanen TA, Keskinen M, Kummu M (2012) Future changes in Mekong river hydrology: impact of climate change and reservoir operation on discharge. Hydrol Earth Syst Sci 16:4603–4619

Lenderink G (2010) Exploring metrics of extreme daily precipitation in a large ensemble of regional climate model simulations. Clim Res 44:151–166. doi:10.3354/cr00946

Lillesand TM, Kiefer RW, Chipman JW (2004) Remote sensing and image interpretation. Wiley India (P). Ltd, New Delhi

MacQueen J (ed) (1967) Some methods for classification and analysis of multivariate observation. In: Proceedings of the fifth berkeley symposium on mathematical statistics and probability. University of California Press, Berkeley, pp 281–297

Maity R, Nagesh Kumar D (2006) Bayesian dynamic modeling for monthly Indian summer monsoon rainfall using El Niño southern oscillation (ENSO) and equatorial Indian ocean oscillation (EQUINOO). J Geophys Res 111:D07104

Mann HB, Whitney DR (1947) On a test of whether one of two random variables is stochastically larger than the other. Ann Math Stat 18:50–60

Mayer XM, Ruprecht JK, Bari MA (2005) Stream salinity status and trends in south west Western Australia, salinity and land use impacts series, vol 38, Department of Environment, Perth, Western Australia, 188

Mehrotra R, Sharma A (2010) Development and application of a multisite rainfall stochastic downscaling framework for climate change impact assessment. Water Resour Res 46 (W07526). doi:10.1029/2009WR008423

Mehrotra R, Sharma A, Nagesh Kumar D, Reshmidevi TV (2013) Assessing future rainfall projections using multiple GCMsand A multi-site stochastic downscaling model. J Hydrol 488:84–100

Mekong River Commission (2005) Overview of the hydrology of the Mekong Basin, Mekong River Commission, Vientiane, 82

Mekong River Commission (2011) Hydrometeorological database of the Mekong River Commission, Mekong River Commission (MRC). Vientiane, Lao PDR

Mitchell TD, Jones PD (2005) An improved method of constructing a database of monthly climate observations and associated high-resolution grids. Int J Climatol 25:693–712

Morais DC, Almeida AT (2012) Group decision making on water resources based on analysis of individual rankings. Omega 40:42–52

Moriasi DN, Arnold JG, VanLiew MW, Bingner RL, Harmel RD, Veith TL (2007) Model evaluation guidelines for systematic quantification of accuracy in watershed simulations. Trans ASABE 50(3):885–900

Mujumdar PP, Nagesh Kumar D (2012) Floods in a changing climate: hydrologic modeling, international hydrology series. Cambridge University Press, UK

Nagesh Kumar D, Janga Reddy M, Maity R (2007) Regional rainfall forecasting using large scale climate teleconnections and artificial intelligence techniques. J Intell Syst 16:307–322

Nash JE, Sutcliffe JV (1970) River flow forecasting through conceptual models part I—a discussion of principles. J Hydrol 10:282–290
NCDC (2010) Global surface summary of the day (GSOD), US National Climatic Data Center (NCDC)
NCEP (2017). www.cpc.ncep.noaa.gov/. Accessed 31 Jan 2017
NOAA (2011) NCEP-DOE Reanalysis 2, National Oceanic and Athmospheric Admistration, Earth System Research Laboratory, Physical Science Division. http://www.esrl.noaa.gov/psd/data/gridded/data.ncep.reanalysis2.html.
Pearson K (1896) mathematical contributions to the theory of evolution. III. Regres Heredity Panmixia Philos Trans Roy Soc A 187:253–318
Pen LJ, Hutchison J (1999) Managing our rivers: a guide to the nature and management of the streams of south-west Western Australia, water and rivers commission, 382 p
Pennell C, Reichler T (2011) On the Effective Number of Climate Models. J Clim 24:2358–2367
Raju KS, Nagesh Kumar D (2014) Multicriterion analysis in engineering and management. Prentice Hall of India, New Delhi
Raju KS, Nagesh Kumar D (2016) Selection of global climate models for india using cluster analysis. J Water Clim Change 7(4):764–774. doi:10.2166/wcc.2016.112
Raju KS, Sonali P, Nagesh Kumar D (2017) Ranking of CMIP5 based global climate models for india using compromise programming. Theor Appl Climatol Springer 128(3):563–574. doi:10.1007/s00704-015-1721-6
Reichler T, Kim J (2008) How Well Do Coupled Models Simulate Today's Climate? Bull Am Meteorol Soc 89:303–311
Reshmidevi TV, Nagesh Kumar D (2014) Modelling the impact of extensive irrigation on the groundwater resources. Hydrol Process 28(3):628–639
Reshmidevi TV, Nagesh Kumar D, Mehrotra R, Sharma A (2017) Estimation of the climate change impact on a catchment water balance using an ensemble of GCMs. J Hydrol. doi:10.1016/j.jhydrol.2017.02.016
Sahai AK, Soman MK, Satyan V (2000) All India summer monsoon rainfall prediction using an artificial neural network. Clim Dyn 16:291–302
Sahai AK, Grimm AM, Satyan V, Pant GB (2003) Long-lead Prediction of Indian Summer Monsoon Rainfall from Global SST Evolution. Clim Dyn 20:855–863
Schaefli B, Gupta HV (2007) Do nash values have value? Hydrol Process 21(15):2075–2080
Šercl P (2008) Assessment of methods for area precipitation estimates, Meteorological Bulletin, vol 61
Smola A, Scholkopf B, Muller K (1998) The connection between regularization operators and support vector kernels. Neural Netw 11(4):637–649
Sousa A, García-Murillo P, Morales J, García-Barrón L (2009) Anthropogenic and natural effects on the Coastal Lagoons in the Southwest of Spain (Doñana National Park). ICES J Mar Sci 66:1508–1514
Spearman C (1904a) General intelligence' objectively determined and measured. Am J Psychol 5:201–293
Spearman C (1904b) Proof and measurement of association between two things. Am J Psychol 15:72–101
Srivastava AK, Rajeevan M, Kshirsagar SR (2009) Development of a high resolution daily gridded temperature data set (1969–2005) for the Indian region. Atmos Sci Lett 10:249–254
Stern H, Hoedt GD, Ernst J (2000) Objective classification of Australian climates. Aust Meteorol Magaz 49:87–96
Strahler A (2013) Introducing physical geography. Wiley, 664 p
Sugawara M (1979) Automatic calibration of the tank model. Hydrol Sci Bull 24:375–388
Sunyer MA, Hundecha Y, Lawrence D, Madsen H, Willems P, Martinkova M, Vormoor K, Bürger G, Hanel M, Kriauciuniene J, Loukas A, Osuch M, Yücel I (2015) Inter-comparison of statistical downscaling methods for projectionof extreme precipitation in Europe. Hydrol Earth Syst Sci 19:1827–1847

Taylor KE, Stouffer RJ, Meehl GA (2012) An overview of CMIP5 and the experiment design. Bull Am Meteor Soc 93:485–498

Timbal B, Fernandez E, Li Z (2009) Generalization of a statistical downscaling model to provide local climate change projections for Australia. Environ Model Softw 24:341–358. doi:10.1016/j.envsoft.2008.07.007

Tripathi S, Srinivas VV, Nanjundiah RS (2006) Downscaling of precipitation for climate change scenarios: a support vector machine approach. J Hydrol 330:621–640

Tveito OE, Bjørdal I, Skjelvåg AO, Aune B (2005) A GIS-based agro-ecological decision system based on gridded climatology. Meteorol Appl 12:57–68

van der Linden P, Mitchell JFB (eds) (2009) ENSEMBLES: climate change and its impacts: summary of research and results from the ENSEMBLES project. Technical report. http://ensembles-eu.metoffice.com/docs/Ensembles_final_report_Nov09.pdf. Accessed 3 June 2014. Met Office Hadley Centre, UK, 160 p

van Griensven A (2005) Sensitivity, auto-calibration, uncertainty and model evaluation in SWAT2005 (Unpublished report)

Wang K (2007) Cluster validation toolbox MATLAB central, file exchange, Math works. http://www.mathworks.com/matlabcentral/fileexchange/14620. Last Accessed 31 Jan 2017

Wang K, Wang B, Peng L (2009) CVAP: validation for cluster for analyses. Data Sci J 8 (20):88–93

Weedon GP, Gomes S, Viterbo P, Shuttleworth WJ, Blyth E, Österle H, Adam JC, Bellouin N, Boucher O, Best M (2011) Creation of the WATCH forcing data and its use to assess global and regional reference crop evaporation over land during the twentieth century. J. Hydrometeorol 12:823–848

Westerberg I, Guerrero JL, Seibert J, Beven KJ, Halldin S (2011) Stage-discharge uncertainty derived with a non-stationary rating curve in The Choluteca river. Honduras Hydrol Process 25:603–613

WFD (2011) EU WATCH—Home. http://www.eu-watch.org/templates/dispatcher.asp?page_id=25222705. Accessed 7 Feb 2013

Wilby RL, Charles SP, Zorita E, Timbal B, Whetton P, Mearns LO (2004) The guidelines for use of climate scenarios developed from statistical downscaling methods. In: Supporting material of the intergovernmental panel on climate change (IPCC), prepared on behalf of task group on data and scenario support for impacts and climate analysis (TGICA). www.ipcc-data.org/guidelines/dgm_no2_v1_09_2004.pdf. Accessed 31 0 Jan 2017

Wilcoxon F (1945) Individual comparisons by ranking methods. Biometrics 1:80–83

Winkler J, Palutikof J, Andresen J, Goodess C (1997) The simulation of daily temperature time series from GCM output part ii: sensitivity analysis of an empirical transfer function methodology. J Clim 10(10):2514–2532

Yu PS, Yang TC (2000) Using synthetic flow duration curves for rainfall-runoff model calibration at ungauged Sites. Hydrol Process 14:117–133

Appendix A: Representative Data Sources

A1. India Meteorological Department (IMD) (http://www.imd.gov.in/pages/about_mandate.php)

National Data Centre (NDC), IMD (http://www.imdpune.gov.in) provides Meteorological Data.

Data request format is available at http://www.imdpune.gov.in/ndc_new/Data_Request/DATA_REQUISITON_FORM.pdf.

Based on the request received, NDC would intimate the cost, availability of the data along with a format of undertaking certificate for the data utilization. More details about data requisition are available at http://www.imdpune.gov.in/ndc_new/Request.html.

Email: ndcsupply@imd.gov.in

A2. Water Resources Information System (WRIS) (http://india-wris.nrsc.gov.in/wris.html)

Homepage consists of Accessibility, Tools, Metadata, WRIS Wiki, Publications (includes project documents, Basin reports, River basin Atlas of India, Watershed Atlas of India, Pre Generated Map, Other reports), Gallery, Mobile, FAQ, WRIS Info discovery, WRIS explorer (which includes Geo-visualization, Sub-Info System, Temporal Analyst, PMP module), WRIS connect (which includes Live Telemetry data, Data Download, Reservoir Module, Automatic Map Generation, Advanced Report Generation, Web Map Services), WR Planning and management, and Input data builder. Information on other categories is also available in the homepage.

K. Srinivasa Raju and D. Nagesh Kumar, *Impact of Climate Change on Water Resources*, Springer Climate, https://doi.org/10.1007/978-981-10-6110-3

A3. Bhuvan: Indian Geo Platform of ISRO (http://bhuvan.nrsc.gov.in/bhuvan_links.php#)

ISRO launched web-based GIS tool, Bhuvan which offers detailed imagery of Indian locations compared to other Virtual Globe Software, with spatial resolutions ranging up to 1 meter. At present, 177 cities high-resolution datasets are available, while the other parts of the country are covered by 2.5-m-resolution imagery. Locations can be viewed from different perspectives, and the software also provides functionality for the measurement of distances and other geoprocessing capabilities (https://en.wikipedia.org/wiki/Bhuvan).

Four categories of information are available in the homepage

Visualization and free download: Features include Bhuvan-2D, Bhuvan-3D, Open data archive, Climate and Environment, Thematic Services, Disaster Services, Ocean Services, and Create a Map/GIS.

Governance/Central Ministries: Chaman, Clean Ganga, SAT-AIBP, Flood Warning, Census data, Deltas of India, Environment and Forests, CRIS, Flycatchers distribution, Islands information, School Bhuvan, Toll Information, Groundwater, Pipe grid, Watersheds, Urban survey, and Monuments.

Application Sectors: Agriculture, forestry, e-governance, water, tourism, urban, rural, and tourism.

Special Applications: Data discovery, hydrological products, international disasters, etc.

A4. CLIMWAT (http://www.fao.org/nr/water/infores_databases_climwat.html)

CLIMWAT is a climatic database in combination with the computer program CROPWAT and allows the calculation of crop water requirements, irrigation supply, and irrigation scheduling for various crops for a range of climatological stations worldwide. CLIMWAT provides long-term mean daily maximum and minimum temperature in °C, mean relative humidity in %, mean wind speed in km/day, mean sunshine hours per day, mean solar radiation in $MJ/m^2/day$, monthly rainfall in mm/month, monthly effective rainfall in mm/month, and reference evapotranspiration calculated with the Penman–Monteith method in mm/day.

Select the Climwat 2.0 for Cropwat (See the download section). The data can be extracted for a single or multiple stations in the format suitable for their use in CROPWAT (http://www.fao.org/land-water/databases-and-software/cropwat/en).

A5. GCMs Data from Coupled Model Intercomparison Project 5 (CMIP5)

Coupled Model Intercomparison Project 5 (CMIP5) provides data of General Circulation Models (GCMs) (http://cmip-pcmdi.llnl.gov/cmip5/availability.html).

Home page consists of Home (overview and history), News, CMIP3 (CMIP3 overview, CMIP3 home, Data archive, Variable list, Data availability, and experiment design), CMIP5 (CMIP5 overview, experiment design, modeling info, data access, more info, and contact), Accomplishments (CMIP3 download rate, publications, usage, and awards), Links (exhaustive and very useful), and contacts.

Some of the important and relevant links are as follows:

- Guide to CMIP5: http://cmip-pcmdi.llnl.gov/cmip5/guide_to_cmip5.html
- CMIP5 data access/ availability (including GCMs): http://cmip-pcmdi.llnl.gov/cmip5/availability.html
- CMIP5 publications: http://cmip.llnl.gov/cmip5/publications/allpublications
- CMIP5 citations information: http://cmip-pcmdi.llnl.gov/cmip5/citation.html
- Frequently asked questions: http://cmip-pcmdi.llnl.gov/cmip5/data_faq.html.

The following four links are useful, and by selecting on any one of the links they are diverted to Earth System Grid Federation (ESGF) home page.

Program for Climate Model Diagnosis and Intercomparison PCMDI http://pcmdi9.llnl.gov/.
British Atmospheric Data Centre (BADC) http://esgf-index1.ceda.ac.uk.
Deutsches Klimarechenzentrum (DKRZ) http://esgf-data.dkrz.de.
National Computational Infrastructure (NCI): http://esg2.nci.org.au.

Various steps for registration/acquiring the data

Step 1 ESGF will need details for login and to access data. To create a new account, select create account option.
A new window will be displayed; all the details displayed are mandatory and are to be provided, such as username, first name and last name, email, password, confirm password, institution, department, city, state, country, interest keywords, interest statement, etc.

Step 2 Once the above information is submitted, an account will be created and the corresponding link is sent to the user's registered email id. On enabling the link successfully, the account will be activated and connected to ESGF Login. After the above procedure, account details are summarized on the screen.

Step 3 Select on the search button option which will contain many categories of search.

Step 4 Select project option (for example CMIP5).

Step 5 Select models development centers and selection of the modeling center (institution) option.

Step 6 Select GCM and experiment family options.

Step 7 Select time frequency option.
Step 8 Selection of name of the Realm option (e.g., Atmos).
Step 9 Selection of the name of the variable option.
Step 10 Selection of Summarized Information and Data Cart Option.

Further steps are quite simple and self-explanatory including data downloading.

A6. National Centers for Environmental Prediction (NCEP) and the National Center for Atmospheric Research (NCAR) (http://rda.ucar.edu/)

- It is a joint product from the National Centers for Environmental Prediction (NCEP) and the National Center for Atmospheric Research (NCAR).
- NCEP/NCAR reanalysis data can be downloaded from the CISL (Computational and Information Systems Lab) Research Data Archive managed by NCAR's Data Support Section.
- The downloaded files are in General Regularly-distributed Information in Binary (GRIB) format.
- Data Available: Maximum temperature, precipitation rate, geopotential height, pressure, u and v-components of wind, etc.

Various steps for registration/acquiring the data

Step 1 Go to the website—http://rda.ucar.edu/.
Step 2 Under the Reanalyses section choose NCEP/NCAR Reanalysis Project.
Step 3 Now choose NCEP/DOE Reanalysis II (ds091.0).
Step 4 Select the Data Access tab (You will need to login to access data). If you are not registered, email address, password, title, first name and last name, organization name and type, and country can be provided for registration purpose.
Step 5 In the table of contents, select "Web File Listing" which is against "Union of Available Products".
Step 6 Select the required NCEP/NCAR data along with appropriate time period and click "continue".
Step 7 Select two or more files of the corresponding Months/Years and click "Download" to download them as a single UNIX tar file.

A7. Climate Research Unit (CRU) (http://www.cru.uea.ac.uk/)

The Climatic Research Unit (CRU) was established in the School of Environmental Sciences at the University of East Anglia (UEA), Norwich and home page consists of home, about CRU, data, academic programs, research, staff and students, information sheets, publications, media, and news events.

Procedure for collecting data (http://www.cru.uea.ac.uk/cru/data):

Step 1 For example, select Temperature (5° × 5° gridded versions) (Select the link: Main webpage for the temperature datasets).

Step 2 Proceed to Data for Downloading: Select NetCDF for downloading the temperature data. Similar steps can be followed for other variables such as precipitation, etc.

CRU research publications are available at https://ueaeprints.uea.ac.uk/view/divisions/CRU.default.html.

List of reports published up to 2009 are available at http://www.cru.uea.ac.uk/documents/421974/1245969/Reports.pdf/d9f96b4c-387c-4dba-83fe-7e0911eb73c6.

List of Ph.D. theses are available at http://www.cru.uea.ac.uk/publications/phd-theses.

A8. University of Delaware Air Temperature and Precipitation (UDEL)

Step 1 Log onto Udel (http://www.esrl.noaa.gov/psd/data/gridded/data.UDel_AirT_Precip.html).

Step 2 Select the links for downloading data.

A9. Tropical Rainfall Measuring Mission (TRMM)

The Tropical Rainfall Measuring Mission (TRMM) (http://trmm.gsfc.nasa.gov/) is designed to measure rainfall for weather and climate research.

Section for frequently asked questions is available at http://pmm.nasa.gov/resources/faq.

Precipitation Measuring Stations Glossary is available at http://pmm.nasa.gov/resources/glossary.

Study material is available at http://pmm.nasa.gov/resources/documents/TRMM.

A10. Asian Precipitation-Highly Resolved Observational Data Integration Towards Evaluation (APHRODITE) (http://www.chikyu.ac.jp/precip/index.html)

The APHRODITE (http://www.chikyu.ac.jp/precip/index.html) project develops state-of-the-art daily precipitation datasets with high-resolution grids for Asia. The datasets are created primarily with data obtained from a rain-gauge-observation network.

APHRODITE home page consists of Home, Scope, Products, Download, Project Members, Publication List, and Links.

A11. Koninklijk Nederlands Meteorologisch Instituut KNMI (Dutch: Royal Netherlands Meteorological Institute)

Step 1 Proceed to http://climexp.knmi.nl/start.cgi?id=someone@somewhere.
Step 2 Select the Monthly CMIP5 scenario runs.
Step 3 For example, select the surface variables.

All GCMs data is available in KNMI climate explorer.

Historical data, RCP 2.6 data, RCP 4.5 data, RCP 6.0 data, and RCP 8.5 data along with number of variables are available for number of GCMs (Type the values of latitude and longitude for which region you want, select make a time series, and select the raw data). Data is available in ASCII and NetCDF formats and select on the required formats to download the data (http://climexp.knmi.nl/selectfield_cmip5.cgi?id=someone@somewhere) (Information last accessed on December 30, 2016).

Other Representative-Related Data Sources/Institutions/Home Pages

Description	Homepage
British Atmospheric Data Centre (BADC)	http://badc.nerc.ac.uk/home/index.html
Carbon Dioxide Information Analysis Center (CDIAC)	http://cdiac.ornl.gov/climate/precip/precip_table.html
Centre for International Earth Science Information Network	http://ciesin.columbia.edu/
Climate Diagnostics Centre at NOAA	http://www.cdc.noaa.gov/
Comprehensive Ocean-Atmosphere Data Set (COADS) at NOAA	http://icoads.noaa.gov/Release_1/coads.html
Computational Information Systems Laboratory	http://www2.cisl.ucar.edu/
ECMWF 40 Year Reanalysis (ECMWF ERA-40)	http://apps.ecmwf.int/datasets/data/era40_daily/
Emissions Database for Global Atmospheric Research (EDGAR)	http://edgar.jrc.ec.europa.eu/
Extended Reconstructed Sea Surface Temperature (ERSST)	http://www.ncdc.noaa.gov/data-access/marineocean-data/extended-reconstructed-sea-surface-temperature-ersst-v3b
Food and Agriculture Organization of the United Nations (FAO)	http://www.fao.org/nr/climpag/pub/EN1102_en.asp

(continued)

(continued)

Description	Homepage
GEWEX Asian Monsoon Experiment-Tropics (GAME-T) data center	http://hydro.iis.u-tokyo.ac.jp/GAME-T/GAIN-T/index.html
Global Environment Facility (GEF)	http://www.gefweb.org/
Global Historical Climatology Network (GHCN)	http://www.ncdc.noaa.gov/oa/climate/ghcn-daily/
Global Precipitation Climatology Centre (GPCC)	http://gpcc.dwd.de/
Historical Anthropogenic Sulfur Dioxide Emission (HASO2); assessment	http://sedac.ciesin.columbia.edu/data/set/haso2-anthro-sulfur-dioxide-emissions-1850-25-v2-86
HYDE Land Use	http://themasites.pbl.nl/tridion/en/themasites/hyde/landusedata/index-2.html
IAMC RCP Database	http://www.iiasa.ac.at/web-apps/tnt/RcpDb
India Water Portal	http://www.indiawaterportal.org/
Integrated Assessment Modeling consortium	http://www.iamconsortium.org/
International Comprehensive Ocean-Atmosphere Data Set (ICOADS); assessment	http://icoads.noaa.gov/
International Geosphere-Biosphere Programme	http://www.igbp.net/
International Research Institute for Climate Prediction/Lamont–Doherty Earth Observation at University of Columbia	http://ingrid.ldeo.columbia.edu/
Japanese 55-year ReAnalysis (JRA-55)	http://jra.kishou.go.jp/JRA-55/index_en.html
Land Use Harmonization	http://luh.unh.edu/
Modern Era Retrospective-analysis for Research and Applications (MERRA)	http://disc.sci.gsfc.nasa.gov/daac-bin/DataHoldings.pl
National Aeronautics and Space Administration	http://www.nasa.gov/
National Climatic Data Center (NCDC)	http://www.ncdc.noaa.gov/oa/documentlibrary/ds-doc.html
National Communications Support Programme (UN)	http://ncsp.undp.org/
National Environment Research Council	http://www.nerc.ac.uk/
NOAA CIRES Twentieth Century Global Reanalysis Version (NOAA_CIRES20thC_ReaV2)	http://rda.ucar.edu/datasets/ds131.1/#!access
NOAA Merged Land–Ocean Surface Temperature Analysis (MLOST)	http://www.esrl.noaa.gov/psd/data/gridded/data.mlost.html
UCAR Community Data Portal	http://cdp.ucar.edu/
United Nations Framework Convention on Climate Change (UNFCCC)	http://www.unfccc.int/
United Nations and Climate Change	http://www.un.org/climatechange/
World Climate Research Programme	http://www.wcrp-climate.org/

(continued)

(continued)

Description	Homepage
World Data Centre for Climate	http://www.dkrz.de/daten/wdcc/
World Meteorological Organization (WMO)	http://www.wmo.int/pages/index_en.html
World Data Center for Meteorology	http://www.ncdc.noaa.gov/wdc

Few available data is in NetCDF format. Information about software that process NetCDF and other similar formats are available at http://www.unidata.ucar.edu/software/netcdf/software.html.

Appendix B: Representative List of Journals on Climate and Allied Fields

Name of the journal and publishing house	Relevant link
Advances in Atmospheric Sciences	http://www.springer.com/earth+sciences+and+geography/atmospheric+sciences/journal/376
Advances in Meteorology, Hindwai Publishing Corporation	https://www.hindawi.com/journals/amete/
Advances in Water Resources, Elsevier	https://www.journals.elsevier.com/advances-in-water-resources
Agriculture and Forest Meteorology	https://www.journals.elsevier.com/agricultural-and-forest-meteorology/
Agricultural Systems, Elsevier	https://www.journals.elsevier.com/agricultural-systems
Agricultural Water Management, Elsevier	https://www.journals.elsevier.com/agricultural-water-management
Atmospheric Environment, Elsevier	https://www.journals.elsevier.com/atmospheric-environment/
Atmospheric Research, Elsevier	https://www.journals.elsevier.com/atmospheric-research/
Atmospheric Science Letters, Wiley	http://onlinelibrary.wiley.com/journal/10.1002/(ISSN)1530-261X/homepage/ProductInformation.html
Bulletin of the American Meteorological Society	https://www.ametsoc.org/ams/index.cfm/publications/bulletin-of-the-american-meteorological-society-bams/
Climate and Development, Taylor & Francis	http://www.tandfonline.com/action/journalInformation?show=aimsScope&journalCode=tcld20
Climate Dynamics, Springer	http://www.springer.com/earth+sciences+and+geography/geophysics/journal/382
Climate Policy, Taylor & Francis	http://www.tandfonline.com/action/journalInformation?show=aimsScope&journalCode=tcpo20
Climate Research, Inter-Research Science Center	http://www.int-res.com/journals/cr/about-the-journal/

(continued)

K. Srinivasa Raju and D. Nagesh Kumar, *Impact of Climate Change on Water Resources*, Springer Climate, https://doi.org/10.1007/978-981-10-6110-3

(continued)

Name of the journal and publishing house	Relevant link
Climate Risk Management, Elsevier	https://www.journals.elsevier.com/climate-risk-management/
Climatic Change, Springer	http://www.springer.com/earth+sciences+and+geography/atmospheric+sciences/journal/10584
Current Climate Change Reports	http://www.springer.com/environment/global+change+_+climate+change/journal/40641
Dynamics of Atmospheres and Oceans, Elsevier	https://www.journals.elsevier.com/dynamics-of-atmospheres-and-oceans/
Earth Interactions, American Meteorological Society	https://www.ametsoc.org/ams/index.cfm/publications/journals/earth-interactions/
Earth and Planetary Science Letters	https://www.journals.elsevier.com/earth-and-planetary-science-letters
Earth Sciences Review, Elsevier	https://www.journals.elsevier.com/earth-science-reviews/
Economics of Disasters and Climate Change	http://www.springer.com/economics/environmental/journal/41885
Geophysical Journal International	http://onlinelibrary.wiley.com/journal/10.1111/(ISSN)1365-246X
Geophysical Research Letters, American Geophysical Union, Wiley	http://agupubs.onlinelibrary.wiley.com/hub/journal/10.1002/(ISSN)1944-87/
Global Environmental Change	https://www.journals.elsevier.com/global-environmental-change/
Hydrological Earth System Science, Copernicus Publications	http://www.hydrology-and-earth-system-sciences.net/about/aims_and_scope.html
Hydrological Processes, Wiley	http://onlinelibrary.wiley.com/journal/10.1002/(ISSN)1099-1085/homepage/ProductInformation.html
Hydrological Sciences Journal, Taylor and Francis	http://www.tandfonline.com/action/journalInformation?show=aimsScope&journalCode=thsj20
International Journal of Atmospheric Sciences	https://www.hindawi.com/journals/ijas/
International Journal of Climate Change Strategies and Management	http://www.emeraldgrouppublishing.com/products/journals/journals.htm?id=ijccsm
International Journal of climatology, Wiley	http://onlinelibrary.wiley.com/journal/10.1002/(ISSN)1097-88/homepage/ProductInformation.html
Irrigation and Drainage, Wiley	http://onlinelibrary.wiley.com/journal/10.1002/(ISSN)1531-0361/homepage/ProductInformation.html
Journal of Applied Ecology, Wiley	http://onlinelibrary.wiley.com/journal/10.1111/(ISSN)1365-2664/homepage/ProductInformation.html

(continued)

(continued)

Name of the journal and publishing house	Relevant link
Journal of Applied Meteorology and Climatology, American Meteorological Society	https://www.ametsoc.org/ams/index.cfm/publications/journals/journal-of-applied-meteorology-and-climatology/
Journal of Atmospheric and Oceanic Technology, American Meteorological Society	https://www.ametsoc.org/ams/index.cfm/publications/journals/journal-of-atmospheric-and-oceanic-technology/
Journal of the Atmospheric Sciences	https://www.ametsoc.org/ams/index.cfm/publications/journals/journal-of-the-atmospheric-sciences/
Journal of Climate, American Meteorological Society	https://www.ametsoc.org/ams/index.cfm/publications/journals/journal-of-climate/
Journal of Computing in Civil Engineering, ASCE	http://ascelibrary.org/page/jccee5/editorialboard
Journal of Earth Science & Climatic Change, OMICS	https://www.omicsonline.org/earth-science-climatic-change.php
Journal of Geophysical Research, American Geophysical Union, Wiley	http://agupubs.onlinelibrary.wiley.com/hub/jgr/journal/10.1002/(ISSN)2156-2202/
Journal of Hydroinformatics, IWA Publishing	http://jh.iwaponline.com/content/aims-scope
Journal of Hydrologic Engineering, ASCE	http://ascelibrary.org/page/jhyeff/editorialboard
Journal of Hydrology, Elsevier	https://www.journals.elsevier.com/journal-of-hydrology
Journal of Hydrology Research, IWA Publishing	http://hr.iwaponline.com/content/aims-scope
Journal of Hydrometeorology, American Meteorological Society	https://www.ametsoc.org/ams/index.cfm/publications/journals/journal-of-hydrometeorology/
Journal of Irrigation and Drainage Engineering, ASCE	http://ascelibrary.org/page/jidedh/editorialboard
Journal of Meteorological Research, Springer	http://www.springer.com/earth+sciences+and+geography/atmospheric+sciences/journal/13351
Journal of Water and Climate Change, IWA	http://jwcc.iwaponline.com/content/aims-scope
Journal of Water Resources Planning and Management, ASCE	http://ascelibrary.org/page/jwrmd5/editorialboard
Journal of Southern Hemisphere Earth Systems Sci Australian Meteorological and Oceanographic Society	http://www.bom.gov.au/amm/
Meteorology and Atmospheric Physics	http://www.springer.com/earth+sciences+and+geography/atmospheric+sciences/journal/703/PSE
Mitigation and Adaptation Strategies for Global Change	

(continued)

(continued)

Name of the journal and publishing house	Relevant link
	http://www.springer.com/earth+sciences+and+geography/atmospheric+sciences/journal/11027/PSE
Monthly Weather Review	https://www.ametsoc.org/ams/index.cfm/publications/journals/monthly-weather-review/
Natural Hazards, Springer, Springer	http://www.springer.com/earth+sciences+and+geography/natural+hazards/journal/11069
Nature Climate Change, Springer	http://www.nature.com/nclimate/about/index.html
Open Water Journal, IWA Publishing	http://iwaponline.com/content/iwa-open-water-journal
Quarterly Journal of Royal Meteorological Society	http://onlinelibrary.wiley.com/journal/10.1002/(ISSN)1477-870X/homepage/ProductInformation.html
Regional Environmental Change	http://www.springer.com/environment/global+change+-+climate+change/journal/10113/PSE
Sustainability Science	http://www.springer.com/environment/environmental+management/journal/11625/PSE
Tellus, Series A: Dynamic Meteorology and Oceanography, Swedish Geophysical Society: Munksgaard	http://www.tandfonline.com/action/journalInformation?show=aimsScope&journalCode=zela20
Theoretical and applied climatology, Springer	http://www.springer.com/earth+sciences+and+geography/atmospheric+sciences/journal/704
Urban Climate, Elsevier	https://www.journals.elsevier.com/urban-climate/
Water Resources Management, Springer	http://www.springer.com/earth+sciences+and+geography/hydrogeology/journal/11269
Water Resources Research, American Geophysical Union, Wiley	http://agupubs.onlinelibrary.wiley.com/hub/journal/10.1002/(ISSN)1944-7973/
Weather, Climate, and Society, American Meteorological Society	https://www.ametsoc.org/ams/index.cfm/publications/journals/weather-climate-and-society/
Weather and forecasting, American Meteorological Society	https://www.ametsoc.org/ams/index.cfm/publications/journals/weather-and-forecasting/
Wiley Interdisciplinary Reviews: Climate Change, Wiley	http://onlinelibrary.wiley.com/journal/10.1002/(ISSN)1757-7799/homepage/ProductInformation.html

Appendix C: Representative List of Books on Climate and Allied Fields

Baldassarre GD, Brandimarte L, Popescu I, Neal JC, Bates PD, Fewtrell TJ (2013) Floods in a changing climate: inundation modelling. Cambridge University Press

Barry RG, Hall-McKim EA (2014) Essentials of the Earth's climate system. Cambridge University Press

Beniston M (ed) (2002) Climatic change: implications for the hydrological cycle and for water management. Springer advances in global change research

Bonan G (2015) Ecological climatology: concepts and applications. Cambridge University Press

Bridgman HA, Oliver JE, Allan R, Cerveny R, Glantz M, Mausell P, Liu D, Diaz N, Giles B, Wendler G, Zielinski G, Grimmond S, Changnon S, Lau W (2014) The global climate system patterns, processes, and teleconnections. Cambridge University Press

Bulkeley H, Newell P (2015) Governing climate change. Taylor and Francis

Cowie J (2012) Climate change biological and human aspects. Cambridge University Press

Cracknell AP, Krapivin VF (2009) Global climatology and ecodynamics: anthropogenic changes to planet Earth. Springer environmental sciences

Dam JCV (2003) Impacts of climate change and climate variability on hydrological regimes. Cambridge University Press

Dash SK (2017) Climate change: an Indian perspective, part of environment and development series. Cambridge University Press

Dessler A (2015) Introduction to modern climate change. Cambridge University Press

Diaz HF, Markgraf V (eds) (2000) El Niño and the Southern oscillation multiscale variability and global and regional impacts. Cambridge University Press

Diaz HF, Markgraf V, Kiladis GN, Cayan DR, Webb RH, Meehl GA, Branstator GW, Enfield DB, Quinn WH, Nicholls N, Pulwarty RS, Anderson RY, Cook ER, Lough JM, Meko DM, D'arrigo RD, Jacoby GC, Swetnam TW, Betancourt JL, Cleaveland MK, Cooke ER, Stahle DW, Thompson LG, Mosley-Thompson E, Thompson PA, Michaelsen J, Cole JE, Shen GT, Fairbanks RG, Moore M, Sharp GD, Anderson RY, Soutar A, Johnson TC, McGlone MS, Kershaw AP, Hughes MK (2009) El Niño historical and paleoclimatic aspects of the southern oscillation. Cambridge University Press

Dijkstra HA (2013) Nonlinear climate dynamics. Cambridge University Press

Dong W, Huang J, Guo Y, Ren F (2016) Atlas of climate change: responsibility and obligation of human society. Springer atmospheric sciences

Eggleton T (2012) A short introduction to climate change. Cambridge University Press

Field CB, Barros VV, Stocker TF, Dahe Q (2012) Managing the risks of extreme events and disasters to advance climate change adaptation: special report of the intergovernmental panel on climate change. Cambridge University Press

Fitzroy FR, Papyrakis E (2015) An introduction to climate change economics and policy. Routledge textbooks in environmental and agricultural economics

K. Srinivasa Raju and D. Nagesh Kumar, *Impact of Climate Change on Water Resources*, Springer Climate, https://doi.org/10.1007/978-981-10-6110-3

Glantz MH, Katz RW, Nicholls N (2009) Teleconnections linking worldwide climate anomalies. Cambridge University Press
Goosse H (2015) Climate system dynamics and modelling. Cambridge University Press
Fell H-J (2016) Global cooling: strategies for climate protection. Taylor and Francis
Hill M (2013) Climate change and water governance: adaptive capacity in Chile and Switzerland. Springer advances in global change research
Holton JR, Hakim GJ (2012) Introduction to dynamic meteorology. Academic Press
Houghton J (2015) Global warming: the complete briefing. Cambridge University Press
Incropera FP, Earley T, Peterson B, Majumdar A (2015) Climate change: a wicked problem: complexity and uncertainty at the intersection of science, economics, politics, and human behavior. Cambridge University Press
Intergovernmental Panel on Climate Change (2015) Climate Change 2014: Mitigation of Climate Change Working Group III Contribution to the IPCC Fifth Assessment Report, Cambridge University Press
Intergovernmental Panel on Climate Change (2014) Climate change 2014—impacts, adaptation and vulnerability: part A: global and sectoral aspects working group II contribution to the IPCC fifth assessment report. In: Global and sectoral aspects, vol 1. Cambridge University Press
Intergovernmental Panel on Climate Change (2014) Climate change 2014—impacts, adaptation and vulnerability: Part B: regional aspects working group II contribution to the IPCC fifth assessment report. In: Regional aspects, vol 2. Cambridge University Press
Intergovernmental Panel on Climate Change (2014) Climate change 2013—the physical science basis working group I contribution to the fifth assessment report of the intergovernmental panel on climate change
Kitchen DE (2016) Global climate change: turning knowledge into action. Taylor and Francis
Koppmann R (ed) (2014) Atmospheric research from different perspectives. The reacting atmosphere series. Springer
Krishnamurti TN, Stefanova L, Misra V (2013) Tropical meteorology: an introduction. Springer atmospheric sciences
Lal R, Uphoff N, Stewart BA, Hansen DO (eds) (2005) Climate change and global food security. CRC Press
Latin HA (2012) Climate change policy failures: why conventional mitigation approaches cannot succeed. World Scientific Publishing
Lau WKM, Waliser DE (2005) Intraseasonal variability in the atmosphere-ocean climate system. Springer environmental sciences
Leroux M (2005) Global warming—myth or reality? Springer environmental sciences
Leroux M (2010) Dynamic analysis of weather and climate. Springer environmental sciences
Lininger C (2015) Consumption-based approaches in international climate policy. Springer climate series
Martens P, Rotmans J (eds) (1999) Climate change: an integrated perspective. Springer advances in global change research
Mcllveen R (1991) Fundamentals of weather and climate. Psychology Press
Mujumdar PP, Nagesh Kumar D (2013) Floods in a changing climate: hydrologic modeling. Cambridge University Press
Neelin JD (2011) Climate change and climate modeling. Cambridge University Press
Parikh J (2016) Climate resilient urban development: vulnerable profile of Indian cities. Cambridge University Press
Rapp D (2014) Assessing climate change. Springer environmental sciences
Ravindranath NH, Sathaye JA (2002) Climate change and developing countries. Springer advances in global change research
Sarachik ES, Cane MA (2010) The El Niño-Southern oscillation phenomenon. Cambridge University Press
Simonovic SP (2013) Floods in a changing climate: risk management. Cambridge University Press

Sheppard S (2012) Visualizing climate change: a guide to visual communication of climate change and developing local solutions. Routledge

von Storch H, Zwiers FW (2002) Statistical analysis in climate research. Cambridge University Press

Wheeler SM (2012) Climate change and social ecology: a new perspective on the climate challenge. Routledge

Teegavarapu RSV (2013) Floods in a changing climate: extreme precipitation. Cambridge University Press

Trenberth KE (2010) Climate system modeling. Cambridge University Press

Index

K. Srinivasa Raju and D. Nagesh Kumar, *Impact of Climate Change on Water Resources*, Springer Climate, https://doi.org/10.1007/978-981-10-6110-3

The manufacturer's authorised representative in the EU is Springer Nature Customer Service Centre GmbH, Europaplatz 3, 69115 Heidelberg, Germany. If you have any concerns regarding our products, please contact ProductSafety@springernature.com

Printed and bound by CPI Group (UK) Ltd, Croydon, CR0 4YY
30/09/2026
02225448-0008